Mobile Telecommunications Networks

To Mair and Macsen

A lifetime spent acquiring knowledge awaits you. Enjoy.

Mobile Telecommunications Networks

Restructuring as a Response to a Challenging Environment

Peter Curwen

Visiting Professor of Mobile Communications, Northumbria University, UK

Jason Whalley

Professor of Digital Economy, Northumbria University, UK

Edward Elgar

Cheltenham, UK • Northampton, MA, USA

Published by
Edward Elgar Publishing Limited
The Lypiatts
15 Lansdown Road
Cheltenham
Glos GL50 2JA
UK

Edward Elgar Publishing, Inc.
William Pratt House
9 Dewey Court
Northampton
Massachusetts 01060
USA

A catalogue record for this book
is available from the British Library

Library of Congress Control Number: 2014941552

This book is available electronically in the ElgarOnline.com Business Subject Collection, E-ISBN 978 1 78347 570 4

MIX
Paper from
responsible sources
FSC® C013056
FSC
www.fsc.org

ISBN 978 1 78347 569 8

Typeset by Columns Design XML Ltd, Reading
Printed and bound in Great Britain by T.J. International Ltd, Padstow

Contents

Preface

The mobile communications industry has been the subject of surprisingly little research considering the central part that it plays in so many people's lives. It is not possible to specify exactly why that is, but at least part of the answer must lie in the fact that the popularity of mobile as against fixed-wire connectivity came as something of a surprise, and most researchers whose work was locked into a world of fixed-wire connectivity were reluctant to give up a subject in which they had invested so much intellectual capital.

As fate would have it, the authors of this book had only recently turned their attention to the mobile industry at the turn of the millennium, at the point when mobile communications were on the cusp of switching from a voice-driven to a data-driven basis, and hence had invested relatively little time researching fixed-wire connectivity. The world of mobile beckoned and it rapidly became clear that they could not research both mobile and fixed-wire, so the authors decided to specialise in the former.

The huge advantage of doing so is that they began to compile databases over a decade ago which have been updated and embellished ever since. These databases are unique in the public domain – something similar probably exists in the hands of private consultancies, but given that these charge massive amounts of money to access their data, and given that such data are sold only in an aggregated form that sheds no light on the disaggregated data that underpin them, it is difficult to be certain. There are also some patchy databases issued by international bodies, but these tend to be out of date and based on assumptions that are not necessarily appropriate for the serious researcher – even assuming that they are specified in the first place.

And therein lies the rub. The mobile industry is extremely complicated: there are, to begin with, some 225 countries, each with roughly four operators, each of which may be reported in the media under one of two or three different names. And that is just the part of the industry that deals with service provision; it excludes the equipment vendors and the software providers, not to mention the socio-economic consequences arising from the use of mobile devices.

Hence even a book of some size, such as the one you are about to read, cannot encompass the full extent of the industry. In this particular case, the topic to be discussed involves only the service providers and, even more specifically, only those that have made an attempt to create an international operation of some scale. Its focus is accordingly on the structure of the mobile operator sector and the strategic behaviour of its most notable constituents on a worldwide basis.

The standard approach among scholars is to begin by offering a number of alternative hypotheses about these matters and to use databases to test whether they hold true in the real world. This has not so far been a particularly fruitful approach because, as indicated above, these scholars are forced to use bought-in databases that they must take at face value even though the databases are in practice circumscribed by an array of qualifications that make it very difficult to compare like with like.

The approach taken in this book is somewhat different because it starts with the proposition that if the data are of dubious quality, then no amount of statistical manipulation can reveal what is really going on and is therefore an unhelpful distraction. The first implication of such an approach is that the databases underpinning the analysis must have been compiled by the authors themselves so that they understand exactly what is reliable and what is subject to uncertainty. The second implication is that all the qualifications must be spelt out so that readers can decide for themselves whether or not they trust the analysis offered. Understanding the logic of business decisions at second hand will never be all that scientific and differences of opinion are always permissible provided that they are based on known facts rather than pure conjecture.

Possibly uniquely for a book of this kind, very few secondary sources are replicated and not one single table or figure is extracted from an existing source without at least some manipulation, for which an explanation is provided. For the most part the databases are indeed unique in that they have been compiled independently by the authors. That does not mean they are wholly free of errors – many events in the real world are difficult to pin down in terms of precise dates and precise quantities – but all the data have been subjected to a stringent process of cross-checking and are accompanied by the extensive footnotes that are unavoidable in research on the mobile industry.

In terms of content, this book attempts to achieve a number of objectives. At one level, it seeks to explain how individual international operators develop over time and adjust their strategies in response to events in a rapidly changing environment. To help with this, two detailed case studies are included in separate chapters, covering the Vodafone Group – the most internationalised of all the service providers – and the mobile activities of Hutchison Whampoa – the most disruptive of all the service providers. A particular issue addressed in the former case study is how mobile operators seek to avoid tax, which has become a very contentious subject across industry in general in recent years.

The main purpose of the case studies is to illustrate the complexity of the modern international service provider. It is normal to refer, for example, to Vodafone as if it were simply a singular entity, whereas the reality is that it consists of a complex set of holding companies headquartered in a wide variety of countries. Who owns what and why is accordingly something that takes a forensic examination to untangle.

At the other end of the spectrum, the book seeks to examine the behaviour of the 30 or so largest international mobile operators viewed as a whole, since between them they account for the bulk of such restructuring as has taken – and is currently taking – place. To this end, their activities are dissected on a regional basis before being brought together on a worldwide basis in the two concluding chapters.

In addition, the status of these operators is examined on a comparative basis in the first chapter, which also sets out all the major Merger and Acquisition (M&A) activities in which they have engaged in recent years. This is followed by a discussion of so-called fourth generation mobile technology, since this is a major driving force behind the need to restructure.

One of the main lessons of the book is that restructuring is an ongoing process that is unusually rapid in the mobile industry because the environment in which it exists is constantly changing, driven by such forces as developments in technology, regulations, customer behaviour and socio-economic conditions in general. It is fair to argue that no industry has ever had to develop so quickly – the railway mania of the late nineteenth century is comparable in some respects but the technology aspect was much less important.

It must never be forgotten that rapid change is costly. The book sets out to shed some light on the financial implications, but the main message is that the structure of the markets in many countries was designed to reflect the environment ruling at particular points in time. However, the combination of forces that currently beset the mobile industry indicates strongly that existing structures are unsuitable in many cases. The main impact of this will fall upon the service providers discussed in this book, but for now the outcome remains unclear. In essence, market forces exert pressure for solutions that are not necessarily acceptable to regulators and governments. Sooner or later, one or the other must prevail, but it is most unlikely that the outcome will be uniform across the world.

The book attempts to set out the key pressure points where developments are likely to influence outcomes in other jurisdictions. For better or worse, most significant developments, such as decisions on takeovers, take lengthy periods of time to resolve, so outcomes forecast in the book may or may not come to pass. However, the book provides a snapshot of the mobile industry as it stands in mid-2014, and that is something that is not available elsewhere.

Acknowledgements

Table 11.3 is a modified and updated version of Table 8.1 in P. Curwen and J. Whalley (2013), *Fourth Generation Mobile Communication: The Path to Superfast Connectivity*, London: Springer.

A previous, rather longer, version of Chapter 4 appeared as 'Third Time Lucky? An Exploration of Hutchison Whampoa's Involvement in the Mobile Telecommunications Industry', *info*, **14**(2), 2012, 17–41, while a previous, much longer, version of Chapter 3 appeared as 'Unraveling Complex Organisational Structures among Mobile Operators', *info*, **15**(4), 2013, 3–22. The authors acknowledge the blanket permission regime of *info* to re-publish new versions of their work.

Abbreviations

AMR-WB	adaptive multi rate wideband
APT	Asia-Pacific Telecommunity
ARPU	average revenue per user
ASA	authorised shared access
CEO	chief executive officer
CIS	Commonwealth of Independent States
CoMP	coordinated multi point
CWW	Cable & Wireless Worldwide
DC-HSPA+	dual-carrier high-speed packet access+
DoJ	Department of Justice
DoT	Department of Telecommunications
EBITDA	earnings before interest, tax, depreciation and amortisation
eICIC	enhanced inter-cell interference coordination
EU	European Union
EV-DO	evolution-data optimised
FCC	Federal Communications Commission
FDD	frequency division duplex
GDP	gross domestic product
GHz	gigahertz
GSM	global system for mobile
GSMA	Global mobile Suppliers Association
HAT	Hutchison Asia Telecommunications
HD Voice	high definition voice
HSPA	high-speed packet access
HTIL	Hutchison Telecommunications International Limited
IPO	initial public offer
ITU	International Telecommunication Union
ITU-R	ITU Radiocommunications
KIA	Kuwait Investment Authority
LTE	Long Term Evolution
LTE-A	Long Term Evolution – Advanced
M2M	machine to machine
M&A	Mergers and Acquisitions

MAN	metropolitan area network
MAN-A	metropolitan area network – advanced
Mbps	megabits per second
MHz	megahertz
MIMO	multiple input multiple output
MNE	multinational enterprise
MTC	Mobile Telecommunications Company
MVNO	mobile virtual network operator
NSN	Nokia Solutions and Networks
PCS	personal communications services
RAN	radio access network
RCS	rich communication services
RCS-e	RCS-enhanced
RSPG	Radio Spectrum Policy Group
SEC	Securities and Exchange Commission
SIM	subscriber identity module
TDD	time division duplex
TD-LTE	time division long term evolution
TD-SCDMA	time division synchronous code division multiple access
TMT	technology, media and telecommunications
TNI	transnationality index
UMB	ultra mobile broadband
UMTS	universal mobile telecommunications system
VEH	Vodafone Europe Holdings
VHA	Vodafone Hutchison Australia
VIH	Vodafone International Holdings
VIL	Vodafone Investments Luxembourg
VLR	visitor location register
VoLTE	voice services over LTE
W-CDMA	wide-band code division multiple access
WCIS	World Cellular Information Service
WiMAX	worldwide interoperability for microwave access
WRC	World Radiocommunication Conference
2G	second generation
3G	third generation
3GPP	Third Generation Partnership Project
4G	fourth generation

1. Restructuring among mobile service providers: a ten-year perspective

1.1 INTRODUCTION

This book is concerned primarily with the restructuring among international mobile service providers that has taken place during the ten-year period beginning on 1 January 2004. However, a given structure does not appear fully fledged at a particular point in time, so some attention needs to be paid to how the structure existing in early 2004 came into being. Equally, one of the objects of the book is to look forward to how the structure might evolve in the future.

The structure of this industry is of interest for three reasons in particular. In the first place, mobile communication plays an ever-increasing role in modern society and hence the issue of which companies are providing it and the links between them has a geo-political impact. Second, the industry is of huge economic importance as a driver of economic growth. Finally, throughout the entire history of the industrialised world – more than three centuries – no industry has ever evolved so quickly.

Obviously service providers are only part of the story, since this rapid change has encompassed a variety of companies making up the wider technology, media and telecommunications (TMT) sector, which includes anything from software providers to equipment vendors. However, one advantage of concentrating on the service providers is that the sector is relatively homogeneous. Some providers have fixed-wire networks and a small number are conglomerates, but in almost every case their main – and fastest-growing – line of business is currently mobile communication.

Whether in principle that is a good or a bad thing is a moot point. When you provide a service that plays an ever-increasing role in the modern economy, you will be a particular beneficiary during periods of economic growth, but equally in recessionary times you will become in principle highly vulnerable as customers cut back on an ever-rising share of their discretionary spending. What is quite extraordinary, however, given that data networks have been common in most countries for only a decade or so, is the totally unexpected attachment of the general public to staying in contact 24/7. It is not simply that vast numbers of people never switch off their mobile devices, but rather that they cannot bear to ignore a call or text irrespective of the situation or time of day. As a result, social mores have evolved alongside the ever-increasing penetration of mobile devices – penetration levels in all advanced and even many emerging economies commonly exceed 100 per cent, such that if the very young and very old who have no devices are excluded, an average of two devices per person is by no means unusual.

While the sociological implications of recent developments in mobile communication are not the direct concern of this book, it is nevertheless important to remember that what is being described is not taking place in a vacuum, and that service providers are obliged to evolve their structures in line with events outside their direct control. At the very least, they may find that operating in countries somewhat dissimilar to their domestic market is fraught with hazards such as unusual interpretations of the rule of law. Equally, there is still a large number of countries where state-owned service providers are to be found that tend to be treated favourably by governments.

1.2 MEASURING INTERNATIONALISATION

There is no shortage of suggestions as to how internationalisation can be measured. Dörrenbächer (2000) and Sullivan (1994) both suggest three broad categories of possible internationalisation indicators. The first of these is structural. According to Dörrenbächer, structural indicators are those that provide a snapshot of the international engagement of a company at any given point in time. One such indicator is the number of countries in which a company is present, while another is foreign assets as a percentage of total assets or the ratio of domestic to overseas employment.

The second broad category of indicators is performance indicators, which measure how well a company is doing overseas. Dörrenbächer (2000: 120) identifies two such indicators as turnover and operating income, whereas Sullivan (1994: 331) prefers R&D intensity, advertising intensity and export sales as a percentage of total sales. Although Gerpott and Jakopin (2005) prefer to use different terminology, they draw attention to foreign income as a percentage of total income.

Attitudinal indicators form the third broad category of indicators. These measure the relationship between a company's home country and its overseas operations. Sullivan (1994: 332) notes the difficulties inherent in measuring such indicators before suggesting that the cumulative international experience of senior managers could be used as a measure. Recognising these difficulties, Dörrenbächer (2000) suggests drawing on Perlmutter (1969), who identified a range of headquarter/subsidiary relationships. In all, four different relationships are identified – ethnocentric, polycentric, regiocentric and geocentric – and as a company progresses through these sequentially, so internationalisation will increase.

A second attitudinal measure is the psychic dispersion of the international operations of a company. Johanson and Wiedersheim-Paul (1975: 308) define psychic distance as those factors 'preventing or disturbing the flows of information between firm and market', and offer examples that include language, culture and political systems. These create information asymmetries and hence uncertainties and risks for inward investors. With this in mind, Johanson and Vahlne (1977) argue that internationalisation is incremental. Companies initially favour foreign markets that are close in terms of psychic distance to their home market before expanding into more distant ones, and as they enter each market their investment increases over time. While there has been some discussion as to the usefulness of this concept and how it may be measured,[1] one widely used approach was suggested by Kogut and Singh (1988), who draw on Hofstede (1980) to rank countries relative to a specified country. An alternative methodology is to use the ten country clusters identified by Ronan and Shenkar (1985), with a company having a presence in, say, five of the clusters being credited with a score of 50 per cent (Sullivan, 1994).

It is clearly possible to argue that the use of the individual aforementioned indicators provides an incomplete picture of internationalisation. For example, although a company may have a presence in 20 countries, these investments may generate only a small fraction of the company's overall revenues and/or account for a small proportion of its employees. As a consequence, some authors have sought to develop composite indices of internationalisation that combine several hitherto separate indicators. For example, Ietto-Gillies (1998) proposed a composite index that combines the ratio of foreign to total assets, sales and employment with a measure of the number of countries in which the company could be present if it so wished. UNCTAD (2007) uses these measures but in the form of two separate indices: a transnationality index (TNI) and an internationalisation index (II). The former consists of the average of the ratio of foreign to total assets, sales and employment, while the latter is calculated by dividing the number of foreign affiliates by the number of all affiliates. While this approach has its attractions, the inclusion of companies from small countries is likely to distort the indices since, according to Hassel et al. (2003: 721), they have a greater proportion of their operations abroad. In other words, the indices do not control for country size.

Sullivan (1994) combined nine measures – five performance-based, two structural and two attitudinal – into a composite index of internationalisation. However, the identification of relevant measures and their subsequent combination into a composite index is by no means straightforward. By demonstrating the range of measures that are available, Dörrenbächer (2000) also highlights how different indicators have been developed to measure ostensibly the same phenomenon. Even assuming that agreement can eventually be reached on which measures are to be chosen, problems are likely to emerge when attempts are made to combine them into a composite index, as demonstrated in Sullivan. Although Ramaswamy et al. (1996: 176) welcomed the move away from a single measure of internationalisation, they suggested that internationalisation is more complex than indicated by Sullivan. They argued that it was unclear how the various components of the index related to one another, and by combining this point with other criticisms they concluded that attempting to create a composite index was premature. Not surprisingly, Sullivan disputed this, claiming that his original article had been misunderstood.

Although the literature provides a wide array of possible ways through which internationalisation could be measured, some are easier to implement than others. A key factor that determines which measures can, and which cannot, be used is the availability of data. By no means all internationalised service providers describe their operations in detail. As a consequence, it is often difficult to determine, for example, the exact scale of their investments or their subscriber numbers. A company may report an aggregated figure for capital expenditure without making clear where, or in what, it was invested. Furthermore, the data may be inconsistent from year to year. Mobile operators may be obliged to restate their accounts to reflect changes in accounting rules or may switch items within the accounts (as happens, for example, when a network holding is put up for sale). As a result, longitudinal analysis may prove to be unsatisfactory. Analysis may accordingly have to be directed exclusively towards data that are published regularly on a suitably disaggregated basis.

These issues present difficulties when using the standard databases for the sector. For example, UNCTAD (2007) identifies France Télécom as the second-most internationalised service provider, but neglects to make clear how dependent it is on France and a handful of other countries for its subscribers. Moreover, the choice of the parent company as the preferred unit of analysis can potentially obscure how subsidiaries are faring in the international sphere of operations. UNCTAD identifies nine telcos among the top 100 non-financial companies ranked by assets. Not only does this overlook, for example, the mobile activities of Hutchison Whampoa, which is classified as 'diversified', but it also ignores the significant handset business of Samsung Electronics in the vendor sector.

The performance of individual investments is also obscured when partially owned operations are consolidated with those that are wholly owned. Financial data are frequently aggregated and consolidated, with the result that there is insufficient detail to determine the revenues and profits of specific investments. Dassler et al. (2002) and Gerpott and Jakopin (2005) are able to conduct their financial analyses of operators only by limiting their samples to American, European and Japanese companies whose financial reporting requirements are comparatively onerous and hence provide relatively detailed data.

As a consequence of the limited availability of data, there is a tension between the comprehensiveness of the analysis and its level of detail. Being comprehensive will increase the number of operators included in the sample set and hence provide a better picture of internationalisation within the mobile sector, but will do so at the expense of detail. In contrast, upping the level of detail will provide a richer understanding of the internationalisation process but one based on a smaller sample set, and thereby hamper efforts to comment on changes at the sector level.

Overall, the extensive internationalisation of the telecommunications industry that has occurred since the mid-1980s has resulted in a diverse and substantial literature. Part of this literature has examined the internationalisation of equipment vendors (see, for example, Pehrsson, 1996; Steinbock, 2003, 2010) or how technological change within the industry occurs (see, for example, Gao, 2014; Martinelli, 2012). The internationalisation of operators has also been examined (see, for example, Curwen, 2002; Curwen and Whalley, 2004, 2008, 2010, 2013; Eliassen and Sjøvaag, 1999; Sutherland, 2007; Whalley and Curwen, 2003). As part of this, the time at which operators enter a particular market has also been examined (see, for instance, Sung, 2014; Whalley and Curwen, 2012b), as has the organisational structures through which internationalisation is undertaken (see, for example, Luiz and Stephan, 2012; Sutherland, 2014; Whalley and Curwen, 2012b, 2013). How the structure of markets influences internationalisation and the success of operators has also been addressed (see, for example, Pogrebynakov and Maitland, 2011; Sarkar et al., 1999).

In recent years, a significant part of the literature has examined the diffusion of telecommunication services. While this has focused largely on the diffusion of mobile telecommunication services (see, for example, Bohlin et al., 2010; Bijwaard et al., 2008; Gupta and Jain, 2012; Islam and Meade, 2012), the diffusion of specific types of handsets has also been investigated (Lee and Lee, 2014). The influence of regulation on the diffusion of mobile telecommunications has been investigated, to see, for instance, the impact of calling or receiving party pays on penetration rates (Dewenter and Kruse, 2011) or how the lack of competition within a particular country shapes access to mobile services (Thakur, 2012). The literature has also explored the socio-economic opportunities that arise from the widespread diffusion of mobile telecommunications. Not only does the widespread availability of mobile handsets enable individuals to communicate with one another, but such devices also facilitate the development of services in sectors as diverse as banking and healthcare (Curwen and Whalley, 2011).

Although the internationalisation strategy of a company may encounter difficulties, it is generally seen as beneficial. Internationalisation has been integral to the development of competitive markets. Foreign direct investment in the form of new entrants that have competed against incumbent operators has brought about a varied array of socio-economic benefits. This foreign direct investment has enabled the internationalising operator to expand into new markets – it has not only enabled them to generate additional revenues and profits but also to defend their domestic markets. Moreover, internationalisation enables the company to diversify, reducing its risks as well as broadening the sources of its revenues and profits.

Although there are many benefits associated with internationalisation, a recent observation by Obiodu (2012) suggests that the financial benefits that accrue to the operator through internationalisation are not as clear-cut as previously thought. Obiodu (2012) suggests that widening the internationalisation of an operator does not necessarily lead to improved performance. While some aspects of an operator's financial performance appear to be correlated with size – such as net debt, number of employees and EBITDA (earnings before interest, tax, depreciation and amortisation) – other aspects, such as EBITDA margin and return on capital employed, do not. That further internationalisation does not necessarily result in greater financial performance is an interesting suggestion, not least because it challenges one of the major drivers of the telecommunications industry during the past two decades.

Obiodu (2012) is inevitably thin on detail, as is to be expected from a short note that highlights paid consultancy. What this document does, perhaps surprisingly, is to draw attention to the data used in a 'statistical analysis' of the financial results of 20 companies between 2000 and 2011 (Obiodu, 2012: 8). Although the 20 companies included in the analysis are drawn from a wide range of countries, a substantial number of them are based in Europe and quite a few engage in several lines of business while others are largely mobile or fixed-wire orientated. This diversity raises questions as to the appropriateness of the sample. Of greater concern, however, is the source of data, since it is not clear what this is. Consultants could purchase data from other consultants. Alternatively, they could collect the data for themselves from the annual reports of telecommunication companies. The former option is likely to be expensive and the latter time-consuming.

It is perhaps because of these methodological challenges that relatively little attention has been directed towards retreat of telecommunications companies from their overseas forays. Although there is some discussion of de-internationalisation – see, for example, Benito and Welch (1997), Nummela and Saarenketo (2012) and Turcan (2011) – and re-internationalisation (Freeman et al., 2011) in the wider international business literature, these issues have been largely overlooked by the telecommunications community. Having said this, Curwen and Whalley (2004: ch. 3) detail the exit of BT from many of its overseas markets and Curwen and Whalley (2008: ch. 8) outline the retreat of the US-based 'Baby Bells' back to their home market. While discussing the impact of private equity ownership on TDC, the Danish incumbent, Lemstra and Groenewegen (2012) illustrate how it divested itself of its international holdings.

1.3 RANKING THE COMPANIES

As a first step, it is useful to examine the picture presented by the FT500 as set out in Table 1.1. Size is defined here by reference to the market values of the companies listed, the reasons for which are explained in the preamble to the FT500 in the *Financial Times*. Coverage is restricted to companies with at least 15 per cent of their shares in the public domain. Companies are ranked on the final day of each of ten successive years concluding in 2012. As can be seen, once a company leaves the FT500 it is unusual for it to return and quite rare for it to do so after a gap of more than one year, although care must be taken to allow for the consequences of Merger and Acquisition (M&A) activity both in this respect and in terms of sudden and rapid changes in the rank of individual companies.

The first conclusion to be drawn from an examination of Table 1.1 is perhaps somewhat surprising, namely that the mobile sector apparently remained quite stable during the decade end 2003 to end 2012 inclusive. At the end of 2003, 37 companies are listed. This number reached a peak of 38 in 2004 and remained broadly constant until 2008 before dropping sharply to 31 in 2009, subsequent to which there was a minor recovery – but one that did not last through 2013, which witnessed a further net loss of three companies. However, this disguises a significant statistic, which is that whereas there were 25 companies in the top 250 in 2003, there were only 16 such companies in 2012. Furthermore, discounting the minor aberration of América Móvil in 2003, the top 12 companies had remained among the top 250 throughout the entire decade, so the slide down the top 250 rankings must be found elsewhere – France Télécom (now Orange) should also perhaps be added to the 12 as it was only in 2012 that its ranking fell below 250.

Generally speaking, even those numbered among the top 250 throughout the period ended it with a ranking below that held in 2003 – the fact that the only real exception is China Mobile hardly comes as a surprise. So the overall conclusion must be that, taking into account a period covering both good and bad years in terms of economic growth, either the mobile sector has declined in importance relative to other sectors or a significant number of individual companies in other sectors have grown rapidly whereas companies in the mobile sector have not done so.

Now this appears at first sight to be somewhat counter-intuitive, since, as we have previously noted, the mobile sector appears to play an ever-increasing role in modern society. However, a number of factors spring immediately to mind. For example, it may be that there has been a shift in economic power among different parts of the communication industry as a whole, and in particular one whereby service providers have essentially become fat pipes along which data are transported whereas the creators of those data and some of the equipment vendors have grown relatively prosperous. Furthermore, it may be that service providers have been subjected to a barrage of regulations affecting such matters as termination fees – what can be charged to terminate a call originated on another network – or what can be charged for international connectivity (roaming charges).

These matters will be addressed at a later point in the text. Meanwhile, it is necessary to examine Table 1.1 in a little more detail. Of the four new entries in 2012, only two were nationwide mobile incumbents. Whether the other two US-based companies will survive in the FT500 is therefore a moot point. It may be noted that only two nationwide incumbents in the USA made the FT500 in 2012 because Sprint Nextel had long been mired in restructuring problems and T-Mobile US was too small to qualify. This appears, oddly, to be less than is the case for tiny Hong Kong, but the mainland Chinese companies are headquartered there and Hutchison Whampoa is a massive conglomerate that has never relied on mobile communication for a large share of its overall valuation. A notable feature is that all three nationwide incumbents from Japan are present in the FT500.

Altogether, 21 companies survived within the FT500 during the entire period, with two others missing a single year and a further two disappearing only in 2012.[2] It is noteworthy that they are both based in Western Europe and one, Telecom Italia, was once a powerhouse in the mobile sector. It may be argued that European incumbents are mostly located in countries with relatively small GDPs, so one would not expect them to register among the world's largest companies, but the latter are almost all international in operational terms and Telenor and TeliaSonera are clearly not listed by virtue of their modest domestic operations but because of their numerous overseas subsidiaries. Hence the disappearance of Telecom Italia must reflect either a change in its overseas holdings or pressures in the domestic market that have caused a deterioration in its market value, in which case they will be also be occurring elsewhere in Europe – and, indeed, Telenor was (marginally) the only exception to this assertion in 2012.

It is possible to argue that market values by themselves give a distorted view of the health of the sector, but even allowing for the difficulties in comparing revenue streams exclusively generated by the provision of mobile services, and adjusting for a variety of financial year-ends, there will inevitably be a fairly close correlation between the movement of gross revenues and/or net profits because it is precisely to those kinds of financial data that markets respond in terms of the value placed on shares.

For data on a variety of financial and other matters rather than market values it is useful to consult the Global 100 – the latest version published in October 2013 is available at www.totaltele.com. This covers fixed-wire as well as mobile operators and the dates are not quite the same as for Table 1.1, but it is of interest to compare the rank order as follows.

Where gross revenues are concerned, after adjusting for the absence of DoCoMo and omitting Hutchison Whampoa because it is a conglomerate, there are eight entries among both top ten lists that coincide. The exceptions are Telstra, which ranks 17th for gross revenues, and Singapore Telecom, which ranks only 29th. In other words, these are operators that are valued surprisingly highly given their gross revenues. Conversely, France Télécom/Orange has a market value that seems unusually poor

Table 1.1 Companies providing mobile services in the FT500, 31 December 2003 to 31 December 2012

Company	Country	Rank 2012	Rank 2011	Rank 2010	Rank 2009	Rank 2008	Rank 2007	Rank 2006	Rank 2005	Rank 2004	Rank 2003
China Mobile (HK)	Hong Kong	6	10	11	11	4	4	19	44	64	61
AT&T[1]	USA	19	15	19	19	8	9	32	40	42	33
Verizon Communications	USA	40	41	51	56	34	56	46	47	25	29
Vodafone Group	UK	41	27	36	42	29	19	29	25	12	13
NTT DoCoMo	Japan	108	61	76	92	42	96	90	61	39	25
América Móvil (AMX)	Mexico	112	91	65	134	137	107	130	174	167	406
Telefónica	Spain	120	63	47	39	31	37	48	63	38	48
Telstra	Australia	125	153	234	177	136	226	239	173	106	94
NTT	Japan	131	75	101	90	50	93	85	55	49	42
Deutsche Telekom	Germany	150	122	122	87	63	77	81	70	35	44
Hutchison Whampoa[2]	Hong Kong	164	188	167	246	246	188	169	159	134	149
Singapore Telecom	Singapore	170	172	207	190	170	212	235	248	230	246
SoftBank	Japan	185	213	211	301	276	451	425	147	358	–
MTN Group	South Africa	193	206	202	244	236	290	386	430	–	–
China Unicom	Hong Kong	207	123	247	226	168	317	–	–	–	456

BCE	Canada	245	215	318	365	333	292	397	309	260	240
KDDI	Japan	266	256	332	325	148	298	266	266	256	198
Telenor	Norway	270	281	313	335	–	241	247	403	395	445
Vivendi Universal[3]	France	291	272	251	182	115	165	162	177	159	179
BT Group	UK	292	334	414	459	363	216	149	194	163	169
TeliaSonera[4]	Sweden	294	243	233	207	232	232	218	257	193	189
France Télécom	France	302	156	124	83	57	79	96	80	44	55
Chunghwa Telecom	Taiwan	352	291	355	385	281	–	497	419	303	361
CenturyLink	USA	367	–	–	–	–	–	–	–	–	–
Rogers Communications	Canada	383	376	467	410	353	333	453	–	–	–
Saudi Telecom	Saudi Arabia	391	452	388	329	192	205	167	62	105	136
Swisscom	Switzerland	404	396	393	386	316	455	411	356	208	221
Bharti Airtel	India	412	311	278	273	175	193	307	496	–	–
Telus	Canada	425	453	–	–	–	–	–	–	–	–
Crown Castle	USA	428	–	–	–	–	–	–	–	–	–
AIS	Thailand	449	–	–	–	–	–	–	–	–	–
Etisalat	UAE	464	397	382	351	337	316	424	232	288	417
Telekomunikasi Indonesia	Indonesia	476	–	–	377	437	462	–	–	–	–
Telecom Italia	Italy	–	391	367	263	174	151	132	106	77	105

Company	Country	Rank 2012	Rank 2011	Rank 2010	Rank 2009	Rank 2008	Rank 2007	Rank 2006	Rank 2005	Rank 2004	Rank 2003
KPN	Netherlands	–	461	389	262	203	296	300	317	240	253
China United	China	–	463	–	–	–	–	–	–	–	–
Zain Group	Kuwait	–	–	385	–	425	–	–	–	–	–
Belgacom	Belgium	–	–	–	–	430	–	–	–	400	–
SK Telecom	South Korea	–	–	–	–	411	–	–	488	399	375
Teléfonos de Mexico	Mexico	–	–	–	–	393	487	–	397	254	242
VimpelCom	Russia	–	–	–	–	242	469	–	–	–	–
Turkcell	Turkey	–	–	–	–	442	423	–	–	–	–
Reliance Communications	India	–	–	–	–	–	252	405	–	–	–
Sprint Nextel[5]	USA	–	–	–	–	–	270	136	77	156	336
Mobile TeleSystems	Russia	–	–	–	–	–	327	–	–	456	–
Telkom	South Africa	–	–	–	–	–	–	381	–	–	–
Alltel	USA	–	–	–	–	–	–	382	269	335	346
BellSouth	USA	–	–	–	–	–	–	–	122	100	76
Etihad Etisalat	Saudi Arabia	–	–	–	–	–	–	–	365	–	–
Telecom Italia Mobile	Italy	–	–	–	–	–	–	–	–	70	91
Nextel Communications	USA	–	–	–	–	–	–	–	–	166	158

O_2	UK	–	–	–	–	–	–	–	–	282	432
AT&T[1]	USA	–	–	–	–	–	–	–	–	410	317
Portugal Telecom	Portugal	–	–	–	–	–	–	–	–	401	404
AT&T Wireless[1]	USA	–	–	–	–	–	–	–	–	–	227

Notes:

1. AT&T divested AT&T Wireless, which was bought by Cingular Wireless in February 2004. AT&T was taken over by SBC Communications which renamed itself AT&T. AT&T then took over BellSouth in December 2006.
2. Telecoms activities form only a minority part of this conglomerate.
3. A media conglomerate.
4. Formed when Telia acquired Sonera at the end of 2002.
5. Sprint FON was divested from Sprint together with Sprint PCS. Sprint FON and Sprint PCS were recombined in April 2004. Sprint acquired Nextel in August 2005.

Source: Compiled by the authors from data published quarterly by the *Financial Times* as the FT500.

given its gross revenues (equivalent to those of América Móvil), but 2012 does seem to have been something of an aberration in respect of its market value. Other operators simply lag the total revenues of the eight most highly valued by such a margin that it would be most surprising if they had a higher rank in Table 1.1.

Looking below the respective top ten is also instructive. Thus we discover that the likes of MTN, TeliaSonera, Rogers and (especially) Chunghwa are seemingly valued too highly given their gross revenues, whereas the converse is true of the likes of Sprint, VimpelCom, KPN and Telecom Italia, where the latter two are respectively 23rd and 13th in terms of gross revenue but do not even make an appearance in the FT500.

It is possible that these discrepancies can be explained by consulting the Global 100 list in terms of net income. The first thing to note here, however, is the potential for huge annual movements that are relatively rare in the case of market values and gross revenues. During 2012/13, for example, VimpelCom rose by 44 places to 19th and OTE rose by 34 places to 47th, whereas Deutsche Telekom fell by 54 places to 92nd and Vodafone fell by 33 places to 35th. In the case of Deutsche Telekom, the reason was that it had acquired MetroPCS in the USA, which generated an accounting loss of nearly €5 billion on a one-off basis. After adjusting for these notable discrepancies, there appears to be a fairly close correlation between those operators with high rankings in terms of market value and in terms of net income.

The above suggests that, aside from financial data, there is one other factor to which the markets pay close attention that is relevant to the current discussion, namely M&A activity real and/or imagined.

1.4 MERGERS AND ACQUISITIONS

Historically, the structure of the mobile service provider sector was significantly influenced by factors such as privatisation. However, there were relatively few privatisations during the period 2008 to 2013 inclusive, most obviously because so many state-owned operators had already been sold off, and some recent attempts to privatise have been unsuccessful due to the reluctance of prospective private sector buyers to pay what they regard as inflated prices in a competitive market – for example, in respect of 49 per cent of Onatel in Burundi and 79 per cent of Kyrgyz Mobile.

In recent times there has been considerable discussion as to whether a ‘bubble’ is present similar to that which manifested itself during the late 1990s – see, for example, Curwen (2011). In practice, this discussion is concerned not with service operators but with other kinds of companies in the TMT sector such as Facebook. Nevertheless, it is worth reminding ourselves at this point just how high valuations rose among service providers during the period 1999 to 2001 inclusive before the entire TMT sector fell abruptly back to its pre-1999 valuations.

- In October 1999, MCI WorldCom made a takeover bid for Sprint worth $115 billion in cash plus $14 billion of debt. In June 2000, this was vetoed by both the European Commission and the US Department of Justice. BellSouth also made an unspecified approach estimated to have been worth roughly 5 per cent less than that of MCI WorldCom.
- Also in October 1999, Vodafone AirTouch made a takeover bid for Germany's Mannesmann worth $202 billion, the largest ever recorded.
- In May 2000, France Télécom made a takeover bid for Orange worth $46.4 billion.
- In July 2000, Deutsche Telekom made a takeover bid for Voice-Stream Wireless worth $50.7 billion in cash plus $4.2 billion of debt.
- In February 2004, Cingular Wireless and Vodafone both bid for AT&T Wireless, the former raising the latter's offer of $38 billion in cash by a further $3 billion.
- In December 2004, Sprint made a takeover bid for Nextel worth $35 billion in cash.

It is necessary to make some allowances for inflation when viewing these sums in terms of current valuations, which makes the first two entries all the more remarkable even if MCI WorldCom was itself to crash into oblivion at a later date. But the share price of Vodafone remained depressed for roughly a decade after its successful bid for Mannesmann, so the financial markets made their views clear on the (retrospectively) absurd valuations placed on anything connected with the TMT sector at that time.

All the above makes inspection of Table 1.2 especially interesting. As can be seen, there have been some very expensive takeover bids during the period since end 2004 – the term M&A is somewhat misleading because in this sector there are virtually no true mergers as such. However, it should first be noted that the table does not contain the deal completed in September 2013 whereby Verizon Communications agreed to pay Vodafone $130 billion in return for its 45 per cent stake in Verizon Wireless – not listed because Verizon Communications already owned a controlling 55 per cent stake and hence it was not a takeover bid or merger as such. Interestingly, an unofficial bid of $54 billion was allegedly made in May 2006 for the same stake.

Table 1.2 Illustrative major[1] takeovers by/of operators, January 2005 to mid-2014[2]

Bidder	Target	Date	Value $billion
Telecom Italia	Telecom Italia Mobile	January 2005	19.0[3]
Alltel	Western Wireless	January 2005	4.4 + 1.5 debt
SBC Communications	AT&T	January 2005	16.0
TeliaSonera	Turkcell	March 2005	3.1[4]
Alfa Telecom	Turkcell	March 2005	3.2[4]
Blackstone Group et al	Wind	March 2005	n/a[5]
Weather Investments	Wind	March 2005	n/a[5]
MTN	Celtel	March 2005	2.7
MTC	Celtel	March 2005	3.4[6]
Telefónica	Ceský Telecom	March 2005	3.6[7]
Swisscom	Ceský Telecom	March 2005	3.4[7]
Kohlberg Kravis Roberts	Auna	April 2005	13.6
Telekom Austria	MobilTel	June 2005	2.0
France Télécom	Amena	July 2005	12.1[8]
Telefónica	O_2	October 2005	31.3
Apax Partners et al	TDC	November 2005	11.5 + 3.5 debt
Sprint Nextel	Alamosa Holdings	November 2005	3.4 + 0.9 debt
Vodafone	Telsim	December 2005	4.6
Sprint Nextel	Nextel Partners	December 2005	6.5[9]
Temasek Holdings	Shin Corp.	January 2006	3.8[10]
Sonae.com	Portugal Telecom	February 2006	12.8
VimpelCom	Kyivstar	February 2006	5.0 + 0.5 debt
AT&T (SBC)	BellSouth	March 2006	67.0 + 17.0 debt
SoftBank	Vodafone Japan	March 2006	14.8 + 0.8 debt[11]
Cerberus/Providence	Vodafone Japan	March 2006	16.4 + 0.8 debt
Telmex	Verizon Latin America	April 2006	3.7
Babcock & Brown	eircom	April 2006	3.1 + 2.3 debt[12]
América Móvil	Verizon Dominicana	April 2006	2.0
China Mobile	Millicom International	April 2006	5.3
Investcom	Millicom International	April 2006	n/a[13]
MTN	Investcom	May 2006	5.5[14]
Verizon Communications	Vodafone USA	May 2006	48.0 + 6.0 debt[15]
Kohlberg Kravis Roberts	Vivendi Universal	October 2006	51.5[16]
Naguib Sawiris	Weather Investments	December 2006	2.5[17]
Etisalat	TIM Hellas	December 2006	4.9[18]
Turkcell	TIM Hellas	December 2006	4.6[18]
Vodafone	Hutchison Essar	December 2006	17.0[19]
Maxis/Texas Pacific	Hutchison Essar	December 2006	13.5[20]
Essar Group	Hutchison Essar	December 2006	11.0[20]
Swisscom	Swisscom Mobile	December 2006	3.5[21]
Weather Investments	TIM Hellas/Q Telecom	February 2007	4.4
Qatar Telecom	al Wataniya	March 2007	3.7[22]
Providence Equity Ptnrs	BTC	March 2007	2.4[23]
Telco consortium	Olimpia	April 2007	4.5
Binariang	Maxis Communications	May 2007	4.6[24]
Goldman Sachs/TPG	Alltel	May 2007	27.5
AT&T	Dobson Communications	July 2007	2.8 + 2.3 debt
OTPP et al	BCE	July 2007	32.6 + 15.9 debt[25]
MetroPCS	Leap Wireless	September 2007	5.5 + 2.0 debt[26]

Bidder	Target	Date	Value $billion
T-Mobile USA	SunCom Wireless	September 2007	1.6 + 0.8 debt
OTE	CosmOTE	November 2007	2.8[27]
Telemar	Brasil Telecom	March 2008	3.5[28]
France Télécom	TeliaSonera	June 2008	47.0
Verizon Wireless	Alltel	June 2008	5.9 + 22.2 debt
Vivendi Universal	Zain Africa	June 2009	10.5[29]
Consortium	Zain	September 2009	13.7[30]
Telefónica	T-Mobile UK	September 2009	5.6[31]
Vodafone	T-Mobile UK	September 2009	5.6[31]
Bharti Airtel	MTN	September 2009	14.0[32]
Eternad-e-Mobin	TCI (Iran)	September 2009	7.8[33]
Bharti	Zain Africa	March 2010	9.0 +1.7 debt[34]
CenturyTel	Qwest	April 2010	10.6 + 11.8 debt
Telefónica	Portugal Telecom/Vivo	May 2010	7.2[35]
Portugal Telecom	Oi	July 2010	4.8[36]
CVC Capital Partners	Sunrise	September 2010	3.3
Etisalat	'new Zain'	September 2010	11.7[37]
VimpelCom	Weather Investments	October 2010	6.5[38]
Çukurova Holding	'new Zain'	January 2011	7.9[39]
AT&T	T-Mobile USA	March 2011	39.0[40]
Vivendi Universal	Vodafone France	April 2011	11.4[41]
Level 3 Communications	Global Crossing	April 2011	1.9 + 1.1 debt
Zygmunt Solorz-Zak	Polkomtel	July 2011	5.4
Google	Motorola Mobility	August 2011	12.5
Apax Partners	Orange Switzerland	December 2011	2.1
Qatar Telecom	Wataniya Telecom	August 2012	2.2[42]
Dish Network	MetroPCS	August 2012	4.0
T-Mobile	MetroPCS	October 2012	[43]
SoftBank	eAccess	October 2012	2.3
SoftBank	Sprint Nextel	October 2012	20.1[44]
Sprint Nextel	Clearwire	December 2012	2.2[45]
Dish Network	Clearwire	January 2013	4.9
Vodafone	Kabel Deutschland	February 2013	9.3 + 3.6 debt[46]
VTB Group	Tele2 Russia	March 2013	2.4 + 1.2 debt
Dish Network	Sprint Nextel	April 2013	25.5[47]
AT&T	Telefónica	June 2013	95.3 + 69.5 debt[48]
Liberty Media	Kabel Deutschland	June 2013	9.9 + 3.6 debt[46]
AT&T	Leap Wireless	July 2013	1.2 + 2.8 debt
Telefónica	KPN Germany	July 2013	10.7[49]
América Móvil	KPN	August 2013	13.5[50]
Etisalat	Maroc Télécom	October 2013	5.7[51]
PPF	Telefónica CZ	October 2013	3.5[52]
PCCW-HKT	CSL	December 2013	2.3
Liberty Global	Ziggo	January 2014	6.7[53]
Vodafone	Ono	February 2014	4.8 + 4.7 debt[54]
Bouygues Télécom	SFR	March 2014	[55]
Altice Group	SFR	March 2014	[56]
Yahoo! Japan	eAccess	March 2014	3.2[57]

Notes:

1. Valued at over $2 billion including debt taken over. The figures quoted are mostly those that applied at the end of the first day of the offer and hence are often significantly different from the value of the offer when it was finally agreed let alone when the transaction was completed.
2. Not all bids have yet received regulatory approval.
3. Maximum assuming the purchase of two-thirds of the outstanding ordinary shares and all of the outstanding savings shares.
4. For a 27 per cent stake that would give TeliaSonera 64 per cent overall and control.
5. In May 2005, Weather was awarded 62.75 per cent of Wind for €3.1 billion in cash + 26 per cent of Weather, which would be given a 50 per cent + 1 share stake in Orascom Telecom valued at €1.96 billion. In January 2006, the rest of Wind was transferred for a further €328 million in cash and newly issued Weather shares. The total value of the transaction is given as €12.14 billion.
6. An initial $2.84 billion for an 85 per cent stake followed by a further $520 million for the other 15 per cent within two years.
7. Bid for a 51.1 per cent stake. A bid was also made by a consortium including France Télécom.
8. France Télécom won the bidding for Amena, initially buying 80 per cent for $4.1 billion in cash + $3.6 billion in shares + $2 billion debt taken on – $2 billion in tax credits. The fixed-wire assets were then sold to Ono for $2.2 billion.
9. The agreement reached in December was that Sprint Nextel would purchase the outstanding 68 per cent of the shares in Nextel Partners during the first half of 2006.
10. The initial $1.9 billion offer was for the 49.6 per cent stake held by the Thai PM's family, and in March a further 46.7 per cent was acquired via an open offer at the same price per share.
11. The name was subsequently changed to SoftBank Mobile.
12. This would leave Babcock & Brown with 65 per cent and the Employee Share Ownership Trust with 35 per cent. Debt taken on was later quoted as $4.7 billion.
13. Withdrawn after takeover bid by MTN. Other offers for Millicom have been lodged but no details are available.
14. With an irrevocable acceptance of 70.6 per cent.
15. Alleged. For the 45 per cent outstanding stake in Verizon Wireless. Rejected in principle.
16. The price was tabled but the bid was never made formal.
17. Paid to Enel for the outstanding 26.1 per cent of the company.
18. It was subsequently alleged that the highest bid was €3.2 billion. All were rejected by Apax Partners and Texas Pacific Group.
19. Vodafone was bidding for the whole of Hutchison Essar but by law could not end up with more than 52 per cent of the company. It eventually proved victorious with an offer of cash worth $11.1 billion + $2 billion of debt for a 67 per cent stake, but 15 per cent of this had to be held by Indian nationals.
20. The stake under offer elsewhere was 67 per cent of Hutchison Essar. Essar already held the other 33 per cent.
21. Bid for the 25 per cent outstanding stake.
22. Bid for a 51 per cent stake.
23. Bid for a 90 per cent stake.
24. Ananda Krishnan, via Binariang, offered to buy the outstanding 40 per cent of Maxis not owned by his company Usaha Tegas.
25. The technical buyer was Serefina Holdings, owned 71 per cent by BC Partners. The deal collapsed in December 2008.
26. The unsolicited bid was rejected as too low.
27. Bid for the outstanding 32.17 per cent of the shares by way of a share swap for OTE shares.

28. Bid for an initial 60.5 per cent of the voting shares. Altogether, this and related share purchases are expected to cost $7 billion.
29. Bid for a 65 per cent stake. The offer was rejected in July.
30. Bid for a 46 per cent stake which comprised a majority of the voting shares and hence control.
31. Both rejected.
32. Bid for a 49 per cent stake. Rejected.
33. Bid for a 50 per cent + 1 share stake.
34. The deal was technically closed in June 2010 with an upfront payment of $8.3 billion but the ownership of the Nigerian network remained unclear.
35. Bid for PT's 29.7 per cent stake thereby doubling the stake of the acquirer and yielding control. Rejected. Government votes its 'golden share' to veto after raised bid is accepted by majority of shareholders. European Commission declares this to be illegal. Agreement reached in July to sell for $9.8 billion in cash.
36. Bid for a 22.4 per cent stake.
37. Bid for a 46 per cent stake. Did not proceed.
38. A merging of the two companies' holdings with certain exceptions. Weather Investments' owner, Naguib Sawiris, to end up with a 51 per cent stake. Weather Investments was renamed Wind Telecom in January 2011.
39. Bid for a 29.9 per cent stake.
40. Of which $14 billion was initially set aside to provide T-Mobile with an 8 per cent stake in AT&T subject to a provision that the cash element could be enlarged so long as T-Mobile would end up with at least 5 per cent of AT&T.
41. Bid for the outstanding 44 per cent of SFR.
42. Bid for the outstanding 47.5 per cent of Wataniya.
43. After Sprint Nextel pulled out of negotiations to buy MetroPCS for roughly $8 billion including debt taken over, T-Mobile obtained agreement to a deal whereby it would merge MetroPCS with T-Mobile USA and hand over to MetroPCS shareholders 26 per cent of the merged company plus $1.5 billion in cash.
44. Bid for a 71 per cent stake. Successful after bid altered to $21.6 billion (with a larger cash element) for a 78 per cent stake.
45. Bid for the outstanding 51 per cent. Successful.
46. Vodafone responded to Liberty Media by raising its own cash offer to $10.1 billion. It was successful in October and by December had acquired 76.6 per cent of the shares.
47. Bid for a 78 per cent stake, causing SoftBank to raise its own offer to $21.6 billion.
48. A friendly takeover bid blocked by the Spanish government.
49. €5 billion plus a 17.6 per cent stake in Telefónica Deutschland – raised to 20.5 per cent in late August.
50. Value of whole of KPN at price offered for outstanding 70.1 per cent. Subsequently withdrawn.
51. Bid for a 53 per cent stake.
52. Bid for a 65.9 per cent stake which included 65.9 per cent of subsidiary in Slovakia. Successful.
53. For the outstanding 71.5 per cent plus debt taken over.
54. Rejected. Raised in March to $9.9 billion in total which was accepted subject to regulatory approval.
55. Comprising $14.5 billion in cash plus 46 per cent of the merged entity with Bouygues to control 49 per cent of the merged entity. To be followed by an IPO. This was revised upwards to provide $15.5 billion in cash plus 43 per cent of the merged entity, equivalent to $21.4 billion. Rejected. Revised to $18.1 billion in cash plus 21.5 per cent of the merged entity with Bouygues to control 67 per cent. Rejected. Revised to $20.7 billion in cash plus 10 per cent of the merged entity. Rejected. Revised to $21.1 billion in cash plus 5 per cent of the merged entity. Rejected.

56. Acting via subsidiary Numéricable and comprising \$15.3 billion in cash plus 32 per cent of the merged entity. The cash element was raised to \$16.2 billion. This was accepted subject to regulatory approval. Later revised to \$18.6 billion in cash plus 20 per cent of the merged entity. Successful.
57. SoftBank bought eAccess in October 2012 with completion in February 2013. It sold on 11 lots of voting rights, each of roughly 6 per cent, to reduce its own holding to one-third. In March 2014, Yahoo! Japan – in which SoftBank has 42.2 per cent of the voting rights and regards itself as the parent – bought 99.7 per cent of the equity and 33.3 per cent of the voting rights (after the absorption of PHS operator Willcom). The new company will launch as Y!Mobile in June 2014.

Table 1.2 reveals that during the nine-year period 2005 to 2013 inclusive only a few takeover bids (excluding the above) exceeded \$30 billion: Telefónica for O_2 in 2005; AT&T (SBC) for BellSouth and Kohlberg Kravis Roberts for Vivendi Universal in 2006; OTPP et al. for BCE in 2007; France Télécom for TeliaSonera in 2008; AT&T for T-Mobile USA in 2011; and AT&T for Telefónica in 2013. It is notable first that some of these bids were never formally tabled and, second, that only two were successful – Telefónica for O_2 and OTPP et al. for BCE in Canada. As can be seen, not all takeovers of service providers are initiated by other service providers as the likes of private equity and pension funds periodically acquire stakes. However, their main objective in so doing is to restructure a network in decline, restore its fortunes and sell it on at a profit.

A few other things stand out. First, there were no takeover bids of consequence between June 2008 (in the aftermath of France Télécom's unsuccessful bid for TeliaSonera) and June 2009, and two operators, T-Mobile UK and Zain, then dominated proceedings, although in the event the former remained independent. Hence a large European operator and a large European network evaded capture during this period. Subsequently, things went very quiet in Europe (with the exception of Vivendi buying out Vodafone's stake in SFR) until June 2013.

During the same period, the main focus of takeovers involving serious money switched to the USA. Given the small number of nationwide incumbents and the size of the market, this may seem somewhat surprising – and hence is investigated in detail in Chapter 9. Nevertheless, both T-Mobile USA and Sprint Nextel became takeover targets (not to mention smaller operators such as MetroPCS and Leap Wireless), with over \$100 billion being committed in principle. But US operators were not merely involved in taking over each other. In what can only be described as something of a volte face, signs appeared that the period of US isolationism by US operators – the modest stake in América Móvil held by AT&T is treated in this text as an investment because of the absence of any control over operational matters – was about to end. In

June 2013, AT&T made the largest bid recorded in the table, targeting a European operator with a large international portfolio. Not surprisingly, the Spanish government stepped in to squash any thoughts of proceeding to a formal offer – the Italian government had previously reigned supreme as the most xenophobic in Europe and even tried to wrest back the title by interfering with Telefónica's attempt to build up a bigger stake in Telecom Italia.

That the AT&T approach was friendly can be explained fairly easily by reference to the amount of debt to be inherited, and, as will become evident in Chapter 10, Telefónica has subsequently been forced to look for an alternative solution to this problem – for example, it agreed to sell its networks in Ireland (in June 2013) and the Czech Republic (in October 2013). KPN was the next European operator to solicit an approach – in part and as a whole – with the attentions of América Móvil setting off a further bout of xenophobia acting as a deterrent to further activity. However, rumours circulating during October 2013 indicated that a much bigger European prize might again be in the sights of a US invader since, stripped of its stake in Verizon Wireless and with a large part of the proceeds distributed to its shareholders, Vodafone now appeared to be vulnerable to a takeover by AT&T.

1.5 CONCLUSIONS

Restructuring goes in phases that reflect the general economic environment. During the late 1990s, there was a speculative bubble that resulted both in wholly unrealistic market valuations for TMT companies and a surge in M&A activity, which ground to a virtual halt in 2002 when recession struck. In 1999, there were 15 entries in the table in Curwen and Whalley (2008) corresponding to Table 1.2, and that number was not matched again until 2005 after the economic recovery had become established. The year 2006 witnessed the return of the mega-bid, but once again this lasted only as long as the world economy remained buoyant. However, as shown in Table 1.2, after rolling along at roughly the level of six per year for several years, there were 12 entries in 2013, so clearly something significant is taking place, although there is no suggestion this time around that it is a speculative bubble comparable to the late 1990s because, as noted above, most service providers are suffering from relatively modest market valuations.

As Chapter 2 makes clear, technological progress – in this case the move towards so-called 4G – provides part of the driving force underpinning restructuring, but there are reasons to believe that a further significant factor is the realisation among the larger service providers that their portfolios of holdings are in dire need of a dose of rationalisation. The point is made forcefully on a number of occasions in this book that you can only buy what is available at the time, so it is irrational to believe that an operator seeking to build up an international presence can select precisely where and when to do so. Given that roughly a decade has passed since it became clear that data rather than voice would be the driving force in mobile communications, it is not surprising that the time has come for a major dose of restructuring to take place as operators review the structures of their holdings and find them wanting.

In order to assist with the analysis of these and related issues, it is helpful to proceed by splitting up the world into a number of regions, each contained within a separate chapter, before proceeding to bring together the various strands into over-arching concluding chapters. However, it is first necessary to get to grips with technological progress, since this is a significant driver of internationalisation.

NOTES

1. See, for example, Sullivan and Bauerschmidt (1990) or Petersen and Pedersen (1997) for a discussion of Johanson and Vahlne (1977). For a broader discussion of psychic distance see Ellis (2008), O'Grady and Lane (1996) and Shenkar (2001).
2. A comparison with the data in Curwen and Whalley (2008) is not valid as they cover the entire TMT sector rather than simply service providers.

REFERENCES

Benito, G. and L. Welch (1997), 'De-internationalisation', *Management International Review*, **32**, 7–25.

Bijwaard, G., M. Janseen and E. Maasland (2008), 'Early mover advantages: An empirical analysis of European mobile phone markets', *Telecommunications Policy*, **32**(3/4), 246–61.

Bohlin, A., H. Gruber and P. Koutroumpis (2010), 'Diffusion of new technology generations in mobile communications', *Information, Economics & Policy*, **22**(1), 51–60.

Curwen, P. (2002), *The Future of Mobile Communications – Awaiting the Third Generation*, London: Palgrave.

Curwen, P. (2011), 'Bubble trouble: Is history repeating itself?', *info*, **13**(5), 53–71.

Curwen, P. and J. Whalley (2004), *Telecommunications Strategy: Cases, Theory and Applications*, London: Routledge.
Curwen, P. and J. Whalley (2008), *The Internationalisation of Mobile Telecommunications: Strategic Challenges in a Global Market*, Cheltenham, UK and Northampton, MA, USA: Edward Elgar.
Curwen, P. and J. Whalley (2010), *Mobile Telecommunications in a High-speed World*, Cheltenham, UK and Northampton, MA, USA: Edward Elgar.
Curwen, P. and J. Whalley (2011), 'The restructuring of African mobile telecommunications provisions and the prospects for economic development', *info*, **13**(2), 53–71.
Curwen, P. and J. Whalley (2013), 'Mapping worldwide mobile networks: Some problems and indicative solutions', *Telecommunications Policy*, **37**(11), 1150–65.
Dassler, T., D. Parker and D. Saal (2002), 'Economic performance in European telecommunications, 1978–1998: A comparative study', *European Business Review*, **14**(3), 194–209.
Dewenter, R. and J. Kruse (2011), 'Calling party pays or receiving party pays? The diffusion of mobile telephony with endogenous regulation', *Information Economics and Policy*, **23**(1), 107–17.
Dörrenbächer, C. (2000), 'Measuring corporate internationalisation – a review of measurement concepts and their use', *Intereconomics*, **35**(3), 119–26.
Eliassen, K. and M. Sjøvaag (1999), *European Telecommunications Liberalisation*, London: Routledge.
Ellis, P. (2008), 'Does psychic distance moderate the market entry sequence relationship?', *Journal of International Business Studies*, **39**(2), 351–69.
Freeman, S., C. Welch and L. Welch (2011), 'Expanding perspectives on re-internationalisation', AIB UK & Ireland Conference, April, Edinburgh, UK.
Gao, X. (2014), 'A latecomer's strategy to promote a technology standard: the case of Datang and TD-SCDMA', *Research Policy*, **43**(3), 597–607.
Gerpott, T. and N. Jakopin (2005), 'The degree of internationalization and the financial performance of European mobile network operators', *Telecommunications Policy*, **29**, 635–61.
Gupta, R. and K. Jain (2012), 'Diffusion of mobile telephony in India: An empirical study', *Technological Forecasting & Social Change*, **79**(4), 709–15.
Hassel, A., M. Höpner, A. Kurdelbusch, B. Rehder and R. Zugehör (2003), 'Two dimensions of the internationalisation of firms', *Journal of Management Studies*, **40**(3), 705–23.
Hofstede, G. (1980), *Culture's Consequences*: *International Differences in Work-related Values*, Beverly Hills, CA: Sage Publications.
Ietto-Gillies, G. (1998), 'Different conceptual frameworks for the assessment of the degree of internationalization: an empirical analysis of various indices for the top 100 transnational corporations', *Transnational Corporations*, **7**(1), 17–39.
Islam, T. and N. Meade (2012), 'The impact of competition, and economic globalisation on the multinational diffusion of 3G mobile phones', *Technological Forecasting & Social Change*, **79**(5), 843–50.

Johanson, J. and J.-E. Vahlne (1977), 'The internationalisation process of the firm – a model of knowledge development and increasing foreign commitments', *Journal of International Business Studies*, **8**(1), 23–32.

Johanson, J. and F. Wiedersheim-Paul (1975), 'The internationalisation of the firm – four Swedish cases', *Journal of Management Studies*, **12**(October), 305–22.

Kogut, B. and H. Singh (1988), 'The effect of national culture on the choice of entry mode', *Journal of International Business Studies*, **19**(Fall), 411–32.

Lee, S. and S. Lee (2014), 'Early diffusion of smartphones in OECD and BRICS countries: An examination of the effects of platform competition and indirect network effects', *Telematics and Informatics*, **31**(3), 345–55.

Lemstra, W. and J.P.M. Groenewegen (2012), 'Private equity leveraged buyout and public values: the bankruptcy of Eircom, Ireland', *TPRC 40*, ssrn.com.

Luiz, J.M. and H. Stephan (2012), 'The multinationalisation of South African telecommunication firms into Africa', *Telecommunications Policy,* **36**(8), 621–35.

Martinelli, A. (2012), 'An emerging paradigm or just another trajectory? Understanding the nature of technological changes using engineering heuristics in the telecommunications switching industry', *Research Policy*, **41**(2), 414–29.

Nummela, N. and S. Saarenketo (2012), 'Beautiful exit – How to withdraw after rapid internationalisation', AIB UK & Ireland Conference, March, Liverpool, UK.

Obiodu, E. (2012), *Examining the Case for Larger Multinational Telcos*, March, Ovum Consulting.

O'Grady, S. and H. Lane (1996), 'The psychic distance paradox', *Journal of International Business Studies*, **27**(2), 309–33.

Pehrsson, A. (1996), *International Strategies in Telecommunications – Models and Applications*, London: Routledge.

Perlmutter, H. (1969), 'The tortuous evolution of the multinational corporation', *Columbia Journal of World Business*, **IV**(1), 9–18.

Petersen, B. and T. Pedersen (1997), 'Twenty years after – support and critique of the Uppsala internationalisation model', in I. Bjorkman and M. Forsgren (eds), *The Nature of the International Firm. Nordic Contributions to International Business Research*, Copenhagen, Denmark: Handelshojskolens Forlag.

Pogrebynakov, N. and C. Maitland (2011), 'Institutional distance and the internationalisation process: The case of mobile operators', *Journal of International Management*, **17**(1), 68–82.

Ramaswamy, K., K. Kroeck and W. Renforth (1996), 'Measuring the degree of internationalization of a firm: A comment', *Journal of International Business*, **27**(1), 167–77.

Ronan, S. and O. Shenkar (1985), 'Clustering countries on attitudinal dimensions: a review and synthesis', *Academy of Management Review*, **10**(3), 435–54.

Sarkar, M., S. Cavusgil and P. Aulakh (1999), 'International expansion of telecommunication carriers: The influence of market structure, network characteristics and entry imperfections', *Journal of International Business Studies*, **30**(2), 361–82.

Shenkar, O. (2001), 'Cultural distance revisited: towards a more rigorous conceptualisation and measurement of cultural distances', *Journal of International Business Studies*, **32**(3), 519–35.

Steinbock, D. (2003), *Wireless Horizon*, New York: Amacom.

Steinbock, D. (2010), *Winning Across Global Markets*, San Francisco, CA, USA: Jossey-Bass.

Sullivan, D. (1994), 'Measuring the degree of internationalization of a firm: a reply', *Journal of International Business Studies*, **25**(2), 325–42.

Sullivan, D. and A. Bauerschmidt (1990), 'Incremental internationalisation: A test of Johanson and Vahlne's thesis', *Management International Review*, **30**(1), 19–30.

Sung, N. (2014), 'Declining first-mover advantage in mobile telecommunications markets', *The Services Industries Journal*, **34**(2), 89–102.

Sutherland, E. (2007), 'The scramble for Africa', Paper presented at the 35th Telecommunications Policy Research Conference, George Mason University, Arlington, Virginia, 26 September.

Sutherland, E. (2014), 'The MTN Group: Governance of an expanding mobile telecommunication group in Africa and Asia', *Achieving a New Balance*, Association of International Business UK & Ireland conference, 10–12 April, York Management School, University of York, UK.

Thakur, D. (2012), 'Market competition and the distributional consequences of mobile phones in Canada', *Technological Forecasting & Social Change*, **79**(2), 223–30.

Turcan, R. (2011), 'De-internationalisation: A conceptualization', Discussion paper No. 57, May, International Business Economics, Aalborg University, Denmark.

UNCTAD (2007), *World Investment Report 2007 – Transnational Corporations, Extractive Industries and Development*, New York: UNCTAD.

Whalley, J. and P. Curwen (2003), 'Licence acquisition strategy in the European mobile communications industry', *info*, **5**(6), 45–57.

Whalley, J. and P. Curwen (2012a), 'Third time lucky? An exploration of Hutchison Whampoa's involvement in the mobile telecommunications industry', *info*, **14**(2), 17–41.

Whalley, J. and P. Curwen (2012b), 'Incumbency and market share within European mobile telecommunication networks', *Telecommunications Policy*, **36**(3), 222–36.

Whalley, J. and P. Curwen (2013), 'Unravelling complex organisational structures among mobile operators', *info*, **15**(4), 3–22.

2. Mobile technology in the modern era

2.1 INTRODUCTION

One of the key factors underpinning the restructuring of the mobile telecommunications sector is undoubtedly technological change. Naturally, this is a factor in any industry to a greater or lesser degree, but what is unique to the mobile sector is the pace at which change is occurring. Curwen and Whalley (2008) contains a full chapter on the subject of technology, yet little more than one page is given over to a preliminary discussion of what is referred to as '4G'. Five years on, it is necessary in a similar chapter to give over little more than a page to anything other than 4G.

The use of the term 4G results from a tendency to explain technological change as proceeding via a series of 'generations' or 'part-generations'. This is more complicated than it appears at first sight because there is confusion in the public domain as to which technologies fit within which generational numbers. This is particularly true in the case of fourth generation technology (4G), which is widely reported in the media as consisting primarily of Long Term Evolution (LTE), although, in practice, as noted below, LTE should strictly be described as lying somewhere between 3.75G and 4G.

Part of the difficulty resides in the fact that a mobile technology can also be described in terms of the speed at which data are transferred, expressed in megabits per second (Mbps). Needless to say, there is a considerable overlap between the two approaches since it is possible to upgrade a technology used in an earlier generation to achieve speeds at least comparable with the lower range of those available via a subsequent generation.

Technology upgrades are achieved via improved software, hardware or both. An important point is that, whereas it is quite cheap and easy to upgrade a technology (largely via software) provided it remains within the same spectrum band, it is extremely expensive to introduce a new technology in a previously unused spectrum band because a new set of hardware is required. An intermediate step in terms of cost is to open up a different spectrum band for a technology already in use, the reason being that much less new hardware will be needed.

For the purposes of the discussion that follows, the sequence of technological developments based upon the 2G technology known as GSM (global system for mobile) will take pride of place. 2G is a digital technology designed primarily to carry voice, whereas 3G – known in this case either as W-CDMA (wide-band code division multiple access) or UMTS (universal mobile telecommunications system) – is designed to cope with the transmission of modest amounts of data. Retrospectively, it is difficult to appreciate that the number of 3G network launches reached double figures only in 2003 and triple figures in 2006, and that there are still a handful of countries that have yet to license the technology. W-CDMA has been superseded by HSPA (high-speed packet access), which has the advantage that it can be upgraded successively from a downlink of 7.2 Mbps either by doubling up the number of channels (dual-carrier) or through the use of MIMO (multiple input multiple output) antennas – see 3G.co.uk (2009), Wikipedia (2010a)[1] and 4G Americas (2012b). Adding MIMO to HSPA helps to convert it to HSPA+, which is capable of yet higher speeds – dual-carrier HSPA+ (DC-HSPA+), for example, operates with a downlink of 42 Mbps or better and with an uplink of at least 5.7 Mbps, and even these speeds can be doubled through the introduction of 64 QAM modulation.[2]

However, readers should be aware that both equipment vendors and network operators are inclined to hype their equipment and services – for example, data transfer speeds quoted by vendors are normally theoretical maxima available, in practice, only if you are a single customer isolated in a laboratory.

The other 3G technology that is in reasonably active use, especially in the USA, is known as cdma2000 1×EV-DO (evolution-data optimised). This was expected to develop through a sequence of upgrades to what became known as UMB (ultra mobile broadband), which it was claimed would operate with a maximum downlink of 288 Mbps in a 20 MHz (megahertz or million hertz) bandwidth channel. However, as a consequence of the widespread commitment by 1×EV-DO operators to move towards the adoption of LTE during 2008, UMB was effectively abandoned at the year-end. Like W-CDMA, 1×EV-DO uses FDD (frequency division duplex), which means that a channel is devoted exclusively to either a downlink or an uplink, and hence the two channels can be situated in widely separated spectrum bands. It is also necessary to refer here to TD-SCDMA – the 'TD' refers to TDD (time division duplex), which means that the signal travels in both directions over the same bandwidth. This was developed by China in an attempt to prove that it was not dependent on Western technology. TDD is also known as unpaired spectrum, while paired spectrum is used for FDD. With unpaired spectrum, the uplink and downlink can be asymmetrical, with the downlink using up to nine times more spectrum than the uplink.

The Third Generation Partnership Project (3GPP) has long played a very significant role in defining technical standards based upon GSM (Wikipedia, 2010c, 2010d), while its counterpart, the Third Generation Partnership Project 2 (3GPP2), serves the same purpose for systems based on cdma2000. 3GPP is usually identified via a series of Releases, which were denoted by dates until 2000 and numbered consecutively starting with Release 4 in 2001 (Wikipedia, 2010c). Each Release incorporates hundreds of individual standard documents that undergo a continuous state of revision. Release 7 in 2007 is primarily concerned with HSPA+ (3G Americas, 2007), while those starting with Release 8 are concerned with the route to 4G.

The great majority of media reporting treats 4G and LTE as synonymous even though this is not technically correct – for those with a deep interest in the technical aspects, see, for example, Sesia et al. (2011) or Dahlman et al. (2011). There is, accordingly, a difficulty in ascribing a generational number to LTE. However, it is an unavoidable fact of life that 4G and LTE are now synonymous in the eyes of the general public, and hence there is little choice but to persevere with the term in what follows on those few occasions where it is necessary to refer inclusively to LTE and equivalent technologies.

2.2 LTE USING PAIRED SPECTRUM

LTE in the Digital Dividend Bands

Certain spectrum bands are popularly referred to as 'digital dividend' because they are bands where spectrum can be released by switching from analogue to much more efficient digital signals. The analogue signals are predominantly used for broadcast TV, and it is the ambition of all advanced (and many emerging) countries to switch over to wholly digital broadcasting by the mid-decade. However, the process will not take place in one fell swoop, nor will there necessarily be consistent use of spectrum within a given region or even country because the spectrum involved spans a wide range of (possibly overlapping) frequencies. For example, the complete 700 MHz band is 698–806 MHz in the USA, but this is divided into several sub-bands and, in effect, the main incumbents have sought to annex different sub-bands, thereby creating difficulties in roaming across networks. Meanwhile, 800 MHz translates as the 790–862 MHz band in Africa, Europe and the Middle East. In contrast, the digital dividend band in most of Asia spans 470–960 MHz, although many individual countries (most recently Brunei, Indonesia, Malaysia and Singapore) have opted for the 700 MHz band, specified by the Asia-Pacific Telecommunity (APT) as 703–748 MHz paired with 758–803 MHz – see Global mobile Suppliers Association (2010) and 4G Americas (2011, 64–9).

As per usual, the European Union seeks to impose harmonised use of digital dividend spectrum across all member states. The European Commission, together with the Council and Parliament, agreed a text requiring national regulators to authorise the use of digital dividend spectrum by January 2013. This formed part of the Radio Spectrum Policy Programme, which set out the target of making at least 1200 MHz of spectrum available for high-speed broadband by 2015. Naturally, it is one thing to authorise the use of a band and another to clear it of existing users, a problem that is sensitive in this case because, first, there may be military usage and, second, it means switching off the analogue signal for TV and forcing consumers to purchase a new digital television, which they cannot necessarily afford.

As of end 2012, a sufficient number of 800 MHz spectrum auctions had taken place in Europe to throw up some interesting insights (Aetha, 2012). In particular, it is noteworthy that by far the most common outcome was for there to be three winning bids even though there were generally four 2G incumbents, and that in half the auctions the fourth-largest operator declined even to register a bid – which was probably a sensible strategy as it failed to win a licence on every occasion that a bid was registered, despite having a shortage of spectrum overall at the time. However, progress is not all it should be and the European Commission was obliged to allow nine member states to delay well beyond the original January 2013 deadline for auctioning spectrum in the band with a further five registering applications to that end (Hibberd, 2013).

Use of the 800 MHz band is more problematic in the case of Africa because there is a relatively wide provision of fixed wireless in the band and it is not intended that this should be replaced wholesale with fully mobile provision. Hence only the 790–806 MHz band is strictly available as a digital dividend. The situation in the Middle East is not much better.

This partly explains why attention has recently switched to the potential of the 700 MHz band, which spans 694–790 MHz in ITU Region 1 – Europe, the Middle East, Africa and Northern Asia. This band was expected to come up for discussion at the World Radiocommunication Conference (WRC) in 2015, but the preliminary groundwork has been brought forward in the hope that all regions will be able to agree on harmonised use of the band at WRC-15. At WRC-12, the decision was taken to co-allocate it by 2015 for mobile and broadcasting services, but it was not established whether it would be used for FDD or simply as a supplemental downlink.

The European Commission has been turning its attention to the 700 MHz band via the Radio Spectrum Policy Group (RSPG). The Digital Agenda Commissioner of the EU is very keen to harmonise the band across Europe, while giving all interested parties a fair allocation. Not surprisingly, European broadcasters using the band are less than enthusiastic about encroachment by mobile operators, but the significant overlap with the APT plan – the favoured plan is to combine a 703–733 MHz uplink with a 758–788 MHz downlink – is a major incentive with a view to achieving economies of scale in equipment provision. It may be noted that the USA plan for the band is being studiously avoided in order to prevent incumbents there from dictating equipment availability. However, given that many European countries are grappling with problems in developing the 800 MHz band, it is unlikely that they will be rushing to open up the 700 MHz band as well.

LTE in the 1800 MHz Band

The growing importance of the 1800 MHz band partly reflects delays in licensing bands set aside primarily for LTE and also delays in the availability of spectrum in the previously favoured digital dividend band. Finland was the first country to permit the use of the 1800 MHz band in 2009, since which time there have been numerous other launches – see Global mobile Suppliers Association (2012b).

Aside from the coverage issue, the 1800 MHz band has rapidly become favoured for LTE because it is used for 2G throughout Europe and in sufficient other parts of the world to foster international roaming. Furthermore, most operators have spare capacity in the band – they often rely mainly on the 900 MHz band for 2G.

In April 2011, the European Commission approved, by way of a Decision that extended the remit of the GSM Directive, technical rules on how the 900 MHz and 1800 MHz bands should be opened up for LTE and WiMAX (worldwide interoperability for microwave access). National regulators were given until 31 December 2011 to implement the Decision (Telecom.paper, 2011a). However, the 900 MHz band holds little appeal, given the paucity of spare capacity to re-farm and the proximity of the digital dividend band in Europe.

LTE in the 2.6 GHz Band

Finland was the first European country specifically to allocate new spectrum for 4G to incumbents in the GSM band without the need for new licences (Global Insight, 2009), although it also assigned licensed spectrum in the 2500–2690 MHz (2.6 GHz) band in November 2009. LTE was expected primarily to utilise existing GSM spectrum together with the 2.6 GHz band, although, in the USA, Verizon Wireless (and to a lesser extent AT&T) intended to use the 700 MHz spectrum won in the course of Auction 73 (Federal Communications Commission, 2007).

The dominant role currently being played by the 2.6 GHz band reflects the fact that it is being used throughout the European Union (EU) in accordance with ECC Decision (05)05 of 18 March 2005, 'On harmonized utilization of spectrum for IMT-2000/UMTS systems operating within the band 2500–2690 MHz'. The EU, unlike the USA, had also mandated common spectrum for all 3G launches, and its action in respect of 4G spectrum reinforces the point that, although one regional group cannot enforce common spectrum use upon other such groups, if a sufficient number of large countries are early adopters of a spectrum band then other countries are under pressure to follow suit because it facilitates international roaming.

2.3 LTE USING UNPAIRED SPECTRUM (TD-LTE)

In 2010, 3GPP was asked to define 2500–2690 MHz as a TDD band for LTE. However, the largest suitable contiguous bandwidth (Band 40 comprising 100 MHz) lies within the 2.3 GHz band, which can be divided efficiently into five 20 MHz channels. Whereas LTE supports channel bandwidths from 1.4 MHz to 20 MHz, a wide channel of between 10 MHz and 20 MHz is preferable because it allows for higher speeds. The fact that developments are largely taking place within Bands 38 and 40 is a significant advantage compared to FDD versions of LTE that are taking place within a multiplicity of bands – for a list of actual and proposed launches in all relevant bands see Global mobile Suppliers Association (2012d). According to the GSMA in April 2014, Band 40 had witnessed the most launches (16), while Bands 38 and 41 (2.6 GHz) had 16 between them, Band 39 (1900 MHz) had only one and Band 42 (3.5 GHz) had three.

Overall, there is some dispute about the prospects for TD-LTE, although increasing numbers of operators are indicating that they intend to proceed on a multi-mode basis – that is, FDD combined with TDD.

However, until such time as suitable smartphones reach the market, there will only be limited interest from consumers.

2.4 WIMAX

Another way forward for 4G is encapsulated in worldwide interoperability for microwave access (WiMAX) technology. The mobile variant of WiMAX in current use is known as 802.16e-2005, and was approved in December 2005 (Wikipedia, 2010b).

With a peak range of at least 50 kilometres – and hence suitable for metropolitan area networks (MANs) – and a peak shared downlink of 70 Mbps, 802.16e certainly appeared to offer an attractive prospect for areas currently either not covered at all, or poorly covered, by other types of network. However, this speed reflects the optimum channel size, power and best-case modulation and takes no account of the network overheads (Wikipedia, 2010d). In practice, as noted, no technology delivers maximum cited speeds, so the more important issue is whether WiMAX can be considered as a 4G technology. For its part, the WiMAX Forum has sanctioned licensing in the 2.3 GHz, 2.5 GHz and 3.5 GHz bands. Unfortunately, there is no agreement on a single-spectrum band to be used on a worldwide basis – the USA favours 2.5 GHz, Asia-Pacific countries favour 2.3 GHz and 2.5 GHz, and Europe and Latin America favour 3.4–3.6 GHz.

Whereas the WiMAX Forum was initially disposed to see TD-LTE as a rival to fight off, it is now resigned to the fact that the battle is already largely lost and hence that its best interests lie in integrating TD-LTE into the specifications for WiMAX2 – see Gabriel (2012d).

2.5 WI-FI

A brief mention of Wi-Fi is warranted at this point – for a fuller discussion see http.//en.wikipedia.org/wiki/Wi-Fi. In essence, the crucial point is that Wi-Fi, which has a limited range and operates over unlicensed spectrum and hence is extremely cheap to use compared to mobile connectivity using licensed spectrum, has long been a means for members of the public to secure an Internet connection without any payment (often in coffee shops and the like). Wi-Fi operates via a wireless network access point or 'hotspot'.

However, Wi-Fi can also be utilised by mobile operators. Normally, an operator pushes the bulk of its traffic along its own network while attempting to off-load its least desirable subscribers – those that provide very little revenue or who 'over-use' their access – onto Wi-Fi connections. But in recent times some MVNOs (mobile virtual network operators) have appeared, primarily in the USA, that send most of their data via Wi-Fi, thereby keeping down costs, and Wi-Fi is used by cable operators that provide mobile services via their own networks.

The problem for mobile operators is that data transfers are growing very rapidly and swamping their networks. Hence it increasingly makes sense for them to reserve as much capacity as they can for the services with high profit margins while off-loading basic services, which account for most of their business, onto Wi-Fi. However, Wi-Fi by its very nature can only provide a best-effort connection, so it is unclear for now how popular its use will become with mobile operators.

2.6 RE-FARMING SPECTRUM FOR LTE

It is worth noting that almost nothing was written into the original contracts for 2G licences about what would happen when they expired, presumably because it was assumed that they would simply be rolled over (as was explicitly made possible in many cases). Where spectrum rights are simply renewed, the term 'grandfathering' is used, but there is nothing in principle to prevent the regulator from re-assigning by administrative fiat or re-selling expiring licences (Vesterdorf, 2012). At the same time, the regulator may try to alter the competitive landscape by imposing spectrum caps, taking into account the way in which the spectrum has been used (or not) during the original licence period.

When detailed discussion about spectrum use for 4G became necessary, most regulators were inclined to assume that, as with 3G in Europe, new spectrum bands would be opened up either in lower bands such as 700 MHz or higher bands such as 2.6 GHz. In practice, however, it rapidly became clear that, because these bands would have to be cleared and subsequently sold, this could turn out to be a long-winded process. Hence operators exerted pressure for existing bands – and in particular the 1800 MHz band – to be re-farmed for LTE. For the most part this was uncontroversial, but it obviously created a difficulty in that many of the relevant licences were near, or at, their expiry date. Hence simply re-assigning them at the going rate to existing licence holders would provide them with a competitive advantage.

2.7 NETWORK SHARING

There is nothing new about this topic, which was regularly discussed in the context of 3G, but the financial problems that beset a significant number of 3G licensees when attempting to roll out their networks in isolation have meant that there has been far greater readiness to address – and indeed to resolve – this issue in the context of LTE.

There are obvious reasons to justify network sharing: it reduces costs for each partner, speeds up network coverage and reduces environmental side-effects. At the end of the day, if a common network has sufficient capacity to serve all the operators that wish to access it, it is seemingly a 'no-brainer' to build it. However, there is an inevitable fly in the ointment, namely how to maintain inter-operator competition.

To understand this, it is first necessary to distinguish between 'passive' and 'active' elements of a network. The passive elements are primarily the sites themselves, the masts and antennae erected on them and the power supplies and other equipment needed for their optimum functioning. The active elements include the radio access network (RAN), including base stations, radio network controllers and backhaul. It is immediately evident that sharing passive elements is much less of a concern than sharing active elements. At the end of the day, operators need to distinguish the services that they sell, and if the signal to the consumer arrives along an entirely common route, then competition may well end up in the form of cut-throat pricing, which is clearly not the object of the exercise since not all operators will survive.

It is also worth bearing in mind that MVNOs have to use the infrastructure of the network that hosts them. A competitive market must therefore provide the requisite spare capacity on such networks as exist and must not allow the network operator to charge fees that will make it impossible for MVNOs to compete.

2.8 'TRUE' 4G/IMT-ADVANCED

As indicated above, LTE has been described as 4G for marketing purposes, but in reality it falls short of the technical specifications in important respects (Wikipedia, 2010e). In practice, there are two main contenders for the right to be called 'true' 4G, namely LTE-Advanced (LTE-A) and WiMAX2. The term IMT-advanced (IMT-A) is often used as a synonym for 'true' 4G.

Both 3GPP and the ITU are keen to have a say in what is meant by 'true' 4G. 3GPP, which covers the development only of GSM-based technology, released the specifications for 'true' 4G within Release 10 & Beyond. In essence, for the 3GPP to sanction it, 'true' 4G must deliver a 100 Mbps downlink with high mobility and wide area coverage, a 1 Gbps downlink and a 500 Mbps uplink when stationary, low latency of under 10 milliseconds round-trip delay and use wide spectrum bands of up to 100 MHz (Global mobile Suppliers Association, 2012c; see also www.radio-electronics.com). Such a wide band can only be achieved via LTE carrier aggregation – that is, by combining carriers/channels in order to build up the bandwidth along which signals can be carried (Arora, 2013). Release 10 specified 'component carriers' of up to 20 MHz, so the best way to achieve a 100 MHz bandwidth is to aggregate five 20 MHz carriers (4G Americas, 2012a). The above requirements are not met by LTE (which meets the specifications of Release 8 and Release 9) but they are met by LTE-A.[3]

For its part, the ITU operates through its radiocommunications sector (ITU-R), which, in late October 2010, chose LTE-A and WiMAX2 as its official candidates for 'true' 4G (Gabriel, 2010c). Final ratification was pencilled in for the ITU-R Study Group 5 meeting in late November.[4] However, the list of potential qualifying technologies was not pre-judged, and in practice six were proposed for consideration in early 2011, including WirelessMAN-A and LTE-A using both FDD and TDD (Morris, 2010). WiMAX2 was the first to be ratified in early April (Gabriel, 2011). In January 2012, the agreed technical specifications for IMT-A were finally ratified by the ITU Radiocommunications Assembly (for details go to www.itu.int), and LTE-A and WirelessMAN-A were accepted as meeting those specifications (Telecom.paper, 2012).

The WiMAX2 Collaboration Initiative was set up in April 2010. It consisted of ten major companies involved in 802.16, including Intel, Samsung, Motorola, Alvarion and ZTE. Its purpose was essentially to support the move up from 802.16e to 802.16m, otherwise known as WiMAX2, via a collaboration with the WiMAX Forum, which announced that it expected WiMAX2 to be completed before the year-end with device certification coming in late 2011 (Gabriel, 2010a). 802.16m is capable, in theory, of a 300 Mbps downlink (Gabriel, 2010b) – a comparison with LTE-A is to be found in Dailywireless (2010). The term WirelessMAN-A is often used as a synonym for WiMAX2 and is favoured by the ITU.

As usual, Japan was at the forefront of developments. In late January 2011, the Ministry of Internal Affairs and Communications issued a licence to DoCoMo to begin pilot trials of LTE-A, the company having by this time demonstrated a 1 Gbps downlink and a 200 Mbps uplink in a laboratory setting (Global mobile Suppliers Association, 2012c). Predictably, South Korea was not far behind, with governmental agencies demonstrating a network capable of a 400 Mbps downlink in February 2011. Two Asian vendors, ZTE and Huawei, both claimed to have been the first to have implemented inter-band carrier aggregation combined with MIMO in February 2012, thereby permitting a theoretical maximum downstream speed of 270 Mbps, in the case of ZTE using 20 MHz channels in the 1800 MHz and 2.6 GHz bands and in the case of Huawei using the 800 MHz and 2.6 GHz bands (Gabriel, 2012a). In Europe, also predictably, Ericsson of Sweden was at the forefront of developments, demonstrating LTE-A using a 3×20 MHz aggregated carrier and 8×8 MIMO in June 2011 (Telecom.paper, 2011b).

Japan is also leading the way with WiMAX2, since in October 2013 WiMAX operator UQ Communications introduced TD-LTE-compatible WiMAX2+ services in central Tokyo (Telecom.paper, 2013c).

In March 2012, Huawei introduced 'Beyond LTE', its self-explanatory term for its latest batch of patents underpinning a prototype network capable of reaching a peak data download rate of 30 Gbps (Gabriel, 2012b). It followed up in May by announcing successful tests over the 2.6 GHz band LTE networks of Tele2 and Telenor in Sweden (acting jointly as NetMobility (TeleGeography, 2012a)) which had achieved a downlink of 290 Mbps. It also announced that it had conducted a field test, using an unspecified commercial European network, which for the first time involved a Category 4 device – one capable of a downlink of at least 150 Mbps as against 100 Mbps in the case of a Category 3 device. This incorporated the first HiSilicon LTE multi-mode chipset to support Release 9 (Gabriel, 2012c).

As well as considering what the future holds, it is perhaps useful to note where precisely things stood as of end 2013. The first Category 4 devices were launched towards the end of 2012. However, the first one of these, in the form of a handset, the Samsung Galaxy S4 LTE-A, was used in June 2013 when SK Telecom launched the first LTE-A service in central Seoul and 42 other cities, providing a maximum downlink of 150 Mbps, twice what was being provided previously. The technology combined carrier aggregation with coordinated multi point (CoMP) software, alternatively known as cooperative MIMO – see, for example, Irmer at al. (2011) – and is to be upgraded in 2014 with enhanced inter-cell interference coordination (eICIC) – see, for example, Kimura and Seki (2012). It should be noted that this is in essence a process of adding technological advancements as part of a rolling programme. The switch to LTE from HSPA was a step-change in technology, but the upgrade from LTE to LTE-A does not follow the same pattern. There is no obligation on an operator to adopt a particular pathway, but, as indicated, almost (if not) all operators are expected to use carrier aggregation as the first step (Gabriel, 2013b; TeleGeography, 2013).

The next development beyond LTE-A – covered by LTE Release 10 and Release 11, which largely performs the role of tweaking what is contained in Release 10 – is commonly referred to as LTE-B and encompasses what is to be contained in LTE Release 12 & Beyond, on which work has begun in 3GPP – see www.3gpp.org/Release-12. Release 12 is discussed in Ericsson (2013), which is in general too technical to warrant detailed analysis at this point. The Ericsson paper goes on to examine 'Small-cell and local area deployment', where it is noted that Release 12 seeks to achieve an even higher degree of inter-working between the macro and low-power layers (such as 'hotspots') including dual-layer connectivity. In essence, this means that a device connects simultaneously to both layers, with, for example, the downlink arriving via one layer while the uplink is sent via the other layer, or with control signalling being provided by the macro layer while connectivity is achieved by the low-power layer. This, in turn, raises issues about the role to be played by Wi-Fi, which currently provides a signal in certain locations provided by an array of (often retail) sources. However, a device will not necessarily pick up a stronger signal than can be provided by an LTE network when it locks on to a Wi-Fi signal.

In February 2013, the European Commission provided €50 million to fund research with a view to delivering so-called 5G technology by 2020 (Cellular-news, 2013a). Subsequently, Broadcom and SK Telecom claimed to have launched a 5G Wi-Fi router (Telecom.paper, 2013a), but there must be a suspicion that if the term '4G' has been open to abuse, this will be even more the case with '5G', particularly during the period before specifications are agreed by 3GPP and the ITU. One problem, as ever, is that various vendors are cracking on regardless. Of particular interest is Samsung's experimental system, announced in May 2013 (Gabriel, 2013a), which concentrates on high-frequency spectrum. This is normally used for access purposes – that is, backhaul or fixed-wire connectivity – but Samsung has managed to achieve a 1 Gbps transmission using the 28 GHz band using adaptive array transceiver technology. In essence, the technology uses 64 antenna elements to reduce the radio propagation loss at millimetre-wave bands, thereby improving both range and indoor penetration. However, this technology will not be commercially available until roughly 2020.

South Korea is unquestionably trying to lead the way with Samsung, LG and the three mobile incumbents all intent upon playing some part. The government has come up with a 'Creative 5G Mobile Strategy', which is focused on practical issues as against some of the theoretical matters of concern to the likes of Samsung. A downlink of 1 Gbps is the target for 2017 and to that end SK Telecom announced in late February 2014 that it had developed the world's first LTE-A tri-band carrier aggregation technology – enabling a maximum downlink of 300 Mbps – which it hoped to launch during 2015. The plan is to aggregate 10 MHz in the 800 MHz band with 20 MHz in the 1800 MHz band and 20 MHz in the 2.1 GHz band. It was given permission to re-farm the latter (used for 3G) earlier in the month. As an intermediate step, it would be expanding the 1800 MHz bandwidth from 10 MHz to 20 MHz in order to raise the downlink to 225 Mbps. The requisite equipment for commercial exploitation – that is, with in-built Category 6 capability (Abazovic, 2014) – is expected to become available around the end of 2014. Subsequently, by combining three 20 MHz carriers, a theoretical maximum of 450 Mbps will become possible. The bands selected for this by SK Telecom comprise the 800 MHz, 2.1 GHz and 2.6 GHz bands.

However, SK Telecom has competition, since, for its part, LG Uplus claimed in February 2014 to have combined 20 MHz in the 800 MHz (mainly for voice), 20 MHz in the 2.1 GHz band and 40 MHz in the 2.6 GHz band (for the bulk of data) via tri-band carrier aggregation – branded as LTE-A 3band – in order to provide an upgraded service in Seoul. The nationwide roll-out was pencilled in for July 2014 (TeleGeography, 2014).

However, even the speeds claimed in South Korea have been exceeded in a trial conducted by Telefónica and Huawei in February 2014, which brought together 70 MHz of both paired and unpaired bandwidth via carrier aggregation so as to achieve a downlink of 550 Mbps.

A further development is associated with Nokia Solutions and Networks (NSN), which in September 2013 demonstrated authorised shared access (ASA) via a trial on a live 2.3 GHz TD-LTE network. A commercial operator and a spectrum-owning local agency (the Finnish research consortium CORE+) were also involved. The concept involves spectrum sharing incorporating bands under the control of the likes of government agencies or TV networks without the need to compromise quality of service on any of the networks involved (Telecom.paper, 2013b).

2.9 LONG TERM HSPA+ EVOLUTION

At this point, it is worth reminding ourselves that it is possible, in principle, for a technology that fails to meet the requirements of 'true' 4G to operate at equivalent speeds. As noted, LTE-A was required to sustain a downlink of 100 Mbps, but during 2010 vendor Huawei claimed that it would be able to upgrade HSPA+ to provide a downlink of 300 Mbps. Not to be outdone, Nokia Siemens, acting in conjunction with operator T-Mobile USA, floated the possibility in December 2010 of a downlink of 672 Mbps, to be achieved by combining up to eight carriers and improving spectrum utilisation. This was described as long term HSPA+ evolution, and it was suggested that it could come on stream commercially in 2013 or 2014 in parallel with LTE-A. It also happened to provide T-Mobile USA with an excuse to describe its HSPA+ network as '4G' even though, at the time, the network was operating at nowhere near 'true' 4G speed.

2.10 VOLTE AND RICH COMMUNICATION SERVICES

Given that LTE is a data-only technology, among the outstanding technical issues to be fully resolved is how to run voice services over LTE (VoLTE) – for the initial discussion see, for example, Poikselkä et al. (2012), GSMA (2010) and Cellular-news (2011) – and how to improve the signal-to-noise ratio. 2G and 3G were designed primarily to carry voice rather than data, but 4G is the reverse and based upon Internet protocol standards. At present, most LTE networks where the operator also provides HSPA use the interim circuit switched fallback solution whereby LTE is used for data and the signal is switched to HSPA for voice – although for cdma2000 networks both radio interfaces must be used simultaneously, which is a heavy drain on handset batteries (Khan, 2012) – a problem that is being addressed with some success (Cellular-news, 2013b).

A big advantage of VoLTE is its ability to handle high definition voice (HD voice) and to provide operators with an in-built alternative to 'over-the-top' services such as Skype. HD voice is the term commonly used in Europe for a version of wideband audio. This has its origins in a decision by the ITU to standardise a version of wideband audio as G.722 in 1987. In effect this uses a speech compression algorithm to extend the range of audio signals from a narrowband range of 300 Hz to 3.4 kHz, far narrower than the potential range of the human voice, to a range of 50 Hz to 7 kHz. Nokia and VoiceAge developed a version of G.722 known either as adaptive multi-rate-wideband, adaptive multi rate wideband or advanced multirate wideband (AMR-WB or W-AMR) – the use of differing versions for an agreed set of initials or even differing initials is not uncommon in telecommunications – which was subsequently taken up by 3GPP (Wikipedia, 2012). VoLTE is defined in GSMA specification VoLTE IR.92, which is based on 3GPP standards.

HD voice is being introduced fairly rapidly. According to the GSA in September 2012, from an initial launch in 2009 by Orange in Moldova, the number of service launches had reached 45 in 35 countries, primarily in Europe (Global mobile Suppliers Association, 2012a). Orange is the leader in the field, with service launches in double figures, and it is trying to get other European operators to follow suit.

It must be borne in mind that HD voice can be introduced in conjunction with any mobile technology, not just LTE. The first commercial VoLTE launches were by SK Telecom and MetroPCS in early August 2012 and the initial launches and deployments were largely restricted to North America and South-east Asia (Global mobile Suppliers Association, 2012a). In December 2012, Sweden's Tele2 claimed that it was the first operator in Europe to test VoLTE and expressed the hope that it would be able to launch before the end of 2013 (TeleGeography, 2012b). VoLTE is expected to generate revenues in the billions within a few years (Cellular-news, 2012).

It is now commonplace to discuss VoLTE in conjunction with rich communication services (RCS), which are marketed by the GSM Association as 'Joyn'. The idea is to combine VoLTE and RCS in order to create a full rich messaging and multimedia communications service spanning 3G and 4G. The main features of RCS are an enhanced phonebook, enhanced messaging (including messaging history) and an enriched call (multimedia content sharing during a voice call).

The RCS initiative, originally called RC suite, began in 2007 and was taken under the auspices of the GSM Association in 2008. In 2011, RC services became the preferred title for a new specification known as RCS-e (RCS-enhanced) (Wikipedia, 2013). So far, there have been five releases for RCS, with Release 4 dealing with support for 4G.

Nevertheless, sceptical voices are increasingly being raised about the prospects for Joyn. The most common criticism is that the likes of WhatsApp, Skype and Facebook have established such a strong foothold in the market that Joyn is largely chasing shadows, especially given that the established brands will be evolving to fend off the potential competition posed by Joyn (Lennighan, 2013). Evidently, the initial results in Spain have been very discouraging.

2.11 LTE LAUNCHES

It is surprisingly difficult to define what is meant by a launch. Does it take place when a single base station in a city centre emits a signal that can be picked up by a few businessmen using dongles, or must the operation have achieved a larger coverage, and, if so, how large? Is it necessary to wait until the general public can access the network, and, if so, is it acceptable that they can only use dongles or must handsets be available? The latter situation is commonly referred to as a 'commercial' launch, but it is left to operators to decide when to use that term, so there are bound to be inconsistencies. Furthermore, launches using existing bandwidth will be more widespread initially than those using spectrum newly allocated for LTE. Clearly, no two databases will agree on all dates and what follows is based exclusively on the specific assumptions made by the authors in creating their own databases.

As shown in Tables 2.1 and 2.2, these databases indicate that there had been 206 launches by nationwide incumbents and 72 other launches of LTE networks, making a total of 278, by the end of 2013.[5] The launches have been allocated on a regional basis according to the breakdown set out in Table 5.1, although, as indicated below, the raw numbers are rather deceptive. The reason for the second table becomes clear when the column for North America is examined: in effect, a large number of regional and local operators is introducing LTE, and while a very few of these, such as Leap Wireless – now a subsidiary of AT&T – are large enough to bear comparison with nationwide operators elsewhere in the world, it does not seem sensible to lump together nationwide, regional, satellite-based and MVNOs as though comparing like with like.

Table 2.1 LTE network launches by region: nationwide incumbent terrestrial networks

	Western Europe	Eastern Europe/CIS	Middle East	Asia-Pacific	North America	Latin America	Africa
2009	2	0	0	0	0	0	0
2010	11	2	0	2	1	0	0
2011	11	2	5	5	4	1	0
2012	26	11	3	16	5	6	8
2013	29	3	8	16	4	17	8
2014[1]	3	1	2	5	2	5	3
Total	82	19	18	44	16	29	19

Note:

1. Before end April.

Table 2.2 LTE network launches by region: non-nationwide/non-incumbent/non-terrestrial networks

	Western Europe	Eastern Europe/CIS	Middle East	Asia-Pacific	North America	Latin America	Africa
2009	0	0	0	0	0	0	0
2010	0	0	0	0	0	0	0
2011	2	0	0	0	2	1	0
2012	4	2	0	4	17	0	1
2013	8	2	4	7	15	2	2
2014[1]	9	0	0	0	–1[2]	0	1
Total	23	4	4	11	33	3	4

Notes:

1. Before end April.
2. BendBroadband shut down its network. Leap Wireless was acquired by AT&T but is continuing to use its Cricket brand.

It is not particularly surprising that the number of nationwide incumbent providers of LTE in Western Europe is relatively high. These networks often have a common ownership, are constantly being modernised and exist in most cases in a highly competitive market, which, in effect, forces all other networks to follow suit once LTE has initially been introduced. Nevertheless, Western Europe is not leading the way in terms of subscribers. As of end 2011, these were only significant in the USA, Japan and South Korea, with the USA accounting for more than half of the total and, in turn, Verizon Wireless accounting for almost all of the US total. According to 4G-Portal, 38 million connections existed at the end of 2012, with that number rising to 100 million in May 2013. Of these, 57 million were recorded in North America.

According to the World Cellular Information Service (as reported in Mobile Communications International, June 2013: 5), the total as of March 2013 was 88.5 million (up from 17.8 million one year earlier), with 48.9 million attributed to USA/Canada and 35.7 million to Asia-Pacific. The 46.9 million attributed solely to the USA by the WCIS represented a penetration rate of 14.8 per cent, somewhat above the 11 per cent penetration achieved in Japan (equivalent to 13.8 million subscribers). However, by far the greatest penetration rate was measured in South Korea at 40.2 per cent (equivalent to 19.6 million subscribers). Hence these three countries accounted for 80.4 million of the total, or 90.8 per cent.

According to the WCIS, the top ten operators in terms of LTE subscribers in March 2013 comprised four in the USA – Verizon Wireless

(26.3 million), AT&T (10.5 million), Sprint Nextel (6.6 million) and MetroPCS (3.5 million) – three in Korea – SK Telecom (9.3 million), LG Uplus (5.2 million) and Korea Telecom (5.1 million) – two in Japan – NTT DoCoMo (11.6 million) and SoftBank (1.3 million) – and one in Australia – Telstra (2 million). The absence of European operators is worthy of note and such evidence as exists is not particularly favourable. Obviously, there are many possible explanations for different penetration rates in different countries, such as subscription and device prices and the transfer speed of existing networks as well as cultural differences, so it is still too early to draw any general conclusions. However, bearing in mind that Europe led the world in the development of 3G because it had a harmonised approach to the use of spectrum, it may be asked why it has evidently not been possible to repeat this feat in respect of LTE. The answer, unsurprisingly, rather depends on who is being consulted. According to the European Commission, there is a lack of coordination and harmonisation among national regulators (with the solution encompassing, inter alia, more centralised control over spectrum allocation), which are also setting auction prices at too high a level. All regulators are unpopular with operators who claim that the constant downward pressure on roaming fees and call termination rates has left them with insufficient resources to invest in the roll-out of LTE. Operators also note that recessionary conditions in Europe have hit subscriber numbers and impaired cash flow and profitability. Concern is also expressed that Wi-Fi and the likes of Skype have affected the prospects for LTE, although there is some contradictory evidence to the effect that the speeding up of mobile connectivity has caused customers to use their mobile devices in preference to searching out a provider of Wi-Fi.

Obviously, in accounting for the success of LTE in the USA, the existence of so few national networks compared to the whole of Europe, given the respective overall market sizes, is a relevant factor as is the tendency for US citizens to both talk and download data at a significantly greater level (but also at greater cost) than their European counterparts. US operators have responded by investing very heavily in LTE roll-outs, spurring each other on in the process, with subscribers switching readily to LTE data plans – a virtuous circle.

However, one should not assume that advances in mobile technology are a blessing for all concerned. The financial implications are, in practice, far-reaching and underpin much of the discussion in subsequent chapters. The issue is in essence straight forward. Improved technology means that networks can operate at higher data transfer speeds. Subsequently, improved devices permit users to download data much faster, which, in turn, means that they use their devices more intensively. And as users switch progressively to smartphones, a given total user base consumes progressively more data.

The numbers are staggering. According to Ericsson, there were 2.1 billion mobile broadband subscribers at the end of 2013 out of a total of some 6.7 billion mobile subscribers – in other words, the switch-over to smartphones still has a long way to go although it must be recognised that the number of independent users is much smaller than the number of subscriptions and some users with multiple devices may need only one smartphone. Overall, the number of subscriptions rose by 6 per cent in 2013 compared to an increase of 70 per cent in the amount of data traffic (Telecom.paper, 2014).

Obviously, upgrading a network to LTE and LTE-A enables it to cope with some of this additional load, but serves simultaneously to attract additional demand.[6] For network operators it is a constant, very expensive, position to maintain, and that has structural implications for the mobile sector.

NOTES

1. Some readers may be surprised at the use of Wikipedia as a reference source, given its reputation. However, this reputation has nothing to do with the quality of the technical writing, which is sound in respect of telecommunications. Furthermore, Wikipedia is highly accessible.
2. For details, see Wikipedia (2010e). The technical aspects involve the bandwidth of the carrier and the use of multiple input multiple output (MIMO) technology, among other factors.
3. A wide variety of technical matters relating to LTE is available at http://lteworld.org. Carrier aggregation can take three forms: contiguous intra-band, non-contiguous intra-band and inter-band. Release 10 contains two intra-band sets (2.1 GHz and 2.3 GHz) plus one inter-band set (2.1 GHz/800 MHz). Work on adding further sets is proceeding at 3GPP, but some are sceptical about the process, noting that vendors will struggle to satisfy many of the potential combinations and that small cells can provide an alternative means for satisfying consumers' expectations.
4. Information on a number of relevant themes can be found on the ITU website, specifically in this case at www.itu.int/ITU-R.
5. The tables in this chapter are based on independent research. There is also a database of LTE commitments and launches kept by the Global mobile Suppliers Association (GSA) at www.gsacom.com, but this differs slightly in terms of the methodology that it uses for counting from that used in this chapter and there are also some issues over what is meant by a 'commercial' launch. At the end of 2013, the GSA recorded 260 commercial launches in 95 countries with, in total, 456 firm commitments to launch across 135 countries plus a further 52 operators engaged in trials (Cellular-news, 2014). The comparable total of launches for April 2014 was 279.
6. The fact that supply creates a (sometimes more than) matching demand is a problem that is common to many other systems, such as hospital A&E departments. However, a smartphone user is not physically aware of the other users trying simultaneously to access such bandwidth as is available and hence is less likely to give up.

REFERENCES

Abazovic, F. (2014), '300Mbit update LTE Cat 6 is a software one', www.policytracker.com, 3 March.

Aetha (2012), *Spectrum value of 800MHz, 1800MHz and 2.6GHz: A DotEcon and Aetha Report for Ofcom*, July, London: DotEcon.

Arora, N. (2013), 'Carrier aggregation in LTE-Advanced – whitepaper', http://4g-portal.com, 19 March.

Cellular-news (2011), 'ZTE participates in first-ever MSF VoLTE interoperability test', www.cellular-news.com, 25 October.

Cellular-news (2012), 'VoLTE service revenues predicted to reach $2 billion by 2016', www.cellular-news.com, 25 June.

Cellular-news (2013a), 'EC announces EUR 50 mln funds for 5G mobile research', www.cellular-news.com, 27 February.

Cellular-news (2013b), 'Next generation VoLTE smartphones offer significant battery life improvement', www.cellular-news.com, 13 March.

Cellular-news (2014), '2013 ends with 260 LTE networks in service', www.cellular-news.com, 16 January.

Curwen, P. and J. Whalley (2008), *The Internationalisation of Mobile Telecommunications: Strategic Challenges in a Global Market*, Cheltenham, UK and Northampton, MA, USA: Edward Elgar.

Dahlman, E., S. Parkvall and J. Skold (2011), *4G: LTE/LTE Advanced for Mobile Broadband*, Oxford: Academic Press.

Daily Wireless (2010), 'WiMAZ2 Collaboration Initiative', www.dailywireless.org, 15 December.

Ericsson (2013), 'LTE Release 12 – Taking another Step toward the Networked Society. White Paper 284-23-3189 Uen, January', www.ericsson.com, 15 January.

Federal Communications Commission (2007), 'Auction of 700 MHz band licences FCC Public Notice DA 07-4514 of 2 November', www.fcc.gov.

Gabriel, C. (2010a), 'WiMAX giants step up defense against LTE', www.rethink-wireless.com, 13 April.

Gabriel, C. (2010b), 'Samsung demonstrates WiMAX2 at 330Mbps', www.rethink-wireless.com, 5 October.

Gabriel, C. (2010c), 'ITU picks WiMAX2 and LTE-Advanced for "true 4G"', www.rethink-wireless.com, 25 October.

Gabriel, C. (2011), 'First true 4G standard approved with 802.16m', www.rethink-wireless.com, 10 April.

Gabriel, C. (2012a), 'ZTE and Huawei vye for LTE-A firsts', www.rethink-wireless.com, 29 February.

Gabriel, C. (2012b), 'Huawei claims 30 Gbps "Beyond LTE"', www.rethink-wireless.com, 21 March.

Gabriel, C. (2012c), 'Huawei leads charge towards LTE-Advanced', www.rethink-wireless.com, 14 May.

Gabriel, C. (2012d), 'WiMAX opens arms to TD-LTE at last', www.rethink-wireless.com, 31 October.

Gabriel, C. (2013a), 'Samsung claims "5G" breakthrough', www.rethink-wireless.com, 13 May.
Gabriel, C. (2013b), 'SKT and Samsung team on "LTE-A" launch', www.rethink-wireless.com, 26 June.
Global Insight (2009), 'Finland allocates new 4G frequencies to TeliaSonera, Elisa, DNA', https://communicationsdirectnews.com, 25 April.
Global mobile Suppliers Association (2010), 'Digital dividend update', www.gsacom.com, 1 April.
Global mobile Suppliers Association (2012a), 'Mobile HD Voice: Global update report', www.gsacom.com, 23 September.
Global mobile Suppliers Association (2012b), 'Status of the Global LTE 1800 market', www.gsacom.com, 14 November.
Global mobile Suppliers Association (2012c), 'Evolution to LTE Report', www.gsacom.com, 21 November.
Global mobile Suppliers Association (2012d), 'Status of the Global LTE TDD market', www.gsacom.com, 22 November.
GSMA (2010), 'GSMA VoLTE initiative', www.gsma.com, 22 December.
Hibberd, M. (2013), 'EC grants 800MHz delays to nine Member States', www.telecoms.com, 23 July.
Irmer, R., H. Droste, P. Marsch et al. (2011), 'Coordinated multipoint: concepts, performance and field trial results', *IEEE Communications Magazine*, February, 102–11.
Khan, A. (2012), 'VoLTE subscriptions to hit 2 million by 2013', www.cellular-news.com, 16 May.
Kimura, D. and H. Seki (2012), 'Inter-cell interference coordination technology', *Fujitsu Scientific and Technical Journal*, **48**(1), 89–94.
Lennighan, M. (2013), 'It's "too late" for Joyn, mobile operators say', www.telecoms.com, 12 August.
Morris, A. (2010), 'True 4G specs to be released in March 2011', www.totaltele.com, 18 February.
Poikselkä, M. et al. (2012), *Voice over LTE (VoLTE)*, Chichester, UK: John Wiley & Sons.
Sesia, S., I. Toufik and M. Baker (2011), *LTE – The UMTS Long Term Evolution: From Theory to Practice*, Chichester, UK: John Wiley & Sons.
Telecom.paper (2011a), 'EU opens up GSM bands for 4G services', www.telecompaper.com, 20 April.
Telecom.paper (2011b), 'Ericsson demonstrates LTE Advanced in Sweden', www.telecompaper.com, 29 June.
Telecom.paper (2012), 'IMT-Advanced standards agreed at ITU Assembly', www.telecompaper.com, 19 January.
Telecom.paper (2013a), 'Broadcom, SK Telecom launch 5G Wi-Fi router', www.telecompaper.com, 8 April.
Telecom.paper (2013b), 'NSN demonstrates spectrum sharing trial of ASA on TD-LTE', www.telecompaper.com, 26 September.
Telecom.paper (2013c), 'UQ launches TDD-LTE compatible Wimax2+ service in Tokyo', www.telecompaper.com, 1 November.

Telecom.paper (2014), 'Mobile data traffic up 70% annually in Q4 – Ericsson', www.telecompaper.com, 18 February.
TeleGeography (2012a), 'LTE Advanced tests reported by Tele2, Telenor', www.telegeography.com, 14 May.
TeleGeography (2012b), 'Tele2 "first in Europe" to test VoLTE', www.telegeography.com, 21 December.
TeleGeography (2013), 'SK Telecom inaugurates commercial LTE-A network', www.telegeography.com, 26 June.
TeleGeography (2014), 'LG Uplus implements tri-band multicarrier system', www.telegeography.com, 11 February.
Vesterdorf, B. (2012), 'Radio spectrum licence renewals', *The Policy Paper Series*, **14**(May), 16–24.
Wikipedia (2010a), 'Evolved HSPA', http://en.wikipedia.org/wiki/HSPA%2B.
Wikipedia (2010b), 'WiMAX', http://en.wikipedia.org/wiki/Wimax.
Wikipedia (2010c), '3GPP', http://en.wikipedia.org/wiki/3GPP.
Wikipedia (2010d), '3GPP Long Term Evolution', http://en.wikipedia.org/wiki/3GPP_Long_Term_Evolution.
Wikipedia (2010e), '4G', http://en.wikipedia.org/wiki/4G.
Wikipedia (2012), 'Wideband audio', http://en.wikipedia.org/wiki/Wideband_audio.
Wikipedia (2013), 'Rich Communication Services', http://en.wikipedia.org/wiki/Rich_Communication_Services.
3G Americas (2007), 'UMTS Evolution from 3GPP Release 7 to Release 8 HSPA and SAE/LTE', www.3gamericas.com, 17 December.
3G.co.uk (2009), 'HSPA+ solution to double 3G download speeds', www.3g.co.uk, 23 September.
4G Americas (2011), 'The Benefits of Using LTE in Digital Dividend Spectrum', www.4gamericas.com, 20 October.
4G Americas (2012a), 'HSPA+LTE Carrier Aggregation', www.4gamericas.com, 25 June.
4G Americas (2012b), 'MIMO and Smart Antennas for Mobile Broadband Systems', www.4gamericas.com, 27 October.

3. Anatomy of an international operator: Vodafone Group

3.1 INTRODUCTION

A mobile operator begins life within a single country, sometimes as the subsidiary of the fixed-wire incumbent. It is expected to begin the process of internationalisation by acquiring stakes in operators within the same region before branching out into new regions where, it must be said, it will have cultural issues to contend with that may in the course of time prove insuperable. But it cannot really pick and choose. It must acquire either stakes or licences, so its options are to grab whatever is available at the time, wait patiently for the right ones to become available or engage in M&A activity on a friendly or hostile basis.

To build up a decent-sized portfolio of international holdings is accordingly a process that takes a long time and is likely to be fraught with setbacks. And at the end of the initial process an internationalising operator will typically find either that there is little or no synergy between its holdings, that it has little or no control over some of its holdings or that some of its holdings are inherently unprofitable. This explains why even the most voracious of acquirers seeks periodically to off-load some of its holdings, which may, for reasons of synergy, prove to be more attractive to another operator.

In reality, given that there are close to 1000 networks worldwide, one might expect that restructuring takes place on a significant scale even on an annual basis, but there is only a limited number of international operators – no more than 30 – of any real size and they are not free to behave without constraints. Furthermore, it is very unusual for a transfer of mobile licences or assets to take place without the intervention of (possibly multiple) regulatory authorities and/or governments, which in itself makes the process fairly long-winded.

The purpose of this chapter is to give readers a flavour of the processes involved in building up and fine-tuning an international portfolio of holdings. The Vodafone Group has been chosen as the primary exemplar for a variety of reasons, including its overall size, the scope of its portfolio and the fact that it controls virtually all its holdings. But it has also been chosen because its internal structure is extremely complicated and this has a bearing on certain other issues of current concern such as the use of tax avoidance measures via the selective location of holding companies.

The geographical scope of Vodafone Group, which until very recently saw itself essentially as a mobile operator even though it had acquired some limited fixed-wire assets, reflects the opportunities that have emerged over the years and the setbacks that have been encountered, and it shapes how the company is structured. New companies are created as fresh markets are entered, or to effect M&A activity. These factors are reflected to a greater or lesser extent in the organisational complexity of all leading mobile service providers that operate in multiple countries.

3.2 LITERATURE REVIEW

Over the years, various aspects of the structure and scope of MNEs have been examined (Ietto-Gillies, 2001; Rugman, 2005). However, when it comes to mobile telecommunications, this focus on the structure and scope of individual companies is largely missing. Much of the research examining, for example, the diffusion of mobile technologies has adopted an industry-wide perspective. Gruber and Verboven (2001) and Li and Lyons (2011) typify the macro-level approach that has been adopted in many cases to identify the factors driving the adoption of a particular wireless technology. In contrast, others have highlighted individual operators while examining the structure of national or regional telecommunication markets – see, for example, Haucap et al. (2010) and Sutherland (2007).

This is not to suggest that the structure of individual operators has been entirely overlooked in the literature. Gerpott and Jakopin (2005) illustrate the variety of entry methods that mobile operators may adopt when internationalising their operations, but this is largely of secondary interest compared to their primary focus on the relationship between internationalisation and financial performance. Through an analysis of why the proposed merger between Telia and Telenor failed, Fang et al. (2004) shed some light on the structure of the proposed post-merger structure.

Within the context of corruption in telecommunications, Sutherland (2012) alludes to the organisational complexity that exists within the industry. Such complexity is more clearly illustrated by Curwen and Whalley (2009) and Whalley and Curwen (2012). By examining the possible merger combinations of European incumbent mobile operators,

Curwen and Whalley (2009) highlight the complex organisational structures that have been adopted as companies internationalise and markets are liberalised. Organisational complexity is also evident in how Hutchison Whampoa participates in the telecommunications industry (Whalley and Curwen, 2012). Both highlight the use of holding companies. There are, of course, examples of holding companies throughout the history of the telecommunications industry (Rosenberg et al., 1993), but what could be viewed as surprising is their widespread use and the tendency of some parent companies to establish them in specific locations.

3.3 CREATING THE DATABASE

The data for this chapter have been drawn from a variety of sources. Telecommunication companies publish data – on their websites and, where applicable, in their annual reports – that identify their operational scope and performance. Both these sources provide a useful starting point when collecting data on the number of countries in which a mobile operator is actively present, as well as on how each of these units is performing. These sources are complemented by the publications of regulatory authorities, trade journals such as *Global Mobile* and Internet-based newsletters such as TeleGeography. Particularly useful is the coverage of any M&A activity in which a telco is engaged. Not only does a takeover generate widespread coverage, but it is also often accompanied by discussion of the strategy of all of the parties involved. As part of this, the vehicle through which the acquisition is being made is often noted. The creation and use of holding companies and special purpose investment vehicles sheds light on the structure of the mobile operator.

Telcos listed on a US-based stock exchange are required to file regular reports with the Securities & Exchange Commission (SEC). US-listed companies are required to file a range of documents relating to the company's quarterly and annual reports as well as to any event that has a material effect on the operation of the company. Such an event could consist of the sale or purchase of equity by management, M&A activity or a change in the risks that affect the company.

In mid-2011, Action Aid published details of the subsidiaries of the FTSE100 constituent members, forcing those that had not previously made the relevant filings with Companies House in the UK to do so (Action Aid, 2011). Action Aid revealed the country of incorporation of subsidiaries as well as details of the equity stake held by the parent company. It is worth noting, however, that the data are incomplete since the equity stake owned – which is not necessarily the same as the proportion of voting shares held – by the parent company is not stated in some cases. Moreover, Action Aid does not identify whether the subsidiary is an operating company or show how the various subsidiaries relate to one another. Some of the entries identified by Action Aid also relate to subsidiaries that have subsequently been sold by Vodafone Group. To ensure that the subsidiary list of Vodafone Group did not contain references to companies that have been sold, it was compared against the company's operational footprint as determined by Curwen and Whalley (2010) and various press releases, and amended accordingly.

After the Action Aid data had emerged, a website dedicated to providing corporate data on an open-access basis was launched. Open Corporates collates data on companies from various sources and presents them on a single website.[1] In contrast to Action Aid, Open Corporates indicates which of the listed companies are 'active' and which are not. Related companies can also be grouped together, thereby suggesting a tentative structure for Vodafone Group.

No single source sheds full light on the structure of Vodafone Group. As a consequence, cross-checking – triangulation – between the various sources was undertaken to improve the accuracy of the research. Thus comments on the structure that are not supported by multiple sources are treated with caution.

3.4 THE STRUCTURE OF VODAFONE

Since its foundation as a subsidiary of Racal Electronics at the start of the 1980s, Vodafone Group has grown into a highly internationalised operator of mobile networks. Organic growth, the friendly merger with California-based AirTouch Communications and the hostile acquisition of Mannesmann in Germany have all contributed to its transformation. At the start of the company's annual report for the financial year ending 31 March 2013 an operational overview of Vodafone Group is provided that highlights the countries where it is present and the subscribers in each of its main markets. As part of detailing its global footprint, the annual report identifies the following four different categories of interests:

- *Northern and Central Europe*: Czech Republic, Germany, Hungary, Ireland, Netherlands, Romania, Turkey and the UK.
- *Southern Europe*: Albania, Greece, Italy, Malta, Portugal and Spain.
- *Africa, Middle East and Asia-Pacific*: Australia, Egypt, Fiji, Ghana, India, Safaricom, New Zealand, Qatar and Vodacom.
- *Non-controlled interests and common functions*: Verizon Wireless.[2]

However, this should not be taken as implying that the underlying organisational structure is the same. Drawing on the various annual reports of Vodafone Group, it is possible to identify a series of key companies within the Group. The most recent annual report identifies 35 such companies, and while the majority are described as 'network operators' – 23 in all – an additional seven are referred to as 'holding companies' (Vodafone Group, 2013a). The remaining five companies provide some form of group services. So far this is more or less the same as the previous year (Vodafone Group, 2012b). The major difference, however, is the inclusion of Cable & Wireless Worldwide, a fixed-wire operator, in the list of principal subsidiaries (Vodafone Group, 2013a).

While the various companies are presented as a list, the accompanying footnotes allow for the construction of an initial organisational chart. Vodafone Group owns 65 per cent of the South African-based Vodacom Group, which, in turn, owns stakes in five mobile network operators and a holding company that provides business-related services across the continent. The footnotes also shed some light on the organisational means through which Vodafone Group is present in India. Having said this, the relationship between the other network operators and the various holding companies is not elaborated, which begs the question as to how the rest of the company is organised.

Assuming that Vodafone Group is structured as a series of cascading holding companies, it is clear that two holding companies in particular play a key role. Although Vodafone International Holdings ('VIH' hereafter) is based in the Netherlands, it is possible to determine through detailed documentary analysis that it has stakes in mobile operators in Eastern Europe, the Middle East, Africa and Asia-Pacific. These are:

- 20.1 per cent of Vodafone Egypt Telecommunications (Wireless Federation, 2011)
- 49 per cent of Vodafone Fiji (Tabureguci, 2007)
- 50 per cent of Vodafone Hutchison Australia (Australian Competition & Consumer Commission, 2009)
- 64.4 per cent of Vodafone Essar (Vodafone Group, 2013a)
- 70 per cent of Ghana Telecommunications (Vodafone Ghana, 2011)
- 100 per cent of Vodafone Kenya, which, in turn, owns 40 per cent of Safaricom (Vodafone Group, 2009)
- 100 per cent of Vodafone Magyarorszag Mobile (InsideView, 2011)
- 100 per cent of Vodafone Czech Republic (European Commission, 2005; Vodafone Group, 2005b)
- 100 per cent of Vodafone Romania (Vodafone Group, 2005b).

In addition to the above, Vodafone International Holdings owns 100 per cent of two other companies that between them control Vodacom Group Limited (Vodafone Group, 2010). Although various documents from Vodafone Group, such as Vodafone Group (2006a), detail the purchase of Telsim of Turkey by the Group, they do not mention how the transaction was structured.

The second key holding company is Vodafone Europe ('VEH' hereafter),[3] a company with an equally wide geographical footprint comprising stakes in mobile operators in the Middle East, Asia-Pacific and Europe. These are:

- 23 per cent of Vodafone Qatar (Vodafone Group, 2008b)
- 34.8 per cent of Vodafone Egypt (Wireless Federation, 2011)
- 50 per cent of Vodafone Albania (Vodafone Albania, 2011)
- 61.4 per cent of Vodafone Portugal (Vodafone Group, 2011c). The remainder is directly held by Vodafone Group
- 76.9 per cent of Vodafone Omnitel (Vodafone Group, 2013d)
- 99.9 per cent of Vodafone Panafon Hellenic Telecommunications (Alacra Store, 2011). In turn, this company owns 100 per cent of Vodafone Panafon International, which, in turn, owns 50 per cent of Vodafone Albania (Vodafone Albania, 2011)
- 100 per cent of Vodafone New Zealand Limited (Vodafone New Zealand, 2011)
- 100 per cent of Vodafone Malta (Malta Communications Authority, 2001)
- 100 per cent of Vodafone España (European Commission, 2001)[4]
- 100 per cent of Vodafone Libertel (Vodafone Group, 1998)
- 100 per cent of Cable & Wireless Worldwide (Vodafone Europe, 2012).

There is a third identifiable holding company, albeit with a smaller geographical footprint than the other two, namely Vodafone Investments Luxembourg ('VIL' hereafter). One of the companies owned by this holding company is Vodafone Ireland (Mulligan, 2010). VIL also owns Vodafone Holding (Goodall, 2010), which, in turn, owns Vodafone D2 in Germany. A few weeks after it emerged that talks were being held with Kabel Deutschland in Germany, Vodafone launched a formal €7.7 billion bid to acquire the cable operator (Cellular-news, 2013c; Vodafone Group, 2013d). The documentation that accompanied the bid is informative as it reveals that, first, the bid would be made through a vehicle jointly owned by Vodafone Group and Vodafone Holding (Vodafone Group, 2013c). Second, once the purchase had been made, Vodafone Group would transfer its stake in the bidding vehicle to Vodafone Holding. As the bid was successful (Cellular-news, 2013c), we can conclude that Vodafone Holding owns Kabel Deutschland. The bidding document also revealed the series of ten holding companies, all directly or indirectly wholly owned by Vodafone Group, which link the parent company (Vodafone Group) with Vodafone Holding in Germany (Vodafone Group, 2013b). Interestingly, nine of these holding companies are located in the UK, with only one, VIL, registered overseas (in Luxembourg).

If we combine the three aforementioned holding companies with the information revealed in the Kabel Deutschland bid documentation, a tentative structure begins to emerge. Not included in this structure, however, are the mobile operators located in the Netherlands, the UK and the USA. As of the 2013 annual report, Vodafone Group had two joint ventures with Verizon Communications that were sold in early 2014 – Vodafone Italy, which operates in Italy although it is based in the Netherlands, and Verizon Wireless, which operates primarily within the USA.[5] As Verizon Wireless was formed through the merger of various wireless operations within the USA, a reasonable assumption is that the relevant assets of the Vodafone Group were held through Vodafone Americas and, in turn, Cellco Partnership (Cellco Partnership, 2011). Publicly available documents do not, however, clearly show how Vodafone Group owned these operations, although Vodafone AirTouch (2000a) suggests that the operation in Italy was, at least at the time of the acquisition of Mannesmann, held through Mannesmann AG. In this respect, the sale of Vodafone Group's 45 per cent stake in Verizon Wireless to Verizon Communications for $130 billion, which was announced in September 2013, is informative (Vodafone Group, 2013c, 2013d). The announcement document revealed that the Group's majority stake in Vodafone Italy was held via VEH (Vodafone Group, 2013d). Moreover, the same document shows that, while the stake in Verizon Wireless was held through Vodafone Americas Inc., at least one of the intermediate steps was located in the Netherlands. In other words, the stake was owned through companies based in the UK, then the USA, the Netherlands and, once more, the USA.

This leaves one company – Vodafone Limited – unaccounted for. While successive annual reports are unclear as to how Vodafone in the UK is owned, Vodafone Group (2001, F-47) states that the company is 'directly held'. As there is no evidence to the contrary in the annual reports and 20-F filings, it is reasonable to assume that it continues to be directly held by Vodafone Group. Between 1993 and 2000 (inclusive), the annual reports and 20-F filings of Vodafone Group and Vodafone AirTouch identify the principal subsidiaries of the company, and by doing so imply a structure. As no suggestion can be found within this structure that Vodafone Limited is part of a cascading series of holding companies, it is reasonable to assume that it is a fourth 'independent' holding company within the Vodafone Group, along with VIH, VEH and VIL. When these are taken together, a tentative structure of the Vodafone Group emerges, shown in Figure 3.1.

3.5 COUNTRY CASE STUDIES

Space prevents exhaustive detailing of how Vodafone Group has grown organisationally over time. Therefore the following case studies are intended to illustrate the mechanisms it has adopted as it has grown in the various markets in which it operates. The Egyptian example illustrates how the merger with AirTouch affected the ownership structure within the country, as well as how it changes to accommodate local market issues. The Indian example not only shows the organisational complexity that often accompanies foreign investment, but also how that foreign investment interacts with host-country legislation and sensitivities.

Egypt

In 1998, a GSM licence was awarded to Misrfone, a consortium in which Vodafone Group held a 30 per cent stake (Vodafone Group, 1998). AirTouch Communications was also a shareholder, owning a similarly sized stake. Thus the 1999 merger between Vodafone Group and AirTouch Communications resulted in Vodafone AirTouch becoming the majority shareholder in Misrfone. In 2002, the name of the operator changed to Vodafone Egypt Telecommunications (Vodafone Group, 2002).

In May 2002, Vivendi Telecom International exercised a put option requiring Vodafone Group to purchase part of its shareholding. As a result, the stake owned by Vodafone Group increased to 67 per cent. In December 2002, Telecom Egypt acquired Egypt's third mobile licence and sought to launch operations in conjunction with a foreign partner (Balancing Act, 2012). However, the existing duopoly of Vodafone Egypt and MobiNil meant that it would be difficult for Telecom Egypt to attract subscribers once it had launched. In December 2003, it was announced that:

- Telecom Egypt would acquire 16.9 per cent of Vodafone Egypt from the Vodafone Group (Vodafone Group, 2004).[6]

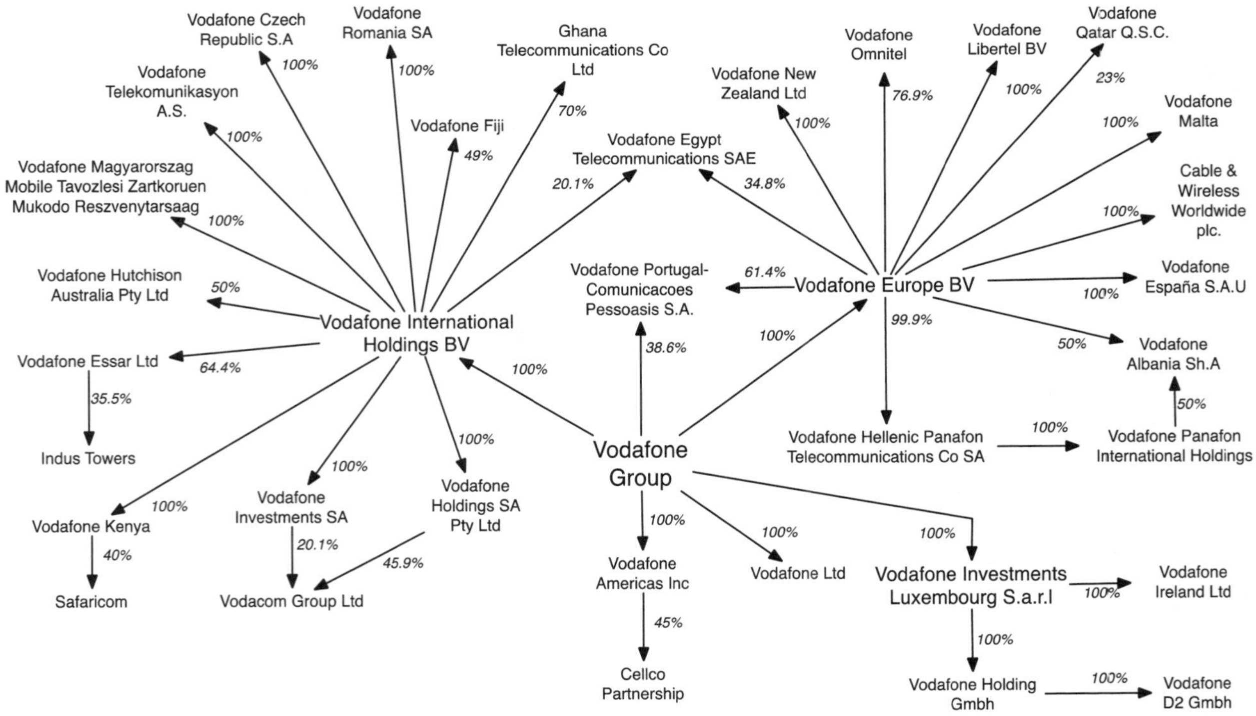

Figure 3.1 The structure of Vodafone, December 2013

- Telecom Egypt would purchase a further 8.6 per cent stake in Vodafone Egypt from its minority shareholders (TeleGeography, 2003).
- Telecom Egypt and Vodafone Group would form Wataniya Consortium, a 50/50 joint venture that would own 51 per cent of Vodafone Egypt (TeleGeography, 2003).
- Telecom Egypt would return its mobile licence (Balancing Act, 2012).
- The National Telecommunications Regulatory Authority would licence 1800 MHz spectrum to Vodafone Egypt (Vodafone Group, 2004).

As a consequence of this deal, Telecom Egypt and Vodafone Group collectively owned 51 per cent of Vodafone Egypt. However, the largest shareholder in Vodafone Egypt remained the Vodafone Group as it did not contribute its entire stake in the operator to the joint venture. When the joint venture stake was combined with this additional equity, Vodafone Group owned 50.1 per cent of Vodafone Egypt (Vodafone Group, 2005a). In other words, Vodafone Group owned 50.1 per cent, Telecom Egypt 25.5 per cent and other shareholders 24.4 per cent of Vodafone Egypt.

The balance between the respective shareholders has changed several times since December 2003. In December 2006, Telecom Egypt further increased its stake in Vodafone Egypt by offering, through a public tender, to buy up to a further 24.4 per cent of the company. When the tender was completed, Telecom Egypt had increased its stake to 49 per cent (Telecom Egypt, 2012). Part of this stake was then sold to the Vodafone Group. At the start of December 2006, Vodafone Group increased its stake in Vodafone Egypt through the purchase of a 4.4 per cent stake (Vodafone Group, 2007a). After Telecom Egypt bought more shares on the open market, its stake increased once more to 44.95 per cent (Telecom Egypt, 2012). As a result of these changes, Vodafone Group owned 54.93 per cent of Vodafone Egypt and Telecom Egypt 44.95 per cent (*Daily News*, 2011).

In mid-2010, it emerged that Telecom Egypt was interested in buying out the Vodafone Group, at a suggested price of around £3 billion. However, it would appear that discussions between the two companies never progressed to the point where they could be considered 'formal' (Parker and Saleh, 2010). In October 2013, Telecom Egypt denied it was considering selling its minority stake to the Vodafone Group (Mansfield, 2013a), though this did not prevent Vodafone Group from expressing an interest in acquiring it (Mansfield, 2013b). Both parties reiterated their desire to buy out the other in November 2013, though it was thought that if Telecom Egypt were to be awarded a mobile licence it would be the selling partner (Morris, 2013). In practice, a licence was duly awarded at a cost of $359 million in April 2014 (Telecom.paper, 2014a). Telecom Egypt will operate as an MVNO for two years, after which it will be able to bid for LTE spectrum, but it must also find a way to dispose of its stake in Vodafone Egypt within one year.

India

As noted above, the Vodafone Group today is partly the product of two mergers with AirTouch in 1999 and Mannesmann in 2000. It is the former of these that provided the company's first exposure to the Indian mobile market. After this, Vodafone Group made two other investments in India. As these investments overlap and have come into conflict with foreign direct investment and tax regulations, the chronology of events is outlined in Table 3.1.

Table 3.1 Vodafone Group presence in India, January 2000 to April 2014

Year	Event
2000	• March: states that it owns 49.0% of RPG Cellcom and 20.6% of RPG Cellular Services Ltd
2001	• March: states that it owns 20.6% of RPG Cellular[3]
2002	• March: states that it owns 20.8% of RPG Cellular[3,4]
2003	• 31 March: agrees to sell its entire stake in RPG Cellular Services[5]
2004	• Sale of RPG Cellular is completed in the financial year ending 31 March 2004[6]
2005	• 18 November: acquires a 10% direct/indirect stake in Bharti Airtel Limited
2007	• 28 February: Telecom Watchdog files an application with the High Court in Delhi questioning whether Hutchison Essar has violated foreign ownership limits • 8 May: completes acquisition of CGP Investment (Holdings) • 9 May: option exercised for Bharti to acquire Vodafone's 5.6% direct stake in Bharti Airtel • August: Vodafone Essar notified by the Indian tax authority of the potential failure to deduct the withholding tax from the sum paid to Hutchison Telecommunications International Ltd • September: Vodafone International Holdings B.V. (VIH) notified by the Indian tax authority of the potential failure to deduct the withholding tax from the sum paid to Hutchison Telecommunications International • September: rebranding of the company to Vodafone • Vodafone Essar and VIH file writs asking for the case against them to be dismissed[1]

Year	Event
2008	• June: final hearing of tax case in Bombay High Court • Indus Towers, infrastructure joint venture with Bharti and Idea, commences operation[2] • December: Bombay High Court dismisses the writs of Vodafone Essar and VIH • December: VIH files a petition with the Supreme Court appealing the dismissal by the Bombay High Court
2009	• 23 January: the Supreme Court refers the jurisdictional tax issue back to the tax authorities, granting VIH permission to appeal back to the High Court if it disagrees with the tax authority's decision • 30 October: VIH asked to explain why it believes that the tax authority does not have jurisdiction
2010	• 29 January: VIH provides the tax authorities with an explanation as to why it believes that they do not have jurisdiction
2012	• 20 January: Supreme Court decides that VIH is not liable to pay tax • March: The government announces that it will change the law to tax mergers undertaken overseas but involving Indian operations
2013	• January: new tax demand presented to VIH • April: Vodafone states its desire for an agreed settlement to the tax dispute • July: FDI threshold raised to 100% • July: Vodafone offers to pay $680 million to extend its existing licences in Mumbai, New Delhi and Kolkata
2014	• January: rumours that Vodafone Group will acquire a majority stake in Tata Teleservices begin to circulate • February: Vodafone fails in its legal attempt to cancel the spectrum auction, with the consequence that it goes ahead • February: Indian government approves Vodafone's purchase of the minority stakes • March: Vodafone raises stake in Vodafone India to 89% • April: Vodafone told by DoT to dispose of residual 4.4% stake in Bharti Airtel held indirectly via a Mittal family holding company • April: Vodafone buys residual 11% of Vodafone India from Piramal Industries

Notes:

1. It is not clear from 2009 annual report when this actually took place.
2. The 2010 annual report states that Vodafone Essar Limited has a 42% interest in Indus Towers.
3. No mention is made of the other company in the relevant annual report.
4. No explanation of the increased shareholding is provided in the annual report.
5. No mention of whom the stake is being sold to is made in the annual report.
6. It is not clear from the 2004 annual report when this actually took place.

Sources: Crabtree (2012, 2013a, 2013b, 2014), Mansfield (2014), Shankar (2013), Vodafone AirTouch (1999, 2000a, 2000b), Vodafone Group (2006b), Vodafone Group (2007a), Vodafone Group (2008a), Vodafone Group (2009), Vodafone Group (2010), Vodafone Group (2011a, 2011b, 2011e) and Vodafone Group (2012a, 2012b).

AirTouch Communications initially invested in India in 1995 (AirTouch Communications, 1996). The first investment was the purchase of a 20 per cent stake in RPG Cellular Services, a company already licensed to provide mobile services in Madras (AirTouch Communications, 1996). Vodafone was also an investor in this company, with a 26 per cent stake, as well as RPG Enterprises with 51 per cent (Mobile Phone News, 1995). The second investment was a 49 per cent stake in RPG Cellcom, a joint venture that subsequently won a mobile licence covering Madhya Pradesh (AirTouch Communications, 1996). Both companies became part of Vodafone AirTouch when the two merged in 1999.

In June 2003, Vodafone Group sold its stake in RPG Cellular Services to IQ Investments (Gairola, 2003). Although RPG Cellular was an Indian mobile operator, the transaction was conducted outside the country; both IQ Investments and AirTouch International (Mauritius) Ltd, the Vodafone Group subsidiary holding its stake in RPG Cellular, were based in Mauritius (Gairola, 2003). IQ Investments acquired AirTouch International (Mauritius) outright, and thus its stake in RPG Cellular. However, another shareholder in RPG Cellular challenged this transaction. Cellfone Ltd, also based in Mauritius, questioned whether the shares could be sold by one shareholder to another without first being offered to all other shareholders. After various appeals and counter-appeals (The Hindu Business Line, 2003a, 2003b), the sale took place in mid-2003.

In late 2005, Vodafone Group acquired a stake totalling 10 per cent in Bharti Tele-Ventures. A direct stake of 5.6 per cent was acquired from Warburg Pincus in November 2005, and a further 4.4 per cent through subscribing to convertible debentures in Bharti Enterprise Private on 22 December. After these purchases, Bharti Tele-Ventures was renamed Bharti Airtel (Vodafone Group, 2006b).

In February 2007, Vodafone Group announced that it would purchase Hutch Essar, the Indian mobile operation of Hutchison Telecommunications International, for $10.9 billion. The acquisition was completed in May 2007 and was followed by the sale of the Group's direct stake in Bharti Airtel to a company within the Bharti Group. After this sale, Vodafone Group continued to own an indirect stake of 4.4 per cent in Bharti Airtel (Vodafone Group, 2007a).

Although Hutch Essar operated in India, it was controlled by a series of cascading holding companies that linked Hong Kong to India via the Caribbean. In part, this reflected the manner in which the mobile industry within India had been liberalised as well as the limits placed on foreign ownership.[7] It also reflected the propensity for Hutchison Whampoa, the ultimate parent company of Hutchison Essar, to favour complex organisational structures – see Chapter 4.

VIH acquired a handful of companies that in turn controlled Hutchison Essar (Vodafone Group, 2007b, 2007c). VIH acquired outright ownership of CGP Investments (Holdings), which owned an indirect 51.95 per cent stake in Hutchison Essar. Through this purchase, minority stakes in two companies – Telecom Investments India and Omega Telecom Holdings – that owned indirect stakes in Hutchison Essar of 19.54 per cent and 5.11 per cent respectively, were also acquired (Vodafone Group, 2007a). As part of the deal, Vodafone Group was granted an option to acquire full ownership of two further companies that owned the outstanding shares in Telecom Investments India. Vodafone Group was also given the option to acquire another company that, in turn, owned the outstanding shares in Omega Telecom Holdings. How these various companies related to one another is shown in Figure 3.2.

Once the various elements of the acquisition were combined, VIH would control 67 per cent of Hutchison Essar. As part of the acquisition of Hutchison Essar, options were granted to VIH enabling it to purchase outstanding shares from companies within the Essar Group. The outright purchase of the equity covered by the options would cost $5 billion, with the options being exercisable between 8 May 2010 and 8 May 2011. The 33 per cent held by the Essar Group was divided between two companies – 11 per cent was owned by Essar Telecommunications Holding and 22 per cent by Essar Telecom India (*The Telegraph*, 2011). The Ruias family, which controlled the Essar Group, proposed to merge the privately owned Essar Telecommunications Holding with India Securities Ltd, a quoted company within the Group.

At the end of March 2011, the Essar Group exercised its put option relating to the 22 per cent held by Essar Telecom India (Vodafone Group, 2011d). In turn, Vodafone Group exercised its call option over the 11 per cent owned by other Essar Group companies. As the purchase would result in a breach of the foreign ownership limit, a small equity stake would be sold on to an Indian investor. However, a few days later it was alleged that too much of Hutchison Essar was foreign-owned. Telecom Watchdog argued that the correct figure was more than 74 per cent and that, as a consequence, the company should be prosecuted and its licences withdrawn (Vodafone Group, 2007a). While this particular complaint seems to have disappeared in subsequent years, the issue itself did not. Following the purchase of shares held by Essar Group, it was alleged that Vodafone Essar had breached the foreign ownership limit by having 75.35 per cent of the company owned by non-Indians. This would require a stake of at least 1.35 per cent to be sold on by VIH to ensure that it was in compliance with the law (Mansfield, 2011).

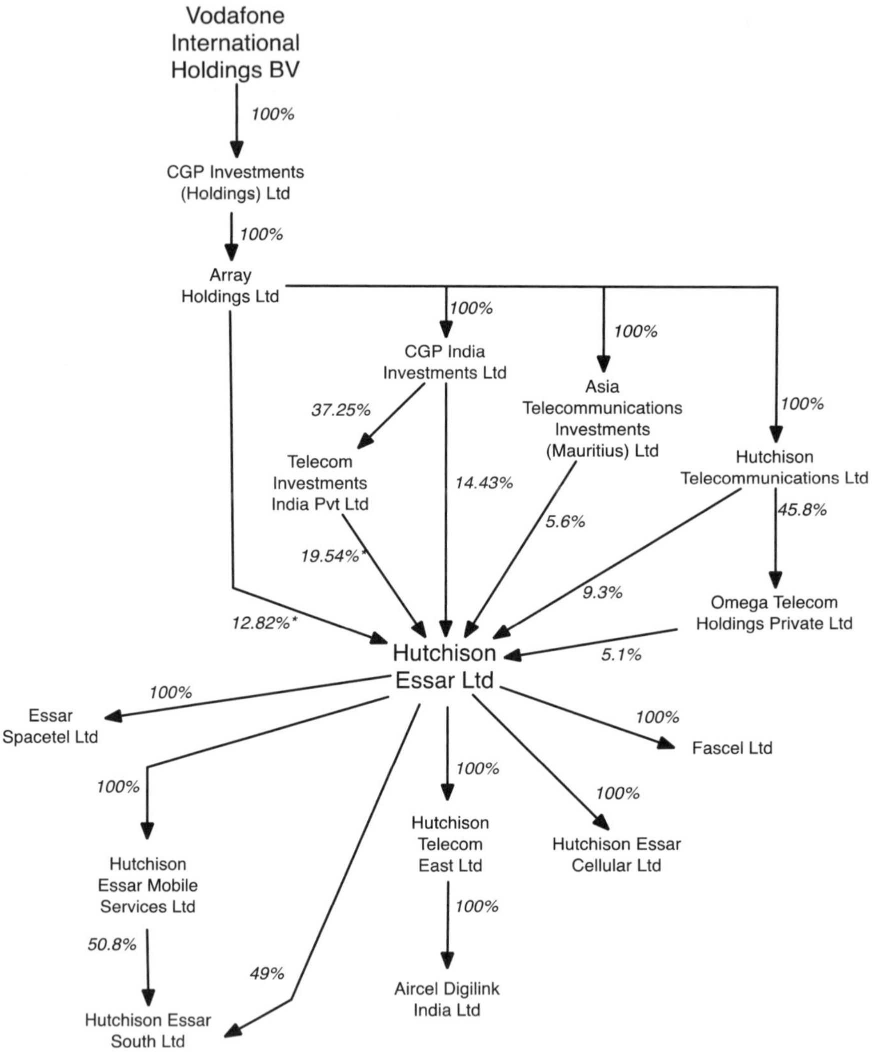

Note: *Combines direct and indirect holdings

Figure 3.2 The purchase of Hutchison Essar by Vodafone International Holdings, February 2007

It was also suggested that, since VIH had options over the indirect equity stakes held by Telecom Investments India and Omega Telecom Holdings, these stakes in Vodafone Essar should in reality be considered to be foreign-owned (*Communications Today*, 2011). The predetermined nature of the amount to be paid to exercise the option meant that any positive or negative returns associated with their equity holdings were actually assumed by indirect equity stakes. In other words, Telecom Investments India and Omega Telecom Holdings enjoyed neither the risks nor rewards associated with holding an equity stake in Vodafone Essar.

The situation was further complicated by changes in the shareholder composition of IDFC, the parent company of Omega Telecom Holdings (Mansfield, 2011). The shareholder composition changed such that the majority of shares were foreign-owned. As a consequence, the 5.11 per cent stake in Vodafone Essar held by Omega Telecom Holdings could also be viewed as foreign-owned.

In August 2011, a new local investor in Vodafone Essar was announced (Parker and Fontanella-Khan, 2011). The Essar Group would sell a 5.5 per cent stake to Piramal Healthcare. Through this sale, the extent to which Vodafone Essar was foreign-owned would decline below the required threshold. The stake held by VIH in Vodafone Essar was further reduced in February 2012 when an additional 5.5 per cent was sold to Piramal Healthcare (Mansfield, 2012). Integral to the sale of 11 per cent to Piramal was a series of exit options that, in essence, enabled Vodafone Group to re-acquire the shares at a later date at a predetermined price (Vodafone Group, 2013a).

In July 2013, India relaxed its foreign direct investment rules, raising the limit on foreign ownership in the telecommunications sector from 74 to 100 per cent (*The New Indian Express*, 2013). Vodafone Group responded by describing itself as a 'natural consolidator' in the market (Crabtree, 2013a). To date, only modest consolidation has occurred – Bharti Airtel has agreed to buy Loop Mobile for roughly $100 million (as it operates only in Mumbai) and Videocon has effectively put itself up for sale, but operators appear to be waiting for clarification of the rules for spectrum transfers before making any major commitments. For its part, Vodafone Group has begun exploratory talks with Tata about acquiring a majority stake in Tata Teleservices (Crabtree, 2014). Vodafone Group has also initiated the process of buying out the minority shareholders from Vodafone Essar – henceforth Vodafone India. In mid-February, the Indian government approved Vodafone's $1.6 billion purchase of the equity held by the minority shareholders in its Indian business (Mansfield, 2014). The stake was raised to 89 per cent in March 2014 and the remaining 11 per cent was bought from Piramal Industries in April (Telecom.paper, 2014b).

Also in April 2014, after the licence awards that saw both Vodafone and Bharti Airtel renew their licences in Delhi, Kolkata and Mumbai,

Vodafone technically fell foul of new regulations forbidding a stakeholder from having a beneficial interest in more than one network within a given circle. As a result, it will be obliged to dispose of its residual stake in Bharti Airtel, quite possibly to Bharti itself. However, whereas Vodafone claims that it has until November 2015 to complete a deal, the Department of Telecommunications claims that a deal must be completed before it is prepared to hand over the operating licences (TeleGeography, 2014).

3.6 DISCUSSION

The previous two sections have illustrated the organisational complexity of the Vodafone Group. It has also been shown that the Vodafone Group is composed of a series of geographically overlapping holding companies. Determining which of the mobile operations that these holding companies own is complicated. One difficulty is that, since Vodafone Group stopped presenting its principal subsidiaries in a format that implied a structure, there have been insufficient data in the company's annual reports and 20-F filings alone to track organisational changes. Another difficulty is that the names of the various companies within the Vodafone Group have changed over time, reflecting mergers on the one hand and new activities on the other. It is possible to obtain data in both areas from Companies House in the UK, but the SIC code used to describe the activities of the company is imprecise and ownership details are not provided. Moreover, as it is an MNE, only some of the companies within the Vodafone Group are covered by Companies House. In other words, to gain a comprehensive picture of Vodafone Group would require detailed data from multiple jurisdictions. In practice, data are more readily available in some countries than in others. The widespread use of, for example, offshore companies registered in the Channel Islands and the Caribbean detracts from our ability to understand the ownership and structure of MNEs like Vodafone Group.

While placing better quality data in the public domain will help further our understanding of MNEs like Vodafone Group, it is only likely to explain in part the complex structures that have been observed. Vodafone Group has established (holding) companies in Luxembourg and the Netherlands, countries with lower tax regimes than the UK. After the UK tax authorities questioned the profitability of VIL, a protracted dispute arose focusing on the applicability of the UK's 'controlled foreign corporations' legislation. After a series of setbacks, the tax authorities won their appeal in 2009 (Dodwell, 2009), leading Vodafone Group to pay £1.25 billion to settle the case in July 2009.

Vodafone Group has also been involved in a tax dispute with the Indian government over its purchase of Hutchison Essar in 2007. Both VIH and CGP Investments were based outside India, with the consequence that no capital gains tax was paid on the transaction. After the transaction was completed, the Indian government asked VIH to explain why no capital gains tax had been paid (KPMG, 2010). A series of court cases resulted that debated whether the transaction was subject to Indian jurisdiction, with the Supreme Court ultimately ruling in January 2012 that no tax should be paid (The Supreme Court of India, 2012). However, this was not the end of the issue. The Indian government then changed the law so that mergers undertaken outside the country but where the companies involved had operations inside it could be taxed. As this change was retrospectively applied, the acquisition of Hutchison Essar by the Vodafone Group technically became taxable. Although it had been stated that the law would be implemented only in the 'rarest of rare cases' (*The Indian Express*, 2012), the fact that its introduction was specifically related to the Vodafone tax case was expected to trigger a coming-together of the parties to the dispute.

Nevertheless, the issue continues to rumble on. In early 2013 it was suggested that no agreement was likely until after the 2014 Indian general elections had been held (Cellular-news, 2013a), while in June 2013 it emerged that Vodafone Group had sought additional time to consider the offer from the government for Indian as against international-based arbitration (Cellular-news, 2013b). The Indian government then walked away from the talks. This was alleged to be because the 2014 auction of spectrum had raised more revenue than had been expected, with the consequence that the government felt that it was in no hurry to settle the dispute for a lower sum. For its part, Vodafone sought to instigate international arbitration in accordance with the terms of the India–Netherlands Bilateral Investment Promotion Agreement, although the Indian government ruled in March 2014 that this did not cover taxation issues. Nevertheless, it agreed once again to pursue a process of conciliation.

More importantly, Vodafone Group's presence in India has been complicated by the emergence of further disputes with the government. In mid-2013, for example, Vodafone Group offered to pay $680 million to renew its soon-to-expire licences in Mumbai, New Delhi and Kolkata (Chilkoti, 2013). This was rejected by the Indian government, which argued that the company should participate in the forthcoming spectrum auction. Subsequent legal attempts by Vodafone Group and other mobile operators in India to have their licences extended and the auction cancelled were unsuccessful (Rautray, 2014), with the consequence that the auction went ahead in February 2014 (Cellular-news, 2014). In the event, Vodafone Group successfully renewed its 900 MHz licences and acquired 900 MHz and/or 1800 MHz spectrum in 11 circles overall at a total cost of $3160 million – the largest sum paid by any operator because of the relatively high cost of spectrum in the big cities.

The other dispute relating specifically to tax matters emerged in December 2013 when Vodafone Group – or, more specifically, Vodafone India Services Private Limited, a subsidiary providing customer support services to the operating companies – was presented with a tax bill for $591 million. This dispute related to transfer pricing – that is, intra-company charges for services – which the government alleges has been organised so as to minimise tax liabilities (Wood, 2014).

These disputes highlight the organisational complexity of Vodafone Group. In the case of the first dispute, it was revealed that a hitherto unknown company – Vodafone 2 – was located between Vodafone Group and VIL. The court cases in India produced a detailed organisational chart, which is presented in a simplified form in Figure 3.2, showing how all the companies related to one another. Clearly, one advantage to be gained from locating companies either in Luxembourg or anywhere else outside India is that of lowering the amount of tax paid by Vodafone Group on either its profits (as in the UK) or a transaction (as in India).[8] Another advantage is that it is sometimes difficult to determine how the various companies involved in a transaction relate to one another. While the UK tax case revealed the presence of Vodafone 2 in the chain of cascading holding companies, the Indian dispute revealed that ownership was affected through a large number of companies and a complex series of options. These options were sufficiently complex to lead some to question whether Vodafone Group actually owned more of its Indian operations than the foreign direct investment regulations permitted.

3.7 CONCLUSIONS

This chapter has sought to explain the organisational structure of Vodafone Group. It probably comes as no great surprise to discover that this structure is complex, although the number of companies within the group and the manner in which they are organised will almost certainly strike the reader as going beyond any reasonable expectation. Although the international operations are largely held through two holding companies (VIH and VEH), while a third holding company (VIL) has a smaller geographical footprint, it can be difficult to understand why newly acquired assets so often need to go hand in hand with the creation of yet another company in a cascading chain of subsidiaries.

While the use of multiple and sometimes cascading holding companies complicates the organisational structure of Vodafone Group between countries, it is also often complex within countries. This inevitably creates a suspicion that the purpose of all this complexity is to avoid individual country regulations on matters such as ownership caps or tax liabilities and, where such a suspicion exists, controversy and recourse to the courts are sure to follow. The latter can be a long-winded process that generally works to the benefit of an operator such as Vodafone – as a debtor it can hold on to the money owed and the creditor may eventually settle for less than it originally wanted – although retrospective legislation such as has been introduced in India will understandably be seen as unfair at best.

NOTES

1. The website can be found at http://opencorporates.com.
2. Safaricom operates in Kenya and Verizon Wireless in the USA. Vodacom, which is based in South Africa, has mobile operations across Africa (Lesotho, Democratic Republic of Congo, Mozambique and Tanzania). Vodacom also owns Vodacom Business Africa Group (PTY) Limited, which is described as a holding company in the 2013 annual report (Vodafone Group, 2013a), and which offers business services to companies across Africa.
3. Close examination of successive annual reports and 20-F filings indicates that the name of the company has changed. Until 1999, the company was known as Vodafone Europe Holdings and from 2002 onwards as Vodafone Europe. In between, the name of the company reflected the merger of AirTouch Communications with Vodafone Group.
4. Nevertheless, some of the earlier sources that are available suggest that Vodafone España was owned, at least in part, by another company within the Vodafone Group, namely Vodafone Europe Investments.
5. Vodafone Omnitel was 76.9 per cent owned by the Vodafone Group and 23.1 per cent owned by Verizon Communications, while Verizon Wireless was owned 45 per cent by Vodafone Group and 55 per cent by Verizon Communications before the transfer of the Vodafone stake to Verizon Communications.
6. Although it was announced in December 2003, it would appear that the transaction actually occurred in January 2005 (Vodafone Group, 2005a).
7. The foreign ownership limit was increased from 49 to 74 per cent in March 2007 (Invest Now India, 2007). A second increase in the foreign ownership limit, to 100 per cent, occurred in July 2013 (*The New Indian Express*, 2013).
8. In April 2013, Vodafone Group declared its tax payments for the financial year ending 31 March 2013. In total, it paid £4.24 billion in direct taxes, up from £3.38 billion the previous year. This reflected a decrease in Europe where revenues had declined but an upturn in emerging markets. In addition, Vodafone Group's non-tax payments, largely for spectrum purchases, rose to £3.23 billion from £1.87 billion the previous year.

REFERENCES

Action Aid (2011), *FTSE 100 Tax Haven Tracker: The Data*, www.actionaid.org.uk.

AirTouch Communications (1996), 'Form 10-K for the year ended December 31 1995, filed on 26 March', www.secinfo.com.

Alacra Store (2011), 'Vodafone-Panafon Hellenic acquires WIND Hellas Telecommunications', *Alacra Store*, www.alacrastore.com, 29 August.

Australian Competition & Consumer Commission (2009), 'Statement of Issues – Vodafone and Hutchison – proposed merger of Australian mobile operations', www.accc.gov.au, 1 April.

Balancing Act (2012), 'Egyptian Telecoms Sector Undergoing Restructuring', www.balancingact-africa.com.

Cellco Partnership (2011), *Report of the Independent Registered Public Accounting Firm*, www.verizon.com, 28 February.

Cellular-news (2013a), 'Vodafone's Indian tax dispute settlement delayed by Indian elections', www.cellular-news.com, 7 May.

Cellular-news (2013b), 'Vodafone sells Verizon Wireless stake for $130 billion in cash and shares', www.cellular-news.com, 2 September.

Cellular-news (2013c), 'Vodafone secures $10.1 billion takeover of Germany's Kabel Deutschland', www.cellular-news.com, 13 September.

Cellular-news (2014), 'Indian spectrum auction gets bids in all circles', www.cellular-news.com, 3 February.

Chilkoti, A. (2013), 'Vodafone makes Indian government an offer for licence extensions', *Financial Times*, www.ft.com, 1 July.

Communications Today (2011), 'Vodafone broke FDI cap in Hutch deal, breach continues, says CBDT', www.communicationstoday.co.in, 1 December.

Crabtree, J. (2012), 'Indian tax reform threatens to reopen $2.9 bn Vodafone case', *Financial Times*, 17/18 March, 13.

Crabtree, J. (2013a), 'Vodafone considers acquisitions in India', *Financial Times*, 15 September.

Crabtree, J. (2013b), 'Vodafone prepares to raise stake in Indian unit', *Financial Times*, 7 October.

Crabtree, J. (2014), 'Vodafone in talks to buy a minority stake in Tata's mobile unit', *Financial Times*, 3 January.

Curwen, P. and J. Whalley (2009), 'Merging incumbent European mobile operators: veni, vidi, non vici', *info*, **11**(4), 34–52.

Curwen, P. and J. Whalley (2010), *Mobile Telecommunications in a High-Speed World*, Farnham, UK: Gower.

Daily News (2011), 'Vodafone Egypt denies Mubaraks' stake in company', www.thedailynewsegypt.com, 6 February.

Dodwell, B. (2009), 'Where next for this tortuous process?', *Tax Adviser*, 9 July.

European Commission (2001), *Case No COMP/M.2469 – Vodafone/Airtel*, Brussels: EC.

European Commission (2005), *Case No COMP/M.3776 – Vodafone/Oskar Mobile*, Brussels: EC.

Fang, T., C. Fridh and S. Schultzberg (2004), 'Why did the Telia–Telenor merger fail?', *International Business Review*, **13**(5), 573–94.

Gairola, M. (2003), 'Sivasankaran acquires Vodafone's RPG Cellular stake', www.articles.economictimes.indiatimes.com, 6 June.

Gerpott, T. and N. Jakopin (2005), 'The degree of internationalisation and the financial performance of European mobile network operators', *Telecommunications Policy*, **29**(8), 635–61.

Goodall, A. (2010), 'Vodafone: HMRC statement fails to halt protest', www.taxjournal.com, 1 November.

Gruber, H. and F. Verboven (2001), 'The diffusion of mobile telecommunication services in the European Union', *European Economic Review*, **45**(3), 577–88.

Haucap, J., U. Heimeshoff and T. Stühmeir (2010), 'Competition on the German Mobile Telecommunications Market', paper presented at 18th Biennial Conference of the International Telecommunications Society, 27–30 June, Tokyo, Japan.

The Hindu Business Line (2003a), 'RPG Cellular: HC direction on Vodafone share transfer to Sterling Co', www.thehindubusinessline.in, 28 August.

The Hindu Business Line (2003b), 'Sterling Infotech buys RPG Cellular', www.thehindubusinessline.in, 19 December.

Ietto-Gillies, G. (2001), *Transnational Corporations: Fragmentation Amidst Integration*, London: Routledge.

The Indian Express (2012), 'Spare Vodafone, go after Hutch if need be: Shome', www.indianexpress.com, 10 October.

InsideView (2011), 'Vodafone Hungary Mobile Telecommunications Company Limited', www.insideview.com.

Invest Now India (2007), 'FDI in telecom raised to 74 per cent', www.ibef.org, 26 March.

KPMG (2010), *Bombay High Court ruling in the case of Vodafone International Holdings B.V*, www.kpmg.com, 8 September.

Li, Y. and B. Lyons (2011), 'Three private firms and an independent regulator are sufficient for rapid mobile network penetration', CCP Working Paper 11-1, Centre for Competition Policy, University of East Anglia, UK.

Malta Communications Authority (2001), *Subsidiary legislation 399.23 – Directives by the Malta Communications Authority*, Malta.

Mansfield, I. (2011), 'Vodafone may have to reduce Indian shareholdings', www.cellular-news.com, 5 April.

Mansfield, I. (2012), 'Piramal Healthcare increases in stake in Vodafone India for $618 million', www.cellular-news.com, 5 February.

Mansfield, I. (2013a), 'Telecom Egypt profits up by nearly 50 per cent', www.cellular-news.com, 14 August.

Mansfield, I. (2013b), 'Vodafone ready to buy out Egyptian partner', www.cellular-news.com, 22 October.

Mansfield, I. (2014), 'Vodafone gains approval for $1.6bn Indian investment', www.cellular-news.com, 10 February.

Mobile Phone News (1995), 'AirTouch expands presence in India', findarticles.com, 25 September.

Morris, A. (2013), 'Telecom Egypt eyes local Vodafone unit', www.fircewireless.com, 13 November.

Mulligan, J. (2010), 'Vodafone pays €400m dividend to sole shareholder as profits plunge', www.independent.ie, 18 January.

The New Indian Express (2013), 'Government raises FDI limit in 12 sectors, 100% in telecom', www.thenewindianexpress.com, 17 July.

Parker, A. and H. Saleh (2010), 'Vodafone rejects sale of Egypt arm', *Financial Times*, www.ft.com, 2 June.

Parker, A. and J. Fontanella-Khan (2011), 'Vodafone Essar to sell stake to Piramal', www.ft.com, 11 August.

Rautray, S. (2014), 'SC rejects bid by Airtel, Vodafone & Idea to cancel spectrum auction', *The Economic Times*, economictimes.com, 3 February.

Rosenberg, E., J. Borrows, C. Hunt, R. Samarajiva and W. Pollard (1993), *Regional Telephone Holding Companies, Affiliate Transactions, and Regulatory Options*, Columbus, OH: The National Regulatory Research Institute.

Rugman, A. (2005), *The Regional Multinationals*, Cambridge, UK: Cambridge University Press.

Shankar, D. (2013), 'I-T dept reminds Vodafone on Rs 11,200 crore tax dues', *Financial Chronicle*, www.mydigitalfc.com, 6 January.

The Supreme Court of India (2012), *Civil Appeal No.733 of 2012, Vodafone International Holdings B.V. versus Union of India & Anr.*, New Delhi, India.

Sutherland, E. (2007), 'The Scramble for Africa', paper presented at 35th Telecommunications Policy Research Conference, 28–30 September, George Mason University, Arlington, VA, USA.

Sutherland, E. (2012), 'Corruption in telecommunications: problems and remedies', *info*, **14**(1), 4–19.

Tabureguci, D. (2007), 'Fiji's sole mobile service fires up', www.islandbusiness.com.

Telecom Egypt (2012), '150 years of telecommunications', www.ir.telecomegypt.com.

Telecom.paper (2014a), 'Telecom Egypt acquires mobile licence for EGP 2.5 bln', www.telecompaper.com, 3 April.

Telecom.paper (2014b), 'Vodafone buys rest of Vodafone India', www.telecompaper.com, 11 April.

TeleGeography (2003), 'Vodafone Egypt lists: gets new shareholders', www.telegeography.com, 23 December.

TeleGeography (2014), 'Vodafone to offload Bharti stake', www.telegeography.com, 3 April.

The Telegraph (2011), 'Vodafone, Essar lock horns', www.telegraphindia.com, 18 January.

Vodafone AirTouch (1999), 'Form 20-F, Annual Report Pursuant to Section 13 or 15(d) of the Securities and Exchange Act of 1934 for the year ended March 31, 1999', www.vodafone.com.

Vodafone AirTouch (2000a), 'Form 20-F, Annual Report Pursuant to Section 13 or 15(d) of the Securities and Exchange Act of 1934 for the year ended March 31, 2000', www.vodafone.com.

Vodafone AirTouch (2000b), 'Annual Report and Accounts for the year ended 31 March 2000', www.vodafone.com.

Vodafone Albania (2011), 'Vodafone Albania to launch 3G mobile broadband', www.vodafone.al, 17 January.

Vodafone Europe (2012), 'Recommended cash offer for Cable & Wireless Worldwide plc. by Vodafone Europe B.V.', www.vodafone.com, 23 April.
Vodafone Ghana (2011), 'About us – company profile', www.vodafone.com.gh.
Vodafone Group (1998), 'Annual Report and Accounts 98', www.vodafone.com.
Vodafone Group (2001), 'Form 20-F, Annual Report Pursuant to Section 13 or 15(d) of the Securities and Exchange Act of 1934 for the year ended March 31, 2001', www.vodafone.com.
Vodafone Group (2002), 'Annual Report and Accounts and Form 20-F for the year ended 31 March 2002', www.vodafone.com.
Vodafone Group (2004), 'Annual Report for the year ended 31 March 2004', www.vodafone.com.
Vodafone Group (2005a), 'Annual Report for the year ended 31 March 2005', www.vodafone.com.
Vodafone Group (2005b), 'Vodafone to acquire control of MobiFon in Romania and Oskar in the Czech Republic', www.vodafone.com, 15 April.
Vodafone Group (2006a), 'Vodafone announces completion of acquisition of the assets of Telsim in Turkey', www.vodafone.com, 24 June.
Vodafone Group (2006b), 'Annual Report for the year ended 31 March 2006', www.vodafone.com.
Vodafone Group (2007a), 'Annual Report for the year ended 31 March 2007', www.vodafone.com.
Vodafone Group (2007b), 'Vodafone agrees to acquire control of Hutch Essar in India', www.vodafone.com, 11 February.
Vodafone Group (2007c), 'Vodafone agrees to acquire control of Hutch Essar in India', Form 6-K, filed with the Securities & Exchange Commission, Washington, DC, USA, 11 February.
Vodafone Group (2008a), 'Annual Report for the year ended 31 March 2008', www.vodafone.com.
Vodafone Group (2008b), 'Vodafone and Qatar Foundation take first steps towards launch of Vodafone Qatar', www.vodafone.com, 23 April.
Vodafone Group (2009), 'Annual Report for the year ended 31 March 2009', www.vodafone.com.
Vodafone Group (2010), 'Annual Report for the year ended 31 March 2010', www.vodafone.com.
Vodafone Group (2011a), 'Annual Report and 20-F Filing', www.vodafone.com.
Vodafone Group (2011b), 'Vodafone and Essar announce Agreements for Sale of Essar stake in Vodafone Essar Limited', www.vodafone.com, 1 July.
Vodafone Group (2011c), 'Annual Report for the year ended 31 March 2011', www.vodafone.com.
Vodafone Group (2011d), 'Exercise of options by Essar and Vodafone', www.vodafone.com, 31 March.
Vodafone Group (2011e), 'Form 20-F, Annual Report Pursuant to Section 13 or 15(d) of the Securities and Exchange Act of 1934 for the year ended March 31, 2011', www.vodafone.com.
Vodafone Group (2012a), 'Statement on Indian tax case judgement', www.vodafone.com, 20 January.
Vodafone Group (2012b), 'Creating a more valuable Vodafone. Annual report for the year ended 31 March 2012', www.vodafone.com.

Vodafone Group (2013a), 'Annual report for the year ended 31 March 2013. The way ahead … introducing Vodafone 2015', www.vodafone.com.
Vodafone Group (2013b), 'Offer document – Voluntary public takeover offer by Vodafone Vierte Werwaltungsgesellshaft mbH to the shareholders of Kabel Deutschland Holding AG', www.vodafone.com, 29 July.
Vodafone Group (2013c), 'Vodafone to realise US$130 billion for its 45% interest in Verizon Wireless', www.vodafone.com, 2 September.
Vodafone Group (2013d), 'News release – group: Vodafone to realise US$130 billion for its 45% interest in Verizon Wireless', www.vodafone.com, 2 September.
Vodafone New Zealand (2011), 'Annual Report for the year ended 31 March 2011', 30 September, Auckland, New Zealand.
Whalley, J. and P. Curwen (2012), 'Third time lucky? An exploration of Hutchison Whampoa's involvement in the mobile telecommunications industry', *info*, **14**(2), 17–41.
Wireless Federation (2011), 'Vodafone Egypt denies Mubarak's stake in company', www.wirelessfederation.com, 2 July.
Wood, N. (2014), 'India set to walk out on Vodafone tax talks', www.totaltele.com, 12 February.

4. Anatomy of a disruptive force: Hutchison Whampoa

4.1 INTRODUCTION

Hutchison Whampoa – which will simply be referred to as Hutchison in what follows where there is no risk of confusion – is normally described as a 'diversified' company and is not readily associated with the telecommunications industry, which is not its main line of business. Nevertheless, it has had a long and significant presence in the industry, particularly in terms of performing the role of a disruptive force in the mobile sector.

Since the late 1980s, Hutchison has invested in several sectors of the industry – mobile, fixed-wire, cable and equipment – and has actively managed its investments, expanding into new markets while divesting itself of others. As a diversified company or conglomerate, it is reasonable to expect that Hutchison would move capital between different industries in the pursuit of maximum returns. However, Hutchison is unusual in that it has always seemed to be willing to continue to pump money into loss-making mobile networks, perhaps because two of its original investments – in Orange plc in the UK and in the Indian mobile market – were highly profitable.

Hutchison continued to invest even though it found that building a third-generation (3G) mobile business from scratch was more challenging and protracted than anticipated (Curwen and Whalley, 2006). It used the proceeds from the sale of Orange and its Indian mobile operations to fund this continued investment, but the ability to invest successfully in one period does not guarantee that this feat can be repeated.

4.2 HUTCHISON AND TELECOMMUNICATIONS

Hutchison Whampoa is a Hong-Kong-based conglomerate. According to the most recent FT Global 500 ranking, which relates to end March 2013, it ranks 180th among the largest companies in the world with sales of \$31.4 billion and net income of \$3.37 billion (*Financial Times*, 2013).

The present size and scope of Hutchison is the result of diversification and significant internationalisation.

As befits a conglomerate, Hutchison Whampoa has a structure that can be described as complex. Hutchison divides its operations into five sectors: ports and related services; property and hotels; retail; telecommunications; and energy, infrastructure, investments and others. These activities are organised into a bewildering array of holding companies, subsidiaries and joint ventures. Some, but by no means all, of these activities are wholly owned and ten subsidiaries and associated companies are listed on stock exchanges.[1] It is also worth noting that Hutchison is itself 49.97 per cent owned by Cheung Kong Holdings.

Due to the small size of its home market and the nature of the industries in which it has invested, Hutchison Whampoa is also highly internationalised. Since 1990, UNCTAD has published details of the largest multinational corporations[2] in terms both of size (revenues, assets, and so on) and the extent to which they are internationalised. Hutchison Whampoa entered the list of the largest 100 multinational corporations by assets in 1999 at number 48 (UNCTAD, 2001). The following year, Hutchison Whampoa rose to be the 14th-largest multinational corporation when measured by foreign assets (UNCTAD, 2002). In subsequent years, Hutchison's position in the rankings fluctuated, so that in 2012 it was the 26th-largest multinational corporation when measured by foreign assets (UNCTAD, 2013).

Involvement in the Telecommunications Industry

Figure 4.1 depicts the current scope of Hutchison Whampoa's involvement in the telecommunications industry. Hutchison's involvement is undertaken through a series of holding companies that have, over the years, experienced significant changes. Companies have not only been bought and sold but have also been floated on stock exchanges and then bought back by Hutchison.

The manner in which Hutchison's involvement in the industry has evolved, coupled with its propensity to move assets between companies and engage in Merger and Acquisition activities, means that disentangling these developments is not straightforward. Each of the holding companies identified in Figure 4.1 could be used to illustrate this complex relationship. However, as the focus of this book is on mobile telecommunications, our discussion here will be limited to Hutchison Asia Telecommunications Ltd and 3 Group.

Until 2010, Hutchison Asia Telecommunications Ltd was known as Hutchison Telecommunications International Ltd (HTIL hereafter). HTIL

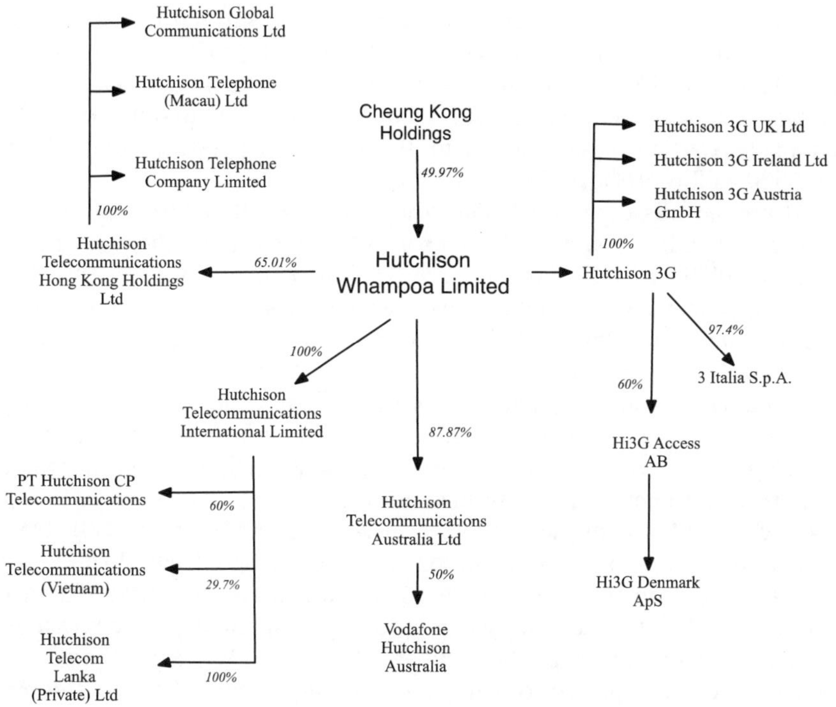

Source: Compiled by the authors from the annual reports and 20-F filings of Hutchison Whampoa, Hutchison Telecommunications (Australia) Ltd, Hutchison Telecommunications International Ltd and Hutchison Telecommunications Hong Kong Holdings Ltd.

Figure 4.1 Structure of Hutchison Whampoa, March 2014

was floated in October 2004 as a holding company encompassing a disparate set of fixed-wire and mobile assets in eight countries: Hong Kong, Israel, India, Macau, Sri Lanka, Ghana, Paraguay and Thailand (Hutchison Whampoa, 2005: 48). In other words, neither the Australian nor West European mobile operations were included in the assets floated as HTIL. The 2004 annual report states that after the flotation Hutchison Whampoa owned 70 per cent of the company.

In August 2005, HTIL purchased the free float of Hutchison Global Crossing Holdings (Hutchison Whampoa, 2006: 47), thereby reducing the percentage of the company owned by Hutchison Whampoa, as did the sale of a 19.3 per cent strategic stake in HTIL to Orascom Telecom. The stake fell to 49.8 per cent, though this changed once more in 2007 when Hutchison acquired shares in the open market to become the majority shareholder in HTIL (Hutchison Whampoa, 2008: 51).

After the flotation of HTIL, the company entered and exited various markets. Its operations in Paraguay (2005), Ghana (2008), Israel (2009)[3] and Thailand (2011) were sold while it entered the Indonesian market (2005) and launched services in Vietnam (2009). In March 2008, the company sold its base stations to PT Profesional Telekomunikasi for US$500 million (HTIL, 2008).

At around this time Hutchison Whampoa decided to 'privatise' HTIL. This privatisation would see Hutchison Whampoa acquire the outstanding stock in HTIL at HK$2.20 per share (Hutchison Whampoa, 2010b). The proposal document highlighted two pertinent issues. The first was that the remaining operations of HTIL lacked scale once those sold in India and Israel and spun off in Hong Kong and Macau were taken into account. The second was that none of the remaining networks was among the largest three in their respective markets, and although Hutchison Whampoa was confident of their long-term potential, achieving it would require investment in the short/medium term.

When coupled with the fall in the market capitalisation of HTIL from HK$95.6 billion to HK$7.9 billion and the limited 'free float' of its shares, Hutchison Whampoa argued that it was no longer appropriate for it to be listed on the stock exchange. Moreover, in removing the pressures associated with being a public company, Hutchison Whampoa stated that it would be better placed to manage the business and make the appropriate investment decisions (Hutchison Whampoa, 2010b). On 24 May 2010, Hutchison Whampoa announced that the Grand Court of the Cayman Islands, where HTIL was incorporated, had approved the privatisation of the company, and that it was expected to become effective on the following day (Hutchison Whampoa, 2010c, 2010d). When the privatisation was effected, HTIL was de-listed from the Hong Kong and New York stock exchanges, and subsequently changed its name to Hutchison Asia Telecommunications (HAT hereafter). With these changes, HAT would become a wholly owned subsidiary of Hutchison Whampoa.

Hutchison Whampoa also has invested in the telecommunications industry through 3 Group. 3 Group was created to encompass the company's 3G investments, primarily in Western Europe but also in Australia. Although Hutchison began to use the term '3 Group' only in its 2004 annual report, the origins of the company's involvement with 3G can be traced back to 2000. Over the course of 2000, Hutchison acquired 3G licences in four countries: Austria, Italy, Sweden and the UK (Hutchison Whampoa, 2001: 28ff). Hutchison explicitly states that it is reinvesting part of the proceeds from the sale of Orange plc in the acquisition of 3G licences. In Austria, the wholly owned Hutchison 3G

Austria paid €139 million for its licence, while in Italy and the UK the sums paid were considerably more, at €3254 million and £4385 million respectively (Hutchison Whampoa, 2001: 29). In contrast, a nominal fee was paid for the Swedish 3G licence.

The licences in Italy, Sweden and the UK were acquired through joint ventures. In November 2000, Andala SpA, a consortium headed by Tiscali, was awarded a 3G licence (Curwen and Whalley, 2006: 629). Three months later, Hutchison Whampoa acquired 78.3 per cent of the company from Tiscali and other shareholders, changing the name of the company to H3G Italy. In the UK it sold 35 per cent of Hutchison 3G UK to NTT DoCoMo and KPN Mobile after it had successfully won the licence – further details of this transaction are set out below. In the case of Sweden, Hutchison Whampoa entered into a joint venture – Hi3G Access – with Investor AB, the investment company of the Wallenberg family. Hutchison Whampoa owns 60 per cent of Hi3G Access, with Investor AB owning the remainder (Hutchison Whampoa, 2001: 29ff).

To offset the costs of rolling out a national infrastructure, Hi3G Access joined with two other 3G licence winners in Sweden, Orange Sverige and Vodafone, to form 3GIS. This company was created to build a 3G infrastructure covering the 70 per cent of the Swedish population living outside the three main urban areas of Gothenburg, Malmö/Lund and Stockholm/Uppsala. Hi3G Access is unusual as it has acquired 3G licences of its own in other Nordic markets. In September 2001, the company successfully bid DKr950 million for a 3G licence in Denmark (Hutchison Whampoa, 2002: 27), and NKr62 million for a 3G licence in Norway in September 2003 (Hutchison Whampoa, 2004: 27).[4] Interestingly, the annual report states that the Norwegian business would be developed as an extension of those in Denmark and Sweden. In other words, Hutchison Whampoa viewed the three markets of Denmark, Norway and Sweden as being closely interlinked.

The final West European market in which 3 Group is present is Ireland. In June 2002, the wholly owned Hutchison 3G Ireland paid $50 million for the largest of the available licences (Curwen and Whalley, 2006: 629). 3 Group also has a history in Australia, where Hutchison Whampoa became the majority shareholder of Hutchison Telecommunications Australia. Hutchison's presence in Australia is outlined in the following section.

4.3 OPERATIONAL PERFORMANCE

Geographical Footprint

The geographical footprint of Hutchison Whampoa is presented in Table 4.1. This table identifies those countries in which Hutchison has invested, either directly or indirectly, through one of its various holding companies. The first observation that can be made is that the number of countries in which Hutchison has been present at one time or another has fluctuated. Broadly speaking, the number of countries in which Hutchison was present increased (1995 to 2002) before stabilising at or near the present total of 12. Second, the main geographical focus of Hutchison's various international investments has been in the Asia-Pacific region and Western Europe. Although the company has invested outside these two regions in countries such as Argentina, Ghana, Paraguay and the USA, these investments have been relatively short-lived. Nevertheless, the shortest time spent by Hutchison in a country was in Western Europe. In the cases of both Belgium and Switzerland, the company entered the market only to exit the following year. As both investments were made by Orange plc, the exit from the market was due to the sale of that company to France Télécom.

A third observation is that, in the case of the UK, Hutchison exited the market in 1999 when it sold Orange only to re-enter it the following year when it successfully participated in the 3G spectrum auction. Shortly after winning the fifth, and largest, 3G licence, Hutchison sold 35 per cent of the company to NTT DoCoMo (20 per cent) and KPN (15 per cent) for a total of £2.1 billion (NTT DoCoMo, 2000). However, the relationship between the three partners was problematic, with KPN refusing to contribute additional funds to Hutchison 3G UK in early 2003 (Leahy and Nutall, 2003: 31). In November of the same year, KPN agreed to sell back its shares to Hutchison Whampoa for £90 million, while in May 2004, NTT DoCoMo also agreed to sell its shares for £120 million. Hutchison Whampoa completed the purchase of these shares in 2005 (Lau, 2005: 28).[5]

The equity held by Hutchison Whampoa has also changed in other countries. Table 4.1 shows that, in the case of India, Hutchison initially increased its ownership before expanding into other parts of the country. Hutchison launched its mobile operations in India in 1995, and was able to become the market leader in the Mumbai 'circle' (licence area) by the end of 1996 with more than 50 000 subscribers (Hutchison Whampoa, 1997). The 1996 annual report states that Hutchison had an effective 29.4 per cent stake in its Indian mobile business. During 1998, Hutchison

Table 4.1 Telecommunication holdings of Hutchison Whampoa, March 2014

	1995	1996	1997	1998	1999	2000	2001	2002	2003	2004	2005	2006	2007	2008	2009	2010	2011	2012	2013	2014
Argentina							Entry[13]				Exit[17]									
Australia[1]	[6]				%↓[10]	%↑	Alliance[14]						%↓		Merger[20]					
Austria[1]						Entry													Merger[24]	
Belgium[2]				Entry	Exit															
Denmark[4]							Entry													
Ghana[1]				Entry							%↑			Exit						
Hong Kong[5]	[7]														Spin-off[21]					
India[4]	[8]			%↑		Expand	Expand					Expand	Exit[18]							
Indonesia[4]											Entry									
Ireland[1]								Entry												[25]
Israel[4]				Entry	%↓[10]								%↑	%↑	Exit					
Italy[1]						Entry							%↑	%↑						
Macau[5]							Entry								Spin-off[21]					
Norway[3]									Entry								[22]			
Paraguay						Entry					Exit									
Sri Lanka[4]			Entry																	
Sweden[1]						Entry														

	1995	1996	1997	1998	1999	2000	2001	2002	2003	2004	2005	2006	2007	2008	2009	2010	2011	2012	2013	2014
Switzerland[2]				Entry	Exit															
Thailand[4]								Entry					Merger[19]				Exit[23]			
UK[1]	[9]				Exit[11]	Entry[12]			%↑[15]											
USA			Entry		Expand	Exit														
Vietnam[4]										Entry[16]										
No. of countries	4	4	6	10	7	11	14	16	17	18	17	17	16	15	14	14	13	12	12	12

Key:
%↑ – Shareholding increased
%↓ – Shareholding decreased

Notes:

1. Part of 3 Group.
2. Held via Orange plc.
3. Wholly owned subsidiaries of 60 per cent Hi3G Access.
4. Held via Hutchison Telecommunications International Limited (HTIL).
5. Part of Hutchison Telecommunications Hong Kong Holdings Limited.
6. Limited information is provided in the earlier annual reports regarding when it entered the Australian market. However, Hutchison Telecommunications (Australia) Ltd (2009) does provide a timeline of the company's presence in Australia, showing that it first entered in 1989.
7. The 1996 annual report mentions rapid growth compared to 1995, thus implying that mobile operations were present in Hong Kong in 1995.
8. The 1996 annual report implies that the Indian joint ventures commenced operations in 1995, though it does not say when the joint ventures were established.
9. Hutchison Whampoa's first entry into the UK telecommunications occurred in 1991, followed by the creation of Orange PCS in 1994.
10. Shares sold as part of IPO.
11. Sale of Orange plc to France Télécom.

12. Shortly after winning a UK 3G licence, Hutchison Whampoa reduced its stake in its UK operations to 65 per cent through selling shares to NTT DoCoMo (20 per cent) and KPN Mobile (15 per cent).
13. According to the 2004 annual report, this investment was not part of either HTIL or 3 Group.
14. Announces strategic alliance with TCNZ.
15. Various annual reports provide contradictory dates regarding when the stakes held by NTT DoCoMo and KPN Mobile were repurchased.
16. The company does not own a network but instead is allocated profits *pro rata*.
17. As no explicit mention of the sale of the Argentina business is made in the annual reports, the 2005 exit date is derived from the footprint maps provided.
18. HTIL sold CGP Investment (Holdings) Limited, the vehicle owning its operations in India.
19. Expands coverage but is not national in scope.
20. Merger with Vodafone Australia with the result that Hutchison Telecommunications Australia Limited owns 50 per cent of Vodafone Hutchison Australia.
21. Although the holding company was spun off, Hutchison Whampoa retained a majority stake in Hutchison Telecommunications Hong Kong (Holdings).
22. Although a licence has been awarded, operations have yet to start in Norway.
23. Sold during January 2011.
24. In January 2013, 3 Austria completed the purchase of Orange Austria.
25. The acquisition of O_2 Ireland has yet to receive regulatory approval.

increased its stake in its Indian mobile business – Hutchison Max Telecommunications – to 49.5 per cent, and subscribed to the preference shares of another company holding a large stake in Hutchison Max (Hutchison Whampoa, 1999: 26).

Purchases of other mobile operators expanded the geographical footprint of Hutchison Max in India. In 2000, it acquired a 49 per cent interest in three GSM operators in the New Delhi, Calcutta and Gujarat state circles (Hutchison Whampoa, 2001: 28), and further expanded in the following year when it paid $99 million for mobile licences in Karnataka, Andhra Pradesh and Chennai (Hutchison Whampoa, 2002: 28). In January 2006, it acquired BPL Mobile Cellular, which, when combined with the pending purchases of BPL Mobile Communications and Essar Spacetel, would expand the company's footprint in India to all 23 circles (Hutchison Whampoa, 2006: 47).

Increasing the geographical footprint over time appears to have been the driving force in Australia. Hutchison has been present in Australia since 1989 when it started offering paging services (Hutchison Telecommunications (Australia) Limited, 2010). At some point after 1989 but before 1997, the company also began to offer GSM-based mobile services (Hutchison Whampoa, 1998).[6] During 1997 Hutchison increased its stake in its Australian joint venture from 56 per cent to 70 per cent (Hutchison Whampoa, 1998). In 1998, the company was awarded CDMA licences covering Sydney and Melbourne, and signed a national roaming agreement with Telstra in the following year (Hutchison Telecommunications Australia, 2010). The year 1999 also saw the initial public offering (IPO) of Hutchison Telecommunications Australia (Hutchison Whampoa, 2000). As a result of the IPO, Hutchison's stake in its Australian business fell to 54 per cent. In 2001, Hutchison Telecommunications Australia won 3G licences covering the country's five major cities (Hutchison Whampoa, 2002). The CDMA network was closed in 2006 (Hutchison Telecommunications Australia, 2010) with services having been switched primarily to the 3G network.

Another joint venture was entered into in 2009, this time with Vodafone when the two companies merged their Australian mobile businesses to create Vodafone Hutchison Australia (VHA). Reflecting the difference in value between the two businesses, Vodafone paid Hutchison Telecommunications Australia A$587 million (Vodafone, 2009). This enabled the joint venture to be owned equally by the two companies. Although the creation of VHA was accompanied by the licensing of the '3' brand, several commentators took the view that the brand would only be used in the short term, unlike that of Vodafone (Sainsbury, 2009; Vodafone, 2009). At the heart of the merger was the desire to improve the performance of both mobile businesses through expanding network coverage, gaining scale economies and combining their different strengths (Middleton, 2009).

In late 2011 it was reported that Hutchison 3G Austria had made a $1.9 billion bid for Orange Austria (Mansfield, 2011). This was confirmed a few months later in February 2012 – Hutchison 3G Austria would acquire Orange Austria for €1.3 billion from Mid Europe Partners and France Télécom (Hutchison 3G Austria, 2012a). This sum, however, would fall to €900 million once the sale of various assets to Telekom Austria to satisfy competition concerns was taken into account (Hutchison 3G Austria, 2012b). Although some doubts about the merger were raised by the European Union (Mansfield, 2012d), concerns that were disputed by Hutchison (Hutchison 3G Austria, 2012b), the merger was eventually approved in December 2012. The merger was formally completed in January 2013 (Hutchison Whampoa, 2013).

Not long after the merger in Austria was finalised, it was confirmed that Telecom Italia and Hutchison 3G Italy were in merger discussions. These discussions did not progress to the satisfaction of Hutchison, and were terminated by them in July 2013 (3 Italia, 2013). In contrast, Hutchison has been engaged in M&A activity in Ireland. In early July 2012 it emerged that Hutchison and Vodafone were discussing an infrastructure-sharing joint venture that would save each company £200 million over five years (Mansfield, 2012b). The joint venture was formally announced in mid-July, though without any details of the cost savings that each party would achieve (Hutchison Whampoa, 2012). Less than a year later, it was reported that Telefónica was on the verge of selling O_2 Ireland to Hutchison. This was subsequently confirmed by Telefónica in late July, with Hutchison paying €850 million for its Irish subsidiary (Mansfield, 2013). Regulatory concerns over the resulting structure of the market where two operators (Vodafone, Hutchison/Telefónica) would be significantly larger than the third (Eircom) have resulted in the acquisition not being finalised (Mansfield, 2014a). Countering these concerns, Hutchison has argued that the merger would provide it with the necessary scale to compete against Vodafone and to fund the necessary infrastructure-related investment (Hutchison Whampoa, 2014).

Geographical expansion within a country played a role in Thailand and the USA, albeit arguably to a lesser extent than in Australia or India. Hutchison Whampoa has been present in Thailand since 2002, with its presence and structure significantly influenced by the unique approach to the telecommunications sector adopted in Thailand. HTIL owned stakes in two companies, one of which – Hutchison CAT Wireless Multimedia

Ltd – provided cdma2000-based mobile services (HTIL, 2006). According to the 2005 annual report of HTIL, two companies owned Hutchison CAT: Hutchison Wireless (73.9 per cent) and CAT Telecom (26 per cent). In turn, 49 per cent of Hutchison Wireless was owned by HTIL and 41 per cent by GMRP (Thailand) Ltd.[7] Not only did HTIL's shares carry more votes, but GMRP held the shares on behalf of HTIL and was constituted through its articles of association so that it could be directed by HTIL to sell the shares it held in Hutchison CAT to another company that it designated.[8] Through this arrangement of cascading stakes, HTIL controlled a 36.2 per cent economic interest in Hutchison CAT.

Within Thailand, the footprint of Hutchison CAT covered 25 out of the country's 76 provinces (HTIL, 2008). Although these provinces accounted for the majority of Thailand's economic output, a significant proportion of the population was obviously unable to access Hutchison CAT services. After technology trials in 2006, Hutchison CAT and CAT Telecom entered into a nationwide roaming agreement in 2007 that expanded the former's scope to encompass the whole of Thailand (HTIL, 2008). While this widening of geographical coverage may have contributed to an increase in the number of Hutchison CAT subscribers and reduced churn between 2006 and 2008, revenues – in total and on a per user basis – fell over the same period (HTIL, 2009). Given this decline in revenues, it is unsurprising that the 2009 annual report stated that the company was in discussions to exit the market (Hutchison Whampoa, 2010a: 50). In April 2010, CAT Telecom was given government approval to purchase Hutchison CAT for THB7.5 billion (TeleGeography, 2010a). However, the appointment of a new board at CAT Telecom resulted in the acquisition initially being questioned before the government ordered CAT Telecom to pay no more than THB4 billion (TeleGeography, 2010b, 2010c, 2010d). With Hutchison refusing to accept such a sum, the sale to CAT Telecom collapsed, allowing True to acquire Hutchison's operations in Thailand for THB4.35 billion (TeleGeography, 2010e). The sale was completed in early 2011 (Hutchison Whampoa, 2011; TeleGeography, 2011).

In 1997, Hutchison acquired minority stakes in two US-based mobile operators: Western Wireless Inc. and Western PCS Corp. (Hutchison Whampoa, 1998). In the following year, Western PCS Corp. was re-branded as VoiceStream Wireless PCS, and expanded its geographical footprint by merging with two other mobile operators – Aerial Communications and Omnipoint – during 1999 and 2000. By the time these mergers had been completed, Hutchison's stake in VoiceStream had risen to 23 per cent (Hutchison Whampoa, 2000).[9] In July 2000, VoiceStream agreed to merge with Deutsche Telekom in a deal worth $50.7 billion plus the assumption of $4.2 billion of debt (Curwen and Whalley, 2004: 16). As a result of this deal, Hutchison received $885 million in cash plus shares in Deutsche Telekom amounting to 4.9 per cent of the company (Hutchison Whampoa, 2002: 29). Hutchison realised a profit of HK$30 000 million on the sale of VoiceStream to Deutsche Telekom, and subsequently sold its stake in Deutsche Telekom over a period of several years (Hutchison Whampoa, 2002; TeleGeography, 2003).

As noted above, the investment made by Hutchison in the USA was one of a handful made outside its main geographical focus of the Asia-Pacific region and Western Europe. Hutchison made just one investment in Africa when, in 1998, it acquired 80 per cent of Kasapa Telecom in Ghana (HTIL, 2007). In December 2004, the company was awarded a 15-year replacement licence enabling CDMA-based services to be delivered. Hutchison acquired the remaining 20 per cent of the company in January 2005. In January 2008, Hutchison agreed to sell Kasapa to EGH International for HK$583.5 million (HTIL, 2008), completing the sale in July 2008 (HTIL, 2009).

Hutchison also invested in two Latin American markets. In the case of Paraguay, the annual reports of both Hutchison Whampoa and HTIL shed little light on the nature and scale of the business in this particular market. The first mention of Hutchison being present in Paraguay occurs in the 2000 annual report, but it is not until the following year that sufficient details are provided to make it possible to state that the business was a 2G mobile operator 60 per cent owned by Hutchison (Hutchison Whampoa, 2002: 27). The network covered the major cities of Paraguay. The business was sold for an undisclosed amount to América Móvil in July 2005 (Hutchison Telecom, 2005).

The second market in which Hutchison invested was Argentina. In Argentina, Hutchison provided mobile services through wireless local loop technologies (HTIL, 2006), and thus did not provide true mobile services as it did in other markets. Hutchison Argentina was a subsidiary of Hutchison Whampoa, so with the creation of HTIL an option to acquire the company was entered into in September 2004 (HTIL, 2006). This option ensured that Hutchison Argentina would continue to operate for three years and could not be sold to another company without both parties agreeing, enabling HTIL to evaluate the viability of its technology and thus determine whether it should purchase the company. This option does not appear to have been exercised, with Hutchison Argentina being removed from the footprint map included in the annual report of Hutchison Whampoa in 2005.

Financial Performance

Determining the financial performance of the telecommunications operations of Hutchison Whampoa is by no means easy. Apart from the seemingly constant restructuring of its various telecommunication operations noted above, determining the financial success of Hutchison is complicated by two other factors: first, the company has frequently restated its revenues and profits; and, second, EBIT/LBIT (earnings/loss before interest and taxation) is used to calculate 3 Group profits. The use of EBIT/LBIT is unusual to say the least, as it leaves out depreciation and amortisation.

The first observation we can make based on earnings as stated in the relevant annual reports from 1995 to 2003 (rather than subsequent restatements) is that the profits from the telecommunications division are rather modest both in absolute terms and when compared to the extraordinary items. It would appear that the telecommunications division produced its highest profit in 1996, though this is slightly misleading as the total for that year, HK$3642 million, included an unspecified profit from the IPO of its UK-based mobile operator, Orange plc. Profits after extraordinary items fall after 1997, then slowly climb back to a broadly similar level over the next six years. Even so, the level of profits reported by the telecommunications division of Hutchison remains relatively modest when compared with the extraordinary items reported. The substantial figure for 1999 is composed of the sale of equity in Orange and the IPO of Partner Communications in Israel. Unravelling the aftermath of the Mannesmann/Vodafone merger contributed to the large level of extraordinary items for 2000, as did the sale of equity in two other Hutchison telecommunication businesses. Together, the extraordinary items for these two years amount to HK$138.4 billion, or more than £11 billion at the then exchange rate.

The picture presented in Table 4.2 is more complicated, reflecting changes to Hutchison's organisational structure and the tendency of the company to report income including extraordinary items. Nevertheless, it appears that HTIL reported profits of almost HK$67 billion, although this is primarily due to the sale of the company's operations in India to Vodafone for HK$69.3 billion (Hutchison Whampoa, 2008). In other words, without this sale HTIL would have been loss-making in 2007. Extraordinary items appear to be determining the profitability of HTIL – the loss incurred on the sale of the Paraguayan operations is only partially offset by the gain from the sale of equity in Hutchison Global Crossing Holdings, resulting in a loss overall for 2005.

Table 4.2 Earnings of the various telecommunication holdings of Hutchison Whampoa, 2004–2012 (in millions)

Year	HTIL HK$	HAT[1] HK$	HTHKH HK$	3 Group HK$[2]	HTAL[3] A$	Extraordinary items HK$
2004	23[5]	–	97[7]	–7 291	–[4]	–
2005	–768	–	–	1 825[8]	–[4]	HTIL – loss on sale of Paraguay, gain from sale of 26 per cent in HGCH
2006	201	–	–	–19 996[8]	–[4]	–
2007	66 884[6]	–	–	–17 938	–[4]	HTIL – gain from sale of Indian operations of 69 343
2008	1 883	–	–	–10 875	–[4]	HTIL – from Indonesian tower lease and buyback
2009	4 940	–	468	–8 922	–[4]	HTIL – from sale of Israeli operations and proceeds from sales in Indonesia 3 Group – gain of 3641 from sale of 3 Australia
2010	–	–2 688	1 090	2 931	476	3 UK – one-off gain from network sharing = 6010. One-off loss against restructuring = 3742 HAT – 699 compensation from suppliers; 251 profit from sale of towers
2011	–	–1 181	1 435	1 481	313	HAT – one-off compensation = 1270 Gain on sale of Thai operations = 463 3 Group – sale of spectrum in Italy, network sharing in UK = 457
2012	–	–846	1 756	3 145	177	HAT – one-off compensation = 1590 3 Group – one-off gain from network sharing in Ireland = 447

Notes:

1. Hutchison Asia Telecommunications, previously Hutchison Telecommunications International Ltd.
2. Calculated using L(E)BITDA (loss [earnings] before interest, tax, depreciation and amortisation) before expensed CAC (customer acquisition costs and retention costs).
3. Hutchison Telecommunications Australia Ltd.
4. Reported as part of 3 Group.
5. Profit attributable to Hutchison Whampoa.
6. Profits before extraordinary items not stated in the 2007 annual report.
7. From Hong Kong and Macau operations.
8. The 2006 annual report restates the 2005 annual report using LBIT. In 2005 LBIT is stated as being HK$ 36 280 million.

Source: Compiled by the authors from the annual reports and 20-F filings of Hutchison Whampoa, Hutchison Telecommunications (Australia) Ltd, Hutchison Telecommunications International Ltd and Hutchison Telecommunications Hong Kong Holdings Ltd.

While the financial performance of HAT appeared to improve between 2010 and 2012 (inclusive), with losses declining over the period, the company reported substantial extraordinary items in each year. That both the losses and the extraordinary items incurred declined between 2011 and 2012 suggests that the underlying financial performance of HAT is improving.

Table 4.2 highlights the gradual financial improvement of the 3 Group. The table suggests that 3 Group was profitable in 2005, albeit at a modest level, before four consecutive years (2006 to 2009 inclusive) of substantial losses. To state that 3 Group was profitable in 2005 is arguably misleading, as the accounts for that year were restated in the following year using a different methodology, resulting in the company reporting a loss (Hutchison Whampoa, 2006). Between 2004 and 2009 (inclusive), the losses incurred by 3 Group amounted to HK$97.7 billion. This is a considerable sum, but it amounts to less than half of the profits Hutchison made from selling assets in just three years (1999, 2000 and 2007). In these three years, extraordinary items totalled HK$207.8 billion. In other words, the losses incurred by 3 Group were less than half of the income gained from the sale of Orange and Hutchison's Indian operations. Interestingly, Hunt (2010) states that Hutchison's 3G operations had accumulated losses of HK$158 billion since 2003. While it is not clear how this has been calculated, it would still be less than the sum raised by the extraordinary items in 1999, 2000 and 2007.

After taking into account the restatement for 2005, 3 Group has been consistently profitable since 2010. While a series of extraordinary items contributed to the group's profitability in 2010, Table 4.2 suggests that collectively the various businesses within 3 Group are performing increasingly strongly. However, the table does not shed any light on the individual performance of these operations. Until recently, the review of Hutchison's telecommunications operations in the annual report did not provide detailed figures regarding the profit or loss of its various businesses, and instead relied on terms such as '151% better than last year' in the case of Italy or, in the case of Sweden and Denmark, 'first full-year positive EBITDA' to describe performance (Hutchison Whampoa, 2007: 55).

By combining qualitative descriptions of the financial performance of the various businesses within 3 Group with the quantitative statements that are available, Table 4.3 highlights two issues. Although at the start of the period the detail provided by Hutchison, with the exception of Australia,[10] is somewhat imprecise, the general trend in the comments indicates a gradual improvement in the operational performance of the various mobile businesses. By the end of 2012, only Ireland continues to be loss-making. There is, however, a degree of volatility in the financial performance of the various businesses. For example, 3 Italia achieves its first positive EBITDA in 2005 before becoming loss-making in 2008 and 2009. From 2010 onwards, only the Irish operations of 3 Group have been loss-making.

The more detailed information that is available from 2010 onwards highlights a second issue, namely, the relatively low profitability of the various businesses within the 3 Group. The most profitable operations would appear to be located in the UK, followed by Italy and Sweden. That UK and Italy are the two most profitable markets is perhaps unsurprising when the large populations, and thus subscriber bases, of Hutchison in these countries are taken into account. That Sweden is the third-most profitable market for 3 Group is somewhat surprising given the relatively small size of the population. While the profits made in Austria reflect the small size of its population, they have been volatile as well: profits more than halved between 2010 and 2011 before increasing in the following year, though not by the same degree.

As noted above, extraordinary items play a significant role in determining the profitability of HTIL. Although Table 4.2 clearly demonstrates that HTIL has been profitable more times than it has made a loss, even when the extraordinary item in 2007 is taken into account, it does not shed light on the performance of individual businesses that are presented in Table 4.4. It is clear from Table 4.4 that the mobile operations in Indonesia have never been profitable, while those in Sri Lanka and Vietnam have been largely loss-making. The operations in Sri Lanka were initially profitable before becoming loss-making, while those in Vietnam became profitable in 2012. In contrast, the mobile operations in India and Israel were profitable throughout the period they were owned by HTIL. It is not clear why HTIL would choose to sell both of its consistently profitable mobile businesses, though the sales would conceivably generate revenue that could sustain losses elsewhere as these businesses edged towards profitability as well as fund new investments.

Table 4.3 The fluctuating financial performance of 3 Group, 2006–2012[1]

Country (currency)	2004	2005	2006	2007	2008	2009	2010[2]	2011[2]	2012[2]
Austria (€ mn)	'… revenues increased 16 times …'	'Total revenue increased by 160 per cent compared to 2004'	'LBITDA before all CACs reduced by 82 per cent to an almost breakeven position'	'LBITDA after all CACs also improved 60 per cent compared to 2006'	'… LBIT, before including the non-recurring foreign exchange gains, improved 24 per cent…'	'… revenue and LBIT … increased 3 per cent and reduced 8 per cent respectively'	77	36	50
Australia (A$ mn)	'… net loss attributable to shareholders of A$552m compared to A$410m last year …'	'… net loss attributable to shareholders of A$547m, a 21 per cent improvement'	'… revenue from its 3G operations of A$849 million, 76 per cent better than in 2005'	'EBITDA after all CACs increased 464 per cent for a second year of positive EBITDA results'	'… revenue of A$1623m, a 23 per cent increase over the previous year and a net loss … of A$163m, a 43 per cent improvement from last year'	'… profit attributable to shareholders of A$468m …'	–	–	–
Denmark (DKK mn)	–	–	–	–	–	–	354	595	663
Italy (€ mn)	'… on target to be EBITDA breakeven, before expensed CAC, monthly from April'	'… reported its first full-year positive EBITDA before all CACs …'	'EBITDA … 151 per cent better than last year'	'… achieved a breakeven EBITDA position for the full year'	'3 Italia's revenue and LBIT … declined 15 per cent and increased 16 per cent respectively'	'… continued to improve, reporting a decrease in LBITDA after all CACs of 70% to €136m …'	263	257	264
Ireland (€ mn)	–	–	–	–	–	–	–56	–28	–19

Sweden and Denmark (SEK mn)	'… revenues increased 27 times …'	'Total revenue grew 135 per cent compared to 2004 …'	'… operation in Sweden achieved its first full-year EBITDA … offset by start-up losses in Denmark'	'Combined revenue … increased by 48 per cent compared to 2006 and LBIDTA after all CACs reduced by 75 per cent as compared to 2006'	'The combined operation achieved a major milestone of positive EBITDA after all CACs … in the second half of the year … , LBIT for 2008 improved 29 per cent compared to 2007'	'… continued to improve, reporting EBITDA after all CACs growth of 1,312%, resulting from a turnaround of LBITDA last year to EBITDA in 2009 …'	–	–	–
Sweden (SEK mn)	–	–	–	–	–	–	616	1 776	1 648
UK and Ireland (£ mn)	–	–	'… an 817 per cent turnaround from the comparable LBITDA last year …'	'3 UK achieved positive EBITDA after all CACs and before including non-recurring foreign exchange gains'	–172	'… combined LBIT … improved 66%…'	–	–	–
UK (£ mn)	'… achieved earnings before EBITDA breakeven before expensed CAC … in Dec 2004 …'	'First full-year positive EBITDA before all CACs …'	–	–	–	'… LBIT totalled £89 million, a 72% reduction from the comparable recurring LBIT last year …'	165	191	281

Notes:

1. EBITDA = earnings before interest, tax, depreciation and amortisation; LBITDA = loss before interest, tax, depreciation and amortisation; CAC = customer acquisition costs.
2. EBITDA.

Source: Compiled by the authors from the relevant reports of Hutchison Whampoa.

Table 4.4 Financial performance of Hutchison Telecommunications International Ltd (2004–2009) and Hutchison Asia Telecommunications (2010–2012)

	2004	2005	2006	2007	2008	2009	2010	2011	2012
Ghana	–	–	[7]	[6]	–	–	–	–	–
India	[3]	EBITDA[1] HK$3237m	EBITDA after CAC[1] HK$4900m	[6]	–	–	–	–	–
Indonesia	–	[4]	'… start–up losses of the Vietnam and Indonesia businesses.'	LBITDA HK$475m	LBITDA HK$1003m	LBITDA HK$1300m	LBIT up 17%	LBIT improved by 3%	LBIT improved by 8%
Israel	US$176m profit	US$77m profit	US$162m profit	US$244m profit	US$277m profit	[6]	–	–	–
Paraguay	–	HK$352m loss[6]	–	–	–	–	–	–	–
Sri Lanka	–	[5]	[7]	EBITDA after CAC HK$93m	EBITDA HK$37m	LBITDA HK$94m	LBIT up 61%	LBIT improved by 41%	LBIT increased by 27%
Thailand	LBITDA[1] HK$233m[2]	LBITDA HK$15m	EBITDA after CAC HK$57m	LBITDA after CAC HK$14m	EBITDA HK$81m	LBITDA HK$83m	[9]	–	–
Vietnam	–	[4]	–	LBITDA HK$228m	–	LBITDA HK$361m	LBIT up 43%[8]	LBIT improved by 29%	EBITDA representing 142% turnaround achieved

Notes:

1. EBITDA = earnings before interest, tax, depreciation and amortisation; LBITDA = loss before interest, tax, depreciation and amortisation; CAC = customer acquisition costs.
2. The reported loss was due to an extraordinary charge of HK$557 million relating to bad debts and customer acquisition costs (Hutchison Whampoa, 2004: 50).
3. The 2004 annual report does not provide specific details regarding the financial performance of the company's operations in India. It does, however, suggest revenue and subscriber growth (Hutchison Whampoa, 2004: 49).
4. Entered the market in this year.
5. No mention is made of Sri Lanka in the annual report (Hutchison Whampoa, 2005).
6. Exited the market in this year.
7. The 2006 annual report does not provide specific details regarding the financial performance in both Ghana and Sri Lanka. Instead, it is noted that in both countries the subscriber base increased as network coverage and quality was improved (Hutchison Whampoa, 2006: 50).
8. The 2010 annual report also states that one-off compensation of HK$669 million resulted in a positive EBIT for the year (Hutchison Whampoa, 2010a: 49).
9. Hutchison exited the Thai market in January 2011.

Source: Compiled by the authors from the relevant annual reports of Hutchison Whampoa.

4.4 DISCUSSION

Sections 4.2 and 4.3 have charted Hutchison Whampoa's involvement in the telecommunications industry. It is clear from both of these sections that the company's involvement is not straightforward. The organisational structure that has been adopted is complex, involving a series of listed holding companies that are connected to the operating companies through another tier of intermediate companies. Hutchison Whampoa has actively engaged in the sale and purchase of companies. Two of these sales have been highly successful – the sale of Orange plc to France Télécom and of the Indian operations to Vodafone. But can Hutchison Whampoa repeat this feat for a third time?

Hutchison Whampoa has used a significant proportion of the proceeds from the sales of Orange and its Indian operations to fund the development of 3 Group. As a result of the acquisition of a series of 3G licences and the launch of mobile services in seven countries, 3 Group now accounts for two-thirds of all mobile subscribers controlled by Hutchison Whampoa. While 3 Group has fared better in some countries than in others, it remains the smallest operator when measured by mobile subscribers in all the markets where it is present. The description of HTIL as 'lacking scale' when it was taken private by Hutchison Whampoa arguably applies to 3 Group. While Hutchison Whampoa could have floated all or part of 3 Group, it is unlikely that such a strategy would have raised the kind of sums that would have enabled it to begin to recoup the investment that it had made. The desire to achieve scale has arguably been acknowledged in respect of the Austrian and Irish markets where it has sought – so far successfully – to acquire another operator. Given the relatively recent nature of these developments, it is not clear whether the consolidation will translate into improved financial performance. Certainly, the concessions attached to both takeover bids will hold back any improvement in the short term.

Since 2006, Hutchison Whampoa has published more detailed financial data regarding the performance of 3 Group. These data are presented in Table 4.5 and enable an initial assessment of the performance of 3 Group to be made. With respect to revenue, Table 4.5 highlights that the decline in Italy that occurred between 2007 and 2010 has almost been offset by the subsequent revenue growth that occurred in 2011 and 2012. Although Table 4.5 shows how revenues in Austria have increased over the period covered by the table, the combination of Denmark with Sweden and of Ireland with the UK between 2006 and 2010 (inclusive) means that calculating whether revenues have increased over the whole period is

Table 4.5 Financial performance of 3 Group, 2006–2012

	2006	2007	2008	2009	2010	2011	2012
				Revenue (million)			
Australia A$	848.9	1156.4	1467.9	1884.5	2201.4	–	–
Austria €	172.9	190.4	169.9	174.2	207.3	321	361
Italy €	2071.8	2018.2	1725.6	1647.0	1705.4	1782	1965
Sweden/Denmark SEK	2632.8	3893.3	4562.3	5552.2	6472.9	–	–
Sweden SEK						5641	5981
Denmark DKK						2355	2098
UK and Ireland £	1519.6	1590.9	1559.3	1548.8	–	–	–
UK £	–	–	–	–	1404.4	1787	1948
Ireland €	–	–	–	–	89.9	150	174
			Average revenue per user per month				
Australia A$	70.5	68.6	66.5	55.9	54.0	–	–
Austria €	51.2	45.3	33.0	23.9	21.8	22.3	22.6
Italy €	33.9	29.3	25.3	23.5	23.6	19.9	18.4
Sweden/ Denmark SEK	404.3	430.8	379.2	347.6	329.0	–	–
Sweden SEK	–	–	–	–	–	307.8	298.9
Denmark DKK	–	–	–	–	–	263.7	216.0
UK and Ireland £	46.6	43.4	33.6	26.5	–	–	–
UK £	–	–	–	–	22.6	21.9	21.2
Ireland €	–	–	–	–	25.4	31.0	32.2
			Average revenue per user per month from non-voice sources, %				
Australia	24.0	27.0	31.2	37.0	40.0	42.0	–
Austria	18.0	26.0	36.3	49.0	52.0	51.0	47.0
Italy	35.0	30.0	31.7	39.0	39.0	42.0	44.0
Sweden/Denmark	21.0	27.0	36.3	42.0	43.0	–	–
Sweden	–	–	–	–	–	41.0	42.0
Denmark	–	–	–	–	–	53.0	54.0
UK and Ireland	29.0	33.0	33.5	37.0	–	–	–
UK	–	–	–	–	41.0	47.0	48.0
Ireland	–	–	–	–	55.0	58.0	56.0

Source: Compiled by the authors from the relevant annual reports of Hutchison Whampoa.

difficult. That aside, the figures for 2010 onwards clearly demonstrate how revenues in three countries – Sweden, Ireland and the UK – have increased while those in Denmark have declined.

Table 4.5 also charts the average revenue per user (ARPU) for 3 Group. In all cases bar Ireland, ARPU has declined. Although a degree of caution has to be exercised because of the manner in which the data were presented for most of the period covered, the picture painted is somewhat bleak in respect of the increase in Ireland and the decreases in Denmark, Sweden and the UK. The widespread nature of the decline suggests a deterioration in 3 Group's financial performance although the pace at which ARPU has fallen has slowed in Austria and Italy. A contributory factor to arresting this decline has been the shifting balance of ARPU composition. The proportion of ARPU accounted for by non-voice products has increased over the period covered by Table 4.5, with the majority of ARPU in Denmark and Ireland coming from non-voice sources.

4.5 CONCLUSIONS

This chapter has outlined the presence of Hutchison Whampoa in the telecommunications industry, and has thereby demonstrated the scope and complexity of Hutchison Whampoa's involvement. Markets have been entered and exited and businesses have been bought and sold around the world. Hutchison Whampoa has also made extensive use of holding companies, some of which have been listed on a variety of stock exchanges, as well as of intermediate companies that link these holding companies to the various operating companies.

The scale and scope of Hutchison Whampoa's involvement in the telecommunications industry provides the context for another question that this chapter has sought to address, namely, whether it has been able to repeat its previous successes of Orange and its Indian mobile operations. Both of these were sold for considerable sums, enabling Hutchison Whampoa to fund the development of both 3 Group and HAT. But has this substantial and continued support resulted in success? The picture is mixed. HAT continues to be loss-making, although over time its financial performance has improved, aided in part by a series of extraordinary items. In contrast, 3 Group is profitable in all bar one of its markets and it has been able to switch its revenues towards non-voice sources. While this suggests that the investment provided by Hutchison Whampoa has begun to bear fruit, the allegation made by the company when HTIL was 'privatised' that it lacked scale arguably continues to apply to its remaining telecommunication operations. This has been acknowledged by the successful consolidation that has occurred in Austria and the proposed takeover in Ireland. If this is successful, then 3 Group will have deepened its financial commitment to these two markets in the hope that the stronger operators that result will be better placed to compete with their larger rivals. That Hutchison Whampoa entered into merger talks, albeit unsuccessfully, with Telecom Italia in Italy also suggests that the need for scale is present in the larger markets where 3 Group operates. Such consolidation would, of course, require yet more investment by Hutchison Whampoa into the telecommunications industry.

New entry into the mobile sector has rarely been successful, and in any event it has largely been the preserve of individual companies that have attempted to take advantage of one-off opportunities. Only one company, Hutchison Whampoa, has pursued new entry across many markets and over a relatively long period. As has been made clear, this is because the parent company was willing and able to withstand the initial losses that are inevitable when setting up a new network. However, as Hutchison Whampoa discovered, incumbents with large market shares are not easy to displace using the same kind of tactics as they themselves employ, and hence Hutchison subsidiaries evolved into an essentially disruptive force, relying increasingly on low prices to buy market share. But this is not the obvious route to profitability and, indeed, it did not tend to produce any.

Hutchison Whampoa was able to withstand the losses but these could not simply be tolerated as an ongoing fact of life, which is why a considerable number of structural changes – sometimes successful and sometimes not – have been attempted and, indeed, are ongoing. However, there was one piece of good news in April 2014 when it was announced that, as a result of a successful appeal to the European Court of Justice against a ruling by the UK tax authorities, 3 UK would be able to offset its tax losses against profits earned by other subsidiaries of the parent group (Mansfield, 2014b).

Whether the current structure of its telecommunications operations is viable is a moot point, and hence Hutchison Whampoa will doubtless remain a disruptive force. However, it does seem fair to conclude that no other company will attempt to become any kind of force in the mobile sector via a strategy of new entry.

NOTES

1. The eight listed subsidiaries and associated companies identified by Hutchison Whampoa on its website are: Hutchison Telecommunications Hong Kong Holdings,

Hutchison Telecommunications Australia, Cheung Kong Infrastructure Holdings Ltd, Hutchison China MediTech, Husky Energy, Hutchison Harbour Ring Ltd, Hutchison Port Holdings and TOM Group Ltd.

2. See www.uncatd.org/wir for further details of the *World Investment Report* published annually by the United Nations Conference on Trade and Development (UNCTAD).
3. In mid-2012 Hutchison Whampoa announced that it was returning to the Israeli market through the purchase of a 75 per cent stake in Scailex Corp, a handset retailer that owns 51 per cent of Partner Communications (Mansfield, 2012a). The purchase would cost $125 million plus the assumption of $760 million of debt. However, two months later Hutchison Whampoa decided against the purchase without stating why (Mansfield, 2012c).
4. Hutchison exited the Norwegian market during 2010 (Hutchison Whampoa, 2011: 65).
5. Interestingly, the various annual reports of Hutchison provide contradictory dates for the completion of the re-purchase of equity in Hutchison 3G UK from NTT DoCoMo and KPN.
6. The 1995 and 1996 annual reports do not describe in detail the businesses operated by Hutchison in Australia, nor do they elaborate on the nature of the investment that the company made.
7. The remaining 10 per cent was owned by three individuals.
8. A similar set of arrangements surrounds BFKT (Thailand) Ltd. HTIL owns 49 per cent of PKNS, with the remaining 51 per cent owned by DPBB (Thailand) Ltd. PKNS, in turn, owns BFKT.
9. However, the following year's annual report stated that Hutchison held an effective interest of 18.4 per cent in VoiceStream (Hutchison Whampoa, 2001: 31).
10. That said, it is worth treating the reported profits and losses for Australia with a degree of caution. In 2004, Hutchison stated that its net loss attributable to shareholders was A$522 million (Hutchison Whampoa, 2004). The following year, the loss of A$547 million was reported as being a 21 per cent improvement on the previous year's results (Hutchison Whampoa, 2005). This is only possible when the restated profits for 2004, published in the 2005 annual report, are used.

REFERENCES

Curwen, P. and J. Whalley (2004), *Telecommunications Strategy – Cases, Theory and Applications*, London: Routledge.

Curwen, P. and J. Whalley (2006), 'The prospects for third generation new entrants in the European mobile telecommunications industry', *Telecommunications Policy*, **30**(10/11), 622–32.

Financial Times (2013), *FT Global 500*, March, www.ft.com.

HTIL (2006), *Form 20-F/A, Amendment No. 1, Annual Report Pursuant to Section 13 or 15(d) of the Securities Exchange Act of 1934, for the fiscal year ended December 31, 2005*, filed 29 June, Washington, DC: Securities and Exchange Commission.

HTIL (2007), *Form 20-F/A, Amendment No. 1, Annual Report Pursuant to Section 13 or 15(d) of the Securities Exchange Act of 1934, for the fiscal year ended December 31, 2006*, filed 27 June, Washington, DC: Securities and Exchange Commission.

HTIL (2008), *Form 20-F/A, Amendment No. 1, Annual Report Pursuant to Section 13 or 15(d) of the Securities Exchange Act of 1934, for the fiscal year ended December 31, 2007*, filed 9 May, Washington, DC: Securities and Exchange Commission.

HTIL (2009), *Form 20-F/A, Amendment No. 1, Annual Report Pursuant to Section 13 or 15(d) of the Securities Exchange Act of 1934, for the fiscal year ended December 31, 2008*, filed 27 May, Washington, DC: Securities and Exchange Commission.

Hunt, K. (2010), 'Li Ka-Shing's Hutchison finally sees 3G profits', www.news.bbc.co.uk, 29 March.

Hutchison Telecommunications (Australia) Limited (2009), *Annual Report 2008*, St Leonards, New South Wales, Australia.

Hutchison Telecommunications (Australia) Limited (2010), *Annual Report 2009*, St Leonards, New South Wales, Australia.

Hutchison Whampoa (1997), 'Annual Report 1996', www.irasia.com.

Hutchison Whampoa (1998), 'Annual Report 1997', www.irasia.com.

Hutchison Whampoa (1999), 'Annual Report 1998', www.hutchison-whampoa.com.

Hutchison Whampoa (2000), 'Annual Report 1999', www.hutchison-whampoa.com.

Hutchison Whampoa (2001), 'Annual Report 2000', www.hutchison-whampoa.com.

Hutchison Whampoa (2002), 'Annual Report 2001', www.hutchison-whampoa.com.

Hutchison Whampoa (2004), 'Annual Report 2003', www.hutchison-whampoa.com.

Hutchison Whampoa (2005), 'Annual Report 2004', www.hutchison-whampoa.com.

Hutchison Whampoa (2006), 'Annual Report 2005', www.hutchison-whampoa.com.

Hutchison Whampoa (2007), 'Annual Report 2006', www.hutchison-whampoa.com.

Hutchison Whampoa (2008), 'Annual Report 2007', www.hutchison-whampoa.com.

Hutchison Whampoa (2010a), 'Annual Report 2009', www.hutchison-whampoa.com.

Hutchison Whampoa (2010b), 'Hutchison Whampoa Group Proposes Hutchison Telecom International Privatisation at HK$2.20 per share', www.hutchison-whampoa.com, 8 January.

Hutchison Whampoa (2010c), 'Proposal to privatise Hutchison Telecom International sanctioned and expected to become effective on 25 May 2010 and AGM on 26 May 2010 to be cancelled', www.hutchison-whampoa.com, 24 May.

Hutchison Whampoa (2010d), 'Privatisation of Hutchison Telecom International becomes effective today', www.hutchison-whampoa.com, 25 May.

Hutchison Whampoa (2011), 'Annual Report 2010', www.hutchison-whampoa.com.

Hutchison Whampoa (2012), 'Vodafone Ireland and Three Ireland announce strategic partnership to share network infrastructure', www.hutchison-whampoa.com, 13 July.
Hutchison Whampoa (2013), 'Annual Report 2012', www.hutchison-whampoa.com, 26 March.
Hutchison Whampoa (2014), 'Three Ireland comments following the issue of the Commission's statement of objectives', www.hutchison-whampoa.com, 3 February.
Hutchison 3G Austria (2012a), 'Hutchison 3G Austria announces acquisition of Orange Austria from Mid Europe Partners and France Telecom combining the third and fourth player in the Austrian mobile telephone market', www.hutchison-whampoa.com, 2 February.
Hutchison 3G Austria (2012b), 'Hutchison 3G Austria welcomes EU approval of the acquisition of Orange Austria', www.hutchison-whampoa.com, 12 December.
Lau, J. (2005), 'Hutchison Whampoa buys out UK 3G partners', *Financial Times*, 11 May, 28.
Leahy, J and C. Nutall (2003), 'Li Ka-Sing's 3G plan hits a snag', *Financial Times*, 13 June, 31.
Mansfield, I. (2011), 'Hutchison 3G in talks to buy Orange Austria', www.cellular-news.com, 28 November.
Mansfield, I. (2012a), 'Hutchison Whampoa returns to Israeli telecoms market', www.cellular-news.com, 8 June.
Mansfield, I. (2012b), 'Vodafone and Three looking to merge Irish network infrastructure', www.cellular-news.com, 3 July.
Mansfield, I. (2012c), 'Hutchison Whampoa walks away from deal to buy control of Israeli network', www.cellular-news.com, 21 August.
Mansfield, I. (2012d), 'Austrian merger of Hutchison 3G and Orange facing competition concerns', www.cellular-news.com, 21 September.
Mansfield, I. (2013), 'Telefonica sells O_2 Ireland to Three to form one', www.cellular-news.com, 24 June.
Mansfield, I. (2014a), 'Hutchison faces regulator hurdles in bid to buy O_2 Ireland', www.cellular-news.com, 29 January.
Mansfield, I. (2014b), 'Three UK wins £1 billion tax case against UK government', www.cellular-news.com, 2 April.
Middleton, J. (2009), 'Vodafone, Hutch merge Australian operations', www.telecoms.com, 9 February.
NTT DoCoMo (2000), 'Hutchison Whampoa, NTT DoCoMo and KPN Mobile announce 3G mobile multimedia strategic co-operation', www.nttdocomo.com, 12 July.
Sainsbury, M. (2009), 'Vodafone Australia and Hutchison "3" mobile merge', *The Australian*, 10 February.
TeleGeography (2003), 'Hutchison bolsters 3G war chest with Vodafone/DT share sales', www.telegeography.com, 23 May.
TeleGeography (2010a), 'CAT gets cabinet approval for Hutch takeover', www.telegeography.com, 8 April.
TeleGeography (2010b), 'CAT's new board queries Hutch acquisition', www.telegeography.com, 13 July.

TeleGeography (2010c), 'CAT Telecom ordered to negotiate lower price for Hutch', www.telegeography.com, 27 September.

TeleGeography (2010d), 'Hutchison stubborn on Hutch CAT sale price', www.telegeography.com, 4 November.

TeleGeography (2010e), 'Can it be True? Thai-owned group to buy Hutch CDMA business', www.telegeography.com, 31 December.

TeleGeography (2011), '4 Real: Cat awards four contracts to True units; Hutch takeover complete', www.telegeography.com, 28 January.

UNCTAD (2001), *World Investment Report – Promoting Linkages*, New York and Geneva: UN.

UNCTAD (2002), *World Investment Report – Transnational Corporations and Export Competitiveness*, New York and Geneva: UN.

UNCTAD (2013), *World Investment Report – Investing in a Low Carbon Economy*, New York and Geneva: UN.

Vodafone (2009), 'Hutchison and Vodafone agree to merge Australian telecom operations', www.vodafone.com, 9 February.

3 Italia (2013), '3 Italia not to continue discussions with TI', www.hutchison-whampoa.com, 4 July.

5. Structural and strategic adjustment among Asia-Pacific mobile operators

5.1 INTRODUCTION TO REGIONS

It is of interest to consider, from a strategic viewpoint, whether internationalising operators concentrate on specific regions of the world. In any event, it would clearly be impossible to analyse structural and strategic changes on a worldwide basis in a single chapter, and hence it has been necessary to divide the world up into a series of manageable regions, with each having a dedicated chapter. Unfortunately, although a number of organisations provide data using a regional framework, they do so in a manner that serves their own particular objectives and, crucially, they disagree significantly with one another.

In the most basic sense, it is possible to use continents as a regional surrogate, but this only really works if there is a clear physical dividing line such as applies to Africa. In contrast, this method does not work at all well either for Europe which, if the evidence of the Eurovision Song Contest is anything to go by, is expanding ever further to the East,[1] or for the Asia-Pacific region.

In any event, the dividing line between North and South America is meaningless for research purposes in mobile telecommunications since the relevant dividing line clearly runs between the USA and Mexico – that is, between North and Latin America. As for Europe, it is, as noted, a moveable feast. Strictly speaking, neither Russia nor Turkey is geographically European, with both consisting of a relatively small Europeanised western section combined with a much larger Asian section to the East – indeed, Turkey in many ways looks to sit most comfortably with neither Europe nor Asia but rather the Middle East. However, the objective is to look at the world through the eyes of an internationalising operator, and this means that both Russia and Turkey are best regarded as part of Europe.

The same ambiguity could be said to arise in the case of Egypt, which is often included under the Middle East heading. Indeed, in some lists, it is unambiguously so included because the Middle East's southern boundary is drawn as an almost straight line below the southern borders of the countries on the northern shores of Africa (with, slightly oddly, the Sudan occasionally thrown in for good measure).

As for the Asia-Pacific region, its sheer diversity leads to it being chopped up in too many ways to mention. A particular issue arises with respect to the CIS countries, which effectively lie on the cusp between Europe and Asia. For our purposes they have been included in with the former Eastern Europe rather than the Asia-Pacific region because that appears to provide a better fit with our internationalisation model.

A further issue relates to the treatment of islands/island groups. There are roughly 100 nations and other entities that comprise islands. Many of these are rather obscure – for example, the Ashmore and Cartier Islands, Svalbard and Wallis and Futuna – and others have fewer than 1000 mobile subscribers, but some have a role to play in the analysis below. Therefore, what appears as Table 5.1 is certainly not intended to be treated as definitive, but it represents a sensible distribution bearing in mind that the focus is on international mobile telecommunications strategy and not, for example, on geo-politics.

It should be borne in mind that the regions in Table 5.1 contain a widely varying number of countries, although this can be somewhat deceptive in the context of the subject matter of this book. In effect, for example, North America contains only two countries of any significance, namely the USA and Canada, but events in the USA by themselves provide sufficient material to fill an entire chapter. In contrast, the countries listed in the Middle East are of little interest in themselves. Rather, it is the operators headquartered within them, which have developed internationally outside the region, that are of interest, and they are accordingly discussed where appropriate in other chapters.

5.2 INTRODUCTION TO THE ASIA-PACIFIC REGION

Curwen and Whalley (2008: 59) made the point that, during the period of adjustment following the market collapse of 2000/2002, relatively little structural change was to be found in the Asia-Pacific region and that the reasons for this were not immediately obvious. It was noted that the region contained large markets, large individual networks and a modest number of established international operators. However, despite the superficially attractive argument that neighbouring countries would have much in common and hence would be quite easy to enter successfully, this did not appear to be happening.

In seeking at least a partial explanation, it may be noted that while Ronen and Shenkar (1985) had demonstrated that some countries in the

Table 5.1 Regional allocation of countries[1]

Region	Countries
Western Europe[2]	Åland Islands, Andorra, Austria, Belgium, Bulgaria, Cyprus (South), Czech Republic, Denmark, Estonia, Faroe Isles, Finland, France, Germany, Gibraltar, Greece, Greenland, Guernsey, Hungary, Iceland, Ireland, Isle of Man, Italy, Jersey, Latvia, Liechtenstein, Lithuania, Luxembourg, Malta, Monaco, Netherlands, Norway, Poland, Portugal, Romania, San Marino, Slovakia, Slovenia, Spain, Sweden, Switzerland, UK
Eastern Europe and CIS[3]	Albania, Armenia, Azerbaijan, Belarus, Bosnia and Herzegovina, Croatia, Georgia, Kazakhstan, Kyrgyzstan, Macedonia, Moldova, Montenegro, Russia, Serbia, Tajikistan, Turkmenistan, Ukraine, Uzbekistan
Middle East	Bahrain, Iran, Iraq, Israel, Jordan, Kuwait, Lebanon, Oman, Palestine, Qatar, Saudi Arabia, Syria, Turkey, UAE, Yemen
Asia-Pacific	Afghanistan, American Samoa, Australia, Bangladesh, Bhutan, Brunei, Cambodia, China, Cook Islands, Fiji, French Polynesia, Guam, Hong Kong, India, Indonesia, Japan, Kiribati, Laos, Macau, Malaysia, Maldives, Marshall Isles, Micronesia, Mongolia, Myanmar, Nauru, Nepal, New Caledonia, New Zealand, Norfolk Island, North Korea, North Marianas, Pakistan, Palau, Papua New Guinea, Philippines, Samoa, Singapore, Solomon Isles, South Korea, Sri Lanka, Taiwan, Thailand, Tonga, Tuvalu, Vanuatu, Vietnam
North America	Bahamas, Bermuda, British Virgin Isles, Canada, Puerto Rico, USA, US Virgin Isles
Latin America	Anguilla, Antigua and Barbuda, Argentina, Aruba, Barbados, Belize, Bolivia, Brazil, Cayman Isles, Chile, Colombia, Costa Rica, Cuba, Dominica, Dominican Republic, Ecuador, El Salvador, Falkland Isles, French Guiana, French West Indies, Grenada, Guatemala, Guyana, Haiti, Honduras, Jamaica, Mexico, Montserrat, Netherlands Antilles, Nicaragua, Panama, Paraguay, Peru, St Kitts and Nevis, St Lucia, St Vincent and Grenadine, Suriname, Trinidad and Tobago, Turks and Caicos, Uruguay, Venezuela
Africa[4]	Algeria, Angola, Benin, Botswana, Burkina Faso, Burundi, Cameroon, Cape Verde Isles, Central African Republic, Chad, Comoros Isles, Congo-Brazzaville, Côte d'Ivoire, Djibouti, DR Congo, Egypt, Equatorial Guinea, Eritrea, Ethiopia, Gabon, Gambia, Ghana, Guinea, Guinea-Bissau, Kenya, La Réunion, Lesotho, Liberia, Libya, Madagascar, Malawi, Mali, Mauritania, Mauritius, Mayotte, Morocco, Mozambique, Namibia, Niger, Nigeria, Rwanda, São Tomé, Sénégal, Seychelles, Sierra Leone, Somalia,[5] South Africa, South Sudan,[6] Sudan,[6] Swaziland, Tanzania, Togo, Tunisia, Uganda, Zambia, Zimbabwe

Notes:

1. Treatment of islands: the status of individual islands or island groups is extremely problematic as different lists – for example those of the OECD, ITU and Wikipedia – use the term 'country' or 'state' in different ways. In essence, there are four considerations: the location of the government, the extent of local autonomy (independent regulation), the independence of the operator and the degree of physical separation. Since these can be combined in various ways, no list can be regarded as definitive. For our purposes, the following factors were most influential:
   Åland Islands: High degree of autonomy and independent networks so included.
   Azores: Governed and regulated by parent country so not counted.
   Canaries: Governed and regulated by parent country so not counted.
   Corsica: A French région but also, in law, a collectivité territoriale; it is not practical to distinguish it from the French mainland so not counted.
   Cocos Islands and Christmas Island: Treated as part of Australia.
   Faroe Islands: Autonomous province of Denmark so counted.
   Greenland: Autonomous province of Denmark so counted, but treated geographically as in Europe.
   Guam: Governed independently so counted.
   Guernsey: Sometimes treated collectively with Jersey, but has different operators so treated separately.
   Isle of Man: A Crown Dependency of Great Britain and self-governing so counted.
   La Réunion: Officially an overseas département of France and one of 26 French régions, but clearly a part of Africa geographically so counted.
   Madeira: Governed and regulated by parent country so not counted.
   Mayotte: Same status as for Corsica, but geographically part of Africa with different set of operators to mainland France so counted.
   North Cyprus: Not recognised internationally so counted as part of Turkey.
   Saipan: Counted as part of North Marianas.
   Sardinia: Part of Italy with some regional autonomy but not warranting a separate inclusion.
   Solomon Islands: Governed independently so counted.
   St Pierre et Miquelon: Orange treated as Canadian regional operator.
2. The Vatican City State is excluded (but counted by the ITU).
3. Abkhazia, Kosovo, Nagorno-Karabakh, South Ossetia and Transnistria are excluded as not internationally recognised on a sufficiently wide basis.
4. Western Sahara is disputed by Morocco and Algeria-backed Polisario Front and not counted separately.
5. Somaliland is not treated as a separate country.
6. The Republic of South Sudan achieved independence from the Republic of Sudan in July 2011. At the time it was claimed to be the world's 196th 'country' and the 55th in Africa. In October 2011, it was accepted as the 193rd member state of the International Telecommunication Union.

region did indeed share cultural affinities, some countries such as China could be said to be less than welcoming to Asia-Pacific inward investment. Furthermore, many countries such as India were strict about the proportion of a domestic operator's equity that could be acquired by a prospective investor, and some countries, including those on the Pacific Rim with its Anglo-Saxon heritage, could be seen only as culturally separated from the rest of the region.

Curwen and Whalley (2008) also observed that, because many countries within the region were emergent rather than developing, their growth potential had yet to be realised. This meant that there was a greater incentive to grow internally rather than to take the risks associated with overseas ventures. However, that is by definition a short-term phenomenon, and should not have applied to anything like the same degree during the period under consideration in this chapter. The other feature that might have changed is the attitude of investors, primarily in Europe, with an ever-increasing need to diversify away from their home markets as they reached saturation point, a possibility that is explored in what follows.

5.3 SETTING THE SCENE

According to Table 5.1, there are 47 countries/islands that can usefully be placed within the Asia-Pacific region for the purposes of this book. However, many of these either contain very small populations or possess a monopoly provider of mobile communications (usually state-owned) and hence they are most unlikely to figure on any list of countries of interest to internationalising operators. For this reason, Table 5.2 is restricted to countries that are either likely to figure on such a list or contain operators that are likely to participate in the internationalising process, and as a result it contains only 27 entries. This alone tends to make the region stand out, as this phenomenon is not replicated in other regions.

There are very substantial variations in the current subscriber bases of Asia-Pacific countries. It is difficult to make precise comparisons because the raw data come with wide variations in the extent to which subscriptions can be said to be 'live'. Overstatement of subscriptions is rife in countries such as India and Vietnam but modest in Japan and South Korea. However, it is evident that China leads the way by a considerable margin as it is the largest market in the world, with comfortably in excess of one billion reported subscribers at the time of writing – although it must be borne in mind that TD-SCDMA customers account (uniquely) for a significant proportion of the total. Although India had nearly 900 million reported subscribers at the end of 2012, it is probable that only 700 million or so were active, but either way it is easily the second-largest market in the world. With 240 million reported subscribers, Indonesia was clearly in third place in the region in 2012, followed by Pakistan, Japan,[2] the Philippines and Bangladesh, each with between 100 million and 130 million reported subscribers. There is then a considerable dropping off in numbers in Vietnam, Thailand and South Korea, with between 50 million and 75 million reported subscribers and then a further sharp fall until one arrives at Malaysia, Taiwan, Australia and Sri Lanka.

Although subscriber growth has been extremely rapid in recent years, taking the region as a whole, there are several individual country exceptions such as Japan and Australia, and to a lesser extent Malaysia, South Korea and Taiwan, which had already reached near-saturation levels five years ago. Although it is well known that penetration levels can soar well above 100 per cent, it is reaching that initial target that accounts for the rapid growth in the likes of Bangladesh, China, India and Indonesia, and overall growth is accordingly bound to slow down progressively as more and more countries in the region approach saturation. In the end, of course, sheer numbers of population are the overriding factor, so China and India will continue to account for a disproportionate share of the region's subscribers. However, it is of interest that the largest markets in the region are structured in fundamentally different ways, so this is a matter to which we must return below.

Meanwhile, we need to take a close look at Table 5.2, which matches the main operators in the region (in terms of networks and subscribers) to the main countries. The table contains data on individual networks without specifying the size of the stake in question or whether it is held directly or indirectly. This is because the networks themselves vary hugely in size and, given that it is the number of proportionate subscribers that matters because that is the source of revenues and profits, it is possible to generate the same number of subscribers from a small stake in a large network as from a large stake in a small network.

Bearing in mind, as noted, that only 27 of the 47 possible countries are listed in the table, it is arguably somewhat surprising that no individual operator can claim to have a presence, whether direct or indirect, in as many as ten countries. Between them, the 11 operators account for 61 entries, implying an average of between five and six, but only DoCoMo and SingTel account for nine and, as noted below, both are characterised by a tendency to hold minority stakes. On the face of it, 61 networks are accounted for out of a possible 122, but that is again rather deceptive for two reasons. First, in Bangladesh, Cambodia, Indonesia, Pakistan and Sri Lanka combined, 26 of the 27 networks are apparently accounted for – in other words, a small sub-set of countries accounts for a disproportionate share of operator entries. Second, however, not all the operator entries are independent – in other words, there may be more than one operator present as an investor in a given network.

Overall, therefore, it is reasonable to conclude that, for whatever reason – and it bears repeating yet again that you can only buy what is put up for sale – empire building has never been a prominent feature of the Asia-Pacific region and that such head-to-head competition between internationalising operators as exists is limited to a small number of countries.

Table 5.2 Location of networks of main Asia-Pacific operators, end 2012

Country	Networks Number	Axiata	Bharti Airtel	Etisalat	Hutchison Whampoa	NTT DoCoMo	Orascom Telecom	Qatar Telecom[1]	SingTel	Telenor	VimpelCom	Vodafone
Afghanistan	4	–	–	Y	–	–	–	–	–	–	–	–
Australia	4	–	–	–	Y	–	–	–	Y	–	–	Y
Bangladesh	6	Y	Y	–	–	Y	Y	–	Y	Y	Y	–
Cambodia	8	Y	–	–	–	–	–	Y	–	Y	Y[7]	–
China	3	–	–	–	–	–	–	–	–	–	–	–
Fiji	2	–	–	–	–	–	–	–	–	–	–	Y
Guam	4	–	–	–	–	Y	–	–	–	–	–	–
Hong Kong	5	–	–	–	Y	Y	–	–	–	–	–	–
India	14[2]	Y	Y	–[2]	–	Y	–	–	Y	Y[2]	–	Y
Indonesia	7	Y	–	Y	Y	–	–	Y	Y	–	–	–
Japan	4	–	–	–	–	Y	–	–	–	–	–	–
Laos	4	–	–	–	–	–	–	Y	–	Y	Y[8]	–
Macau	4	–	–	–	Y	Y	–	–	–	–	–	–
Malaysia	4	Y	–	–	–	–	–	–	–	Y	–	–
Maldives	2	–	–	–	–	–	–	Y	–	–	–	–
Mongolia	4	–	–	–	–	–	–	–	–	–	–	–
Myanmar	1	–	–	–	–	–	–	–[6]	–	–[5]	–	–
New Zealand	3	–	–	–	–	–	–	–	–	–	–	Y
North Korea	1	–	–	–	–	–	Y	–	–	–	–	–
Pakistan	5	–	–	Y	–	–	Y	–	Y	Y	Y	–
Philippines	3	–	–	–	–	Y	–	–	Y	–	–	–
Singapore	4	Y	–	–	–	–[4]	–	Y	Y	–	–	–
South Korea	4	–	–	–	–	Y	–	–	–	–	–	–
Sri Lanka	5	Y	Y[9]	Y[9]	Y[5]	–	–	–	Y	–	–	–
Taiwan	5	–	–	–	–	Y	–	–	–	–	–	–
Thailand	6	–	–	–	–	–	–	–	Y	Y	–	–
Vietnam	6[3]	–	–	–	Y	–	–	–	–	–	–	–
Total	122	7	3	4	6	9	3	5	9	7	4	4

Notes:

1. Renamed Ooredoo in 2013.
2. India is divided into 22 circles and no operator serves every one. 122 licences issued in 2008 were revoked by the Supreme Court in 2012 as a result of which Etisalat withdrew from the market. Telenor is involved in a restructuring exercise. New licences (2G, 3G and 4G) are being issued gradually.
3. In April 2014, the government ordered state-owned VNPT, owner of third-placed (by subscribers) VinaPhone, to sell off its stake in MobiFone (second-placed by subscribers). The number of networks will be unaffected.
4. Parent NTT is present in this market.
5. Sale to be completed during 2014.
6. Awarded new licences in 2013 that were not actually handed over until January 2014.
7. Sold in 2013.
8. Provisionally sold in 2013 but sale vetoed by government.
9. Etisalat agreed to purchase Bharti Airtel Sri Lanka in December 2013.

5.4 UPDATE TO 2012

Curwen and Whalley (2008: 68) contains a table that lists the main mobile operator presence in the Asia-Pacific region during the years 2004 to 2007 inclusive (but excluding China Mobile – see Note 1 to Table 5.3). It is instructive to compare this with Table 5.3, which provides a similar format for the period 2005 to 2012 inclusive – an overlapping and longer period – because by 2005 the immediate consequences of the economic meltdown of 2000/2002 had now passed and the main internationalising operators were in a position to review their strategies in the expectation of better times to come.

The comparison reveals a number of interesting facts once adjustments have been made to align the 15-company sample in the earlier table with the 13 companies listed second time around. In particular:

- Ten of the 13 companies are common to both tables, with Millicom, SK Telecom and Telstra disappearing and being replaced by Bharti Airtel, Qatar Telecom (Ooredoo) and Saudi Telecom. Telekom Malaysia has become Axiata. This replacement makes almost no difference to the number of countries listed.
- The earlier sample's range of proportionate subscribers spans from 4 million to 62 million, whereas the later sample spans from 23 million to 184 million. In other words, there was a massive increase in subscriber numbers between end 2007 and end 2012.

So far as the individual operators are concerned, the biggest changes in country presence apply to VimpelCom (0 to 4) and Orascom (3 to 2), followed by Telenor (4 to 7) and Vodafone (7 to 4). With only two entries, Orascom no longer merits inclusion in Table 5.2 but is included in Table 5.3 to help clarify what has happened.

It is immediately striking that Table 5.3 contains only five operators based in the Asia-Pacific region, although they represent five separate countries and China Mobile is excluded. The remainder (excluding Orascom) comprise three based in Europe (including VimpelCom, although most of Russia is technically in Asia) and three in the Middle East. Predictably, the USA and Africa are not represented.

If one examines the five regionally based operators, it is probably best to treat Hutchison Whampoa as an exception because it is a conglomerate based in Hong Kong for reasons other than its telecommunications activities. The other four obtain at least 75 per cent of their proportionate subscribers from within the region (as does China Mobile) – in other words, they can reasonably be described as uni-regional operators. None of the operators based outside the region obtains a comparable figure above 65 per cent, so there is no overlap. Interestingly, VimpelCom, which might be expected to have the strongest link to the region given its base in Russia, obtains the smallest proportion among the six based elsewhere. The three based in the Middle East average close to 50 per cent, which suggests that they tend to cast their eyes acquisitively to the East whereas, in practice, Etisalat is considerably more involved in Africa so it is important not to jump to conclusions.

The changes in Table 5.3 up to the end of 2007 are set out in Curwen and Whalley (2008: 67, 69) and therefore what follows covers the period beginning 1 January 2008.

- Bharti Airtel entered Sri Lanka in 2009 and Bangladesh in 2010.
- Singapore Telecom entered Sri Lanka in 2009.
- Vodafone left China (and indirectly Hong Kong and Pakistan) in 2010.
- Telenor entered India (and indirectly Cambodia and Vietnam via VimpelCom) in 2009 and exited (indirectly) Vietnam in 2012. In 2012 its licences in India were revoked but subscribers were attributed.
- NTT DoCoMo entered Bangladesh in 2008 and India in 2009 and exited Malaysia in 2010.
- VimpelCom entered Cambodia and Vietnam in 2009. In 2011 it entered Bangladesh, Laos and Pakistan. It exited Vietnam in 2012.
- Hutchison exited Thailand in 2011.
- Etisalat entered Sri Lanka in 2009 and India in 2010 but had its licences revoked in India in 2012 – it shut down its operation in 2013.
- Orascom exited Hong Kong, Indonesia, Macau, Sri Lanka, Thailand and Vietnam when it sold its stake in HTIL in 2008. It entered North Korea in 2009 but exited in 2012.

Table 5.3 Asia-Pacific mobile operator presence: number of countries[1] and subscribers, end year[2]

	2005	2006	2007	2008	2009	2010	2011	2012	Subs (mn)[3]	Asia-Pacific subs as % of total
Bharti Airtel[4]	1	1	1	1	2	3	3	3[10]	184.1[14]	75.8
Singapore Telecom	7	7	8	8	9	9	9	9	150.6	88.8
Vodafone[5]	6	6	7	7	7	4	4	4	115.3[14]	31.7
Telenor	4	4	4	4	7	7	8	7	114.5[14]	65.0
NTT DoCoMo[6]	7	8	8	9	10	9	9	9	97.2[14]	99.6
Axiata[7]	6	7	7	7	7	7	7	7[11]	89.5	99.9
Orascom[8]	7	7	8	2	3	3	3	3	63.3	73.8
Maxis	3	3	3	3	3	3	3	3[11]	60.2	100
Qatar Telecom[9]	0	0	5	5	5	5	5	5	37.6	62.3
VimpelCom[8]	0	0	0	0	2	2	5	4[12]	33.0	19.5
Saudi Telecom	0	0	3	3	3	3	3	3[11]	30.4[14]	54.7
Hutchison	6	6	7	7	7	7	6	6[13]	24.2	54.5
Etisalat	0	1	3	3	4	5	5	4[10]	23.3	36.0

Notes:

1. There is no reference to China Mobile (Hong Kong) Group in the table. This operator had roughly 700 million subscribers at the end of 2012, which renders comparisons with other operators rather pointless. It also operated in Hong Kong (which it regarded as part of China) and in Pakistan, with 2.8 million and 18.3 million subscribers respectively.
2. Numbers at year-end. It is not always possible to be certain of the year in which assets were transferred.
3. As of 31 December 2012. We refer here not to gross subscriber numbers but to 'proportionate' or 'equity' subscribers: that is, the gross figure is weighted according to the proportion of the equity held by the company in question.
4. These are also attributed to Singapore Telecom, which owns a significant minority stake, but not to Vodafone (see note 5).
5. Vodafone was originally in India via its stake in Bharti Airtel. However, after acquiring a large stake in what became Vodafone Essar, the Bharti stake was heavily reduced and hence is treated here as an investment and not counted.
6. NTT mainly operates internationally via majority-owned subsidiary DoCoMo.
7. Before 1 April 2009, the company was known as Telekom Malaysia International.
8. Orascom Telecom Holding was renamed as Global Telecom Holding in September 2013. The 2008 change reflects the sale of a stake in HTIL acquired in 2005. In 2011, a majority stake was acquired by VimpelCom which thereby acquired two new indirect holdings (but not North Korea) in the Asia-Pacific region.
9. Now known as Ooredoo.
10. Etisalat agreed to buy Bharti Airtel Sri Lanka in December 2014.
11. Axis, an Indonesian network with both Saudi Telecom and Maxis as shareholders was sold to Axiata in March 2014.
12. Agreed sale of Laos network in 2013 vetoed by government.
13. Hutch Lanka is being sold in 2014.
14. All these operators are to be found in India where overstatement of subscriber numbers is currently rife. The same is true of Vietnam, where only Hutchison is affected.

5.5 THE LARGEST MARKETS

As noted, one market towers above all others, namely China. But China is not a free market in any meaningful sense even though China Mobile and China Unicom are listed on the Hong Kong Exchange. Technically, China Mobile (Hong Kong) Group Ltd holds 74 per cent of China Mobile Ltd via a wholly owned subsidiary China Mobile HK BVI, so some opportunities arise to take minority stakes, albeit at considerable cost. In 2008, the then China Unicom hived off its CDMA assets to China Telecom and merged with China Netcom, forming an entity that continued to trade as China Unicom and was owned primarily by China Unicom (BVI) with 41 per cent and China Netcom Group (BVI) with 29.5 per cent.

Although a small number of foreign operators have been allowed to build up stakes, these have been kept at a level that provides no meaningful element of control – in other words, they were effectively investments from the buyer's point of view with an added element of joint collaboration as appropriate.

Vodafone held a 3.3 per cent stake in China Mobile in the early 2000s, which appeared to be barely worthwhile although it generated large numbers of proportionate subscribers – over 17 million in 2009. But Vodafone had already withdrawn from Japan and it seemed to be sensible to cash in its stake in China at this time given that it was *de facto* merely an investment and its overall strategy included a desire to dispose of minority holdings and return cash to shareholders.

Telefónica's involvement with China Unicom has been of more recent vintage and somewhat more complicated. In 2009, the companies agreed a cross-shareholding worth $1 billion, to which a further $500 million was added in 2010. Telefónica eventually acquired a 9.7 per cent stake in China Unicom (yielding roughly 18 million proportionate subscribers) and China Unicom ended up with a 1.4 per cent stake in Telefónica. However, China Unicom bought back a 4.6 per cent stake in July 2012, leaving Telefónica with roughly 5 per cent.

In principle, the stake in Telefónica gave China Unicom a foothold in foreign markets, but neither Chinese operator has shown much tangible interest in internationalising – hardly a surprise given the enormous opportunities and even more enormous investment associated with the home market. China Mobile owns the former People's in Hong Kong but tends to regard Hong Kong as part of the home market, as it does in respect of its network in Tibet. The only obviously foreign holding is Paktel (Zong) in Pakistan, which it has owned since 2007. This yielded

19 million subscribers at the end of 2012, but that figure is an insignificant proportion of its total subscribers. Oddly, there was an agreement to buy 12 per cent of Taiwan's FarEasTone in 2009 at a time when inward investment from China was technically forbidden, but aside from brief skirmishes with Reliance Telecom and 3 Italia that came to nothing, there has been no obvious interest other than to widen its operations in Pakistan.

At various times, China Mobile has gone on record with vague expressions of interest in emerging markets in the Middle East and Africa but has traditionally ignored Europe other than possibly to target Chinese inhabitants via MVNOs. In January 2013, general interest was expressed in Brazil, Germany, North Korea, Portugal and South Africa, but it would be unwise to bet on any concrete outcomes in any of these countries even though China Mobile is cash-rich and has very low gearing.

What is particularly intriguing is the claim made in January 2014, immediately after the statement by AT&T that it would not be making a bid for Vodafone Group after the disposal of its stake in Verizon Wireless was finally effected, that China Mobile would make an offer to acquire between 5 per cent and 20 per cent of Vodafone – once it had distributed a significant part of the payment from Verizon Communications in the form of dividends (Cellular-news, 2014). Although the two operators had formed a joint venture to bid (unsuccessfully) for a licence in Myanmar, it seemed unlikely that shareholders – many based in the USA – would welcome a Chinese firm taking even a non-controlling stake, and even if they had no objection it is certain that regulators in many of Vodafone's markets would not be at all happy with such a development.

China Unicom was originally the CDMA operator in China, so it arguably had little incentive to acquire networks running GSM. The small stake in Telefónica was, of course, in a portfolio of GSM networks but was not intended to provide China Unicom with any involvement in operating the networks. That aside, the chairman has made it clear that Unicom is not interested in overseas assets, although it did take part in a consortium bidding for 75 per cent of Nitel/Mtel in Nigeria in 2011. When one takes into account China's aggressive moves into Africa in other industrial sectors, this reluctance to buy mobile assets in Africa is a little surprising, although both Chinese operators are very reluctant to overpay for acquisitions.

Although India is the second-largest mobile market in the world, it operates in a way that could hardly be more different from China. While it is true that China is divided up into many individual regions, each with a superficially separate service provider, these are all direct subsidiaries of the national operators and to all intents and purposes the government calls the shots. In India, the regional breakdown includes the main cities treated separately and altogether there are 22 so-called 'circles', each licensed independently. Historically, no operator could claim to have national coverage as even the two (rather unsuccessful) state-owned operators divided up the country between them. GSM (the dominant technology) coexists with a significant CDMA presence (although all operators are converging on LTE). The private operators tend to be aggressive, which has had the result that markets are typically highly competitive – see http://coai.in for details.

It is beyond the scope of this section to explain in anything like full detail the history of mobile in India over the past five years – but see, for example, Curwen (2009, 2010a, 2010b, 2012a, 2012b) and Whalley and Curwen (2013). Suffice it to say by way of introduction that by the late 2000s the Indian mobile sector was beset by tales of corruption (especially in relation to the award of licences), by internecine strife among various government agencies (of which there are a surprising number with the power to intervene in some shape or form), the sector regulator (the TRAI) and the courts, and by disputes over the permitted scale of foreign ownership of operators.

Before this period of notoriety, the country, despite its huge and growing population, held little appeal for internationalising operators. Indeed, France Télécom sold out in 2004 and the then AT&T Wireless sold out in 2005, discouraged by rules forbidding their holding of a majority stake, a very low average revenue per user (ARPU) and an unpredictable regulatory environment. However, when the government raised the equity cap to 74 per cent for foreign ownership in late 2005, overseas entrants were swift to appear, such as Maxis, Telekom Malaysia (now Axiata) and Vodafone. The latter initially invested in Bharti Airtel, but in 2007 it bought a majority stake in the then Vodafone Essar (Whalley and Curwen, 2013), triggering a series of disputes concerning the size of its stake and the tax implications of its purchase via a company based outside India.

In 2008, Etisalat bought a stake in Swan Telecom and Telenor bought into Unitech Wireless, paying what appeared to be highly inflated prices, prompting NTT DoCoMo in turn to overpay for a minority stake in unprofitable CDMA operator Tata Teleservices – it promised to hold on to this stake until 2015 but in April 2014 announced that it would be selling it to Tata Group later in the year – and Russia's Sistema to acquire a majority stake in Shyam Telelink.

However, this apparent tale of success in attracting inward investment came to an abrupt halt in September 2010 when the Supreme Court notified the Minister of Telecommunications that 122 mobile licences had been issued unlawfully in January 2008 – these related only to 2G as 3G licensing had yet to take place. Possibly tens of billions of dollars in revenue had evidently been forgone – an extraordinary sum that arose from the fact that the Department of Telecommunications (DoT) had chosen to allocate pan-India 2G (Unified Access Services) licences to nine operators for INR16.5 billion apiece – a fee originally set in 2001 when subscriber numbers were much lower – despite demands by parties such as the Ministry of Finance that the licences be sold at auction.

The effects of the Supreme Court ruling were fairly traumatic for overseas operators that it caught out. Etisalat, for example, was forced to shut down its operations and, unable to resolve matters with its joint venture partner, placed its network into bankruptcy in November 2013. Of the biggest three in terms of subscribers – Bharti Airtel, Vodafone and Reliance Communications, with the latter two effectively constituting an inward investor and a CDMA operator – only Bharti (with Vodafone now left with only a very small stake) could be said to be a domestic operator in a position to look beyond the explosively growing internal market, as noted below. The dust has yet to settle as 3G and LTE licensing is of recent vintage and the Supreme Court ruling has effectively instigated a restructuring with as yet unknown consequences, so we must wait to see how many of the seven international operators listed in Table 5.2 survive in the market.

One recent development in early 2014 was the claim that Vodafone would use some of the cash to be received from the sale of its stake in Verizon Wireless to make a bid for Tata Group's 59.5 per cent stake in Tata Teleservices, even though this is a CDMA operator. Such a move became possible only as a result of a loosening of the rules governing the size of an inward investor's stake in an Indian network towards the end of 2013 – Vodafone also intends to raise its stake in its own joint venture. As NTT DoCoMo had an existing option to buy the Tata Group stake before the end of March 2014, this option needed to be extended for one year if Vodafone and DoCoMo were to come to an amicable arrangement, and this duly took place. Tata Group may also buy the state's minority stake in Tata Communications before selling the whole company to Vodafone.

As it happens, Vodafone had simultaneously been engaged in talks with the owners of Loop Mobile with a view to a purchase of the network, contingent upon Loop's renewal of its 900 MHz licence in early 2014. If these two projects had come to fruition, and Vodafone had also proved able to take full control over its own network, then Vodafone would have supplanted Bharti Airtel as easily the largest operator in India. In the event, Loop Mobile failed to renew its licence and was promptly subject to a takeover bid by Bharti Airtel. However, the joint problems of poor profitability and high levels of debt may not be soluble simply by upping the scale of operations.

The third market of interest is Japan – which may appear to be somewhat surprising given that it is no longer a particularly large market, and the only operator listed in the country in Table 5.2 is NTT DoCoMo. However, it was the third-largest market until the late 2000s and in its time the world's most technologically advanced, so the absence of foreign-owned operators is unexpected. In practice, the reality is that Vodafone was present until 2006 but failed to develop a successful strategy in a very different cultural context from Europe and, somewhat chastened as it was not used to failure in its overseas ventures, sold out to SoftBank. SoftBank, like its other domestic rival KDDI, which has never shown any real interest in foreign sorties, appeared initially to be content with expanding internally, but to general surprise it has recently become a significant player in the USA, as noted in Chapter 9, rather than elsewhere in the Asia-Pacific region.

5.6 OPERATOR STRATEGIES

In the light of the above, it is now time to turn our attention to the individual strategies of operators listed in Tables 5.2 and 5.3, in accordance with the rank order shown in Table 5.3. As indicated above, Bharti Airtel was until 2010 simply another domestic operator with one modest network in Sri Lanka (acquired in 2009) as its only other holding in the region, which is why it is absent from Curwen and Whalley (2008). However, in 2010 it suddenly acquired a further 17 networks, all bar one in Africa, with the exception being a 70 per cent stake in Warid Bangladesh. This meant, interestingly, that it was buying into a small operation in a backward country with only 2G licensing where it would face competition from several other networks with foreign investors. It has subsequently shown no interest in buying stakes in other regional networks. Hence, while remaining much the largest operator in the region in terms of subscribers other than those based in China, it is far from being a regional powerhouse.

Singapore Telecom (SingTel) held pride of place in Curwen and Whalley (2008), with half as many again proportionate subscribers in the region as its nearest rival Vodafone. It had as many networks in the region as any other operator (8) and was dependent on the region for all its subscribers. Although it has now dropped down one place in Table 5.3, little else appears to have changed. It still has far more proportionate subscribers in the region than its nearest rival Vodafone, and it has only acquired one new network via a minority stake in Warid in Pakistan. However, it no longer relies on the region for all its subscribers because, interestingly, it holds a 32 per cent stake in Bharti Airtel. Given that its home market is small and static, it is unsurprising that SingTel has sought to internationalise, although it has been unusually aggressive for a network based in such a small country. It was unaware at the time of taking its investment in Bharti Airtel that the latter would spread its wings as it has, but it has been supportive and this has to some extent obviated the need for it to buy on its own behalf. In point of fact it was highly expansionist in the period 2005 to 2007 inclusive (see Curwen and Whalley, 2008: 71) but largely unsuccessful in its endeavours. Since then it has shown no real interest in furthering its empire, preferring to consolidate and, if appropriate, to divest, as in the case of its stake in Warid, which it has sold back to its majority owner as it has been steadily losing subscribers.

As noted in Curwen and Whalley (2008: 74), Vodafone was highly active in the region during the period 2005 to 2007 inclusive, culminating in its acquisition of a majority stake in Vodafone Essar in India – which simultaneously obliged it to sell down its existing stake in Bharti Airtel. However, its departure from Japan, followed by the sale of its small stake in China Mobile (which reduced its number of networks in the region to four), has left it as an ostensible major player in the region, albeit one overdependent on one network for its subscribers, and the only foreign operator heavily involved in Australasia – although it would be perfectly happy to sell the networks at a decent price. It has shown no real interest in the region in recent years as its attention has been focused elsewhere, primarily in Europe and the USA. It seems unlikely that the proceeds from the sale of its stake in Verizon Wireless will be spent in the region.

It is surprising to find Nordic operator Telenor playing a significant role in the region, but this has been the case for a good many years and, as shown in Table 5.3, it is currently heavily dependent on the region for its subscribers – indeed, even more so than in 2007. To some extent this is a consequence of its stake in VimpelCom – and hence indirectly of the latter's purchases from Orascom – but it is its holdings in Bangladesh, India, Pakistan and Thailand that largely account for its status in the region. Interestingly, these are all rather backward countries in the sense of being latecomers to 3G, so in principle they present excellent opportunities for revenue growth, and Telenor has a controlling stake in every case. The downside is that none of the networks has been gaining subscribers at all rapidly, and Telenor has been caught up in the Supreme Court decision in India noted above, so its future has a question mark hanging over it. Not surprisingly, therefore, Telenor is concentrating on its relationship with VimpelCom and consolidating its existing networks in the region rather than seeking new countries to enter.

NTT DoCoMo is arguably the most Asia-Pacific-centric of Asia-Pacific operators. It has for many years invested exclusively in the region – apart from a recent small MVNO operation in the USA for its nationals living there – and held as many individual country stakes as its rivals. However, it has been reluctant to take controlling stakes – outside Japan, only in tiny Guam – so it is not really a regional powerhouse in any meaningful sense of the term. This may partly be due to cultural sensibilities relating to its previous bellicose history and partly to the fact that it is reluctant to enter countries that are technologically immature. Its dependence on the region also reflects the fact that its attempts to invest outside the region have met with failure. Its second-biggest market is India, but that has run foul of the 2G licence revocation and has been put up for sale, so its overall subscriber numbers are unlikely to rise in the near future.

Axiata is another operator that is trading almost entirely within the region – it also has a few subscribers in Iran but wishes to divest its stake. Although it has not managed to increase its portfolio for over five years, it is nothing if not acquisitive in principle, albeit reluctant, perhaps, to pay out on a scale sufficient to snare its targets. Bangladesh, India and Indonesia provide more proportionate subscribers than the home market, and it very much looks as if that will continue to be the case for the foreseeable future.

Orascom, which is now known as Global Telecom Holding, had only three entries in the region in 2012, but because among them were two large wholly owned networks, it occupies a fairly high position in Table 5.3. However, since the sale of its stake in HTIL it has shown no interest at all in acquiring further holdings in the region, preferring to concentrate on Europe and Africa.

Maxis Communications does not appear in Table 5.2 because at the end of 2013 it was left with only two networks. However, it previously did, and still does, account for a large number of proportionate subscribers in the region as a result of its majority stake in Aircel in India.

Qatar Telecom (now known as Ooredoo) is a much smaller regional operation than any of the above but is still fairly heavily dependent on the region for its proportionate subscribers. Notably, it entered the region only in 2007 but did so in a fairly dramatic fashion. However, this considerably disguises the reality that the only stake of any consequence was in Indonesia, which in 2012 provided all bar roughly half a million of its proportionate subscribers in the region. Effectively, therefore, Ooredoo is little more than a single network in a single country.

VimpelCom follows much the same pattern overall, although its entry into the region was even later and in two stages rather than one. It does not depend particularly on the region for its proportionate subscribers and, indeed, it is the indirect stakes in Global Telecom Holding's networks in Bangladesh and Pakistan that are the important elements in its presence in the region, especially in the light of its exit from Vietnam in 2011 and Laos in 2013.

Saudi Telecom does not appear in Table 5.2 because at the end of 2013 it was left with only two networks. However, it was previously of little consequence in relation to the region since, although it has three entries in Table 5.3, the only network in which it had a majority stake, in Indonesia, was sold to Axiata and the Indian network is also for sale in the wake of the Supreme Court ruling. It clearly does not consider itself to have a real future in the region.

Hutchison Whampoa has more network investments but slightly fewer proportionate subscribers (for now). It was described as a declining force in the region in Curwen and Whalley (2008: 73), although, surprisingly, it has significantly increased its dependence on the region for proportionate subscribers since 2007. However, this reflects not simply significant growth in a number of Asia-Pacific countries such as Indonesia, but also virtual stagnation in Europe. Despite this, Hutchison has made no recent efforts to widen its Asia-Pacific portfolio – indeed quite the opposite, since its only moves have been finally to get rid of its pointless investment in Thailand in 2011 and to negotiate the sale of its network in Sri Lanka to state-owned Mobitel at some point during 2014.

The region is moderately important to Etisalat in terms of proportionate subscribers, and this has not been much affected by its travails in India as the network in question was very small. However, it has recently sold most of its modest holding in Indonesia, which was nevertheless one of its biggest sources of subscribers, so it will become increasingly reliant on Pakistan (where its stake is also modest) and, unusually, Afghanistan. It has expressed an interest in entering Bangladesh, either attracted by, or ignoring, the presence of its main competitors in the region.

It is worth noting, finally, that the appearance of Middle Eastern operators for the first time in the mid-2000s was noted in Curwen and Whalley (2008:75), where it was suggested that their presence would increase although they would probably remain focused on Africa. The first suggestion has not, in practice, come to pass as, by the end of 2013, the three companies had exactly the same number of holdings in the region as they had in 2007.

5.7 DISCUSSION

The Asia-Pacific region contains several of the most populous countries in the world and is home to several of the world's largest operators in terms of subscribers. However, it is also characterised by significant differences in income levels within countries and, on average, between countries. It is also characterised, partly as a result, by differences in the sophistication of the mobile technologies in use. On the one hand, there is South Korea, where operators are discussing when to shut down their 3G networks, while, on the other hand, there is Bangladesh, where operators are only just about to launch their 3G services.

It is of interest that, for example, South Korean operators have not invested heavily in the region and that DoCoMo has been extremely cautious outside its home market, all evidently appreciating – partly in the light of experiment – that it is not practicable to transfer experience gained in an advanced country's mobile market to that in an emerging economy. Equally, operators like Bharti Airtel that have developed a strategy to make money in a low-ARPU competitive environment might well consider that this model is easier to export to Africa than to other countries in the Asia-Pacific region.

However, Table 5.2 does reveal one significant factor, which is the manner in which regionally based operators have homed in on the same small sub-set of markets. In practice, there is an unusually large number of markets in the region with more than four licensees. In Europe, this has tended to be a transitional stage because that number has proved to be unsustainable, but many of the countries in Asia have relatively large populations so it is difficult to predict how things will develop. Clearly, the fact that only three operators are present in China represents an anomaly because it is not a market-based solution and there is no prospect of any change in the future – indeed, the first MVNOs are only now being licensed. However, the situation in India is equally atypical even if changes are certain to take place during 2014, driven on the one hand by the Supreme Court decision and on the other by the excessive degree of competition in many circles.

Nevertheless, even if it seems reasonable that Bangladesh should have six networks, the presence of six internationalised operators comes as something of a surprise. After all, it is not as though ARPU is running at an appealingly high level. Equally, the presence of five internationalised operators in Sri Lanka, a market with five networks, does not appear to be a sustainable structure. It is evident that there is little incentive to restructure while the market is growing rapidly overall, but there have recently been some signs that the main internationalised operators are reconsidering their portfolios now that penetration levels have risen to levels that not so long ago would have been regarded as unachievable in emerging economies.

For example, there is no mention above of Millicom, which once had five holdings in the region and figured in Curwen and Whalley (2008: 68). Millicom exited from Cambodia and Sri Lanka in 2009 and Laos in 2011, and is no longer present at all in the region. So far, there are no signs of other operators taking such a drastic view, and Millicom was somewhat exceptional in that it was based in Europe yet had no holdings in that region. Nevertheless, 2012 saw Telenor and VimpelCom exit Vietnam and TeliaSonera exit Cambodia, while 2013 witnessed the departure of Saudi Telecom from Indonesia, Singapore Telecom from Pakistan and VimpelCom from Laos. Hardly a rush for the exit door, it is true, but the only concrete entries in the region relate to the new licences in Myanmar, where the market has been opened up for the first time. Furthermore, only Etisalat has expressed any interest in entering new markets, surprisingly targeting the already crowded market of Bangladesh although the recent award of 3G licences is arguably an incentive to do so.

5.8 CONCLUSIONS

Looked at from a structural perspective, the Asia-Pacific region differs markedly from the other two relatively underdeveloped regions, namely Latin America and Africa. In the former case, which admittedly contains a modest number of countries if the Caribbean section is excluded, the key characteristic is the struggle for dominance between two rivals with Hispanic roots, América Móvil and Telefónica. Africa, in contrast, while containing roughly the same number of countries/islands (but of somewhat less disparate size) as the Asia-Pacific region, is home to five operators with large numbers of holdings, among which only MTN is Africa-based. Furthermore, trading of sizeable portfolios of holdings is by no means unknown, yet this is a phenomenon that is not to be found in the Asia-Pacific region, where it is more usual for the odd network to be transferred. In that region, it is more a question of adjusting portfolios at the margin as circumstances permit.

As has been demonstrated, operators based in the region do not have much of a record of successful investment outside the region even though parts of the region are in the vanguard of technological advance. It has been suggested that such operators may not be comfortable in environments where political, cultural, linguistic and other factors come into play unless, like Bharti Airtel, the driving force is the belief that the operational strategy as such can be readily transferred – which, in essence, means to Africa. However, the likes of Axiata and Singapore Telecom exist in relatively small markets with limited prospects, so, in principle, they are likely to be relatively acquisition-minded. But, as ever, this kind of argument has to take account of such features as the ability of an operator such as Singapore Telecom to use another operator, in this case Bharti Airtel, as an indirect route for the acquisition of a broad portfolio of holdings. Equally, majority state ownership is rarely an attractive feature of potential targets and, more particularly, aggravates the targets' governments.

However, while the indirect route can in principle be replicated, consideration needs to be given to the financial circumstances of the relevant operators and the attitude of the financial markets to the implications of large-scale purchases for the indebtedness of operators, especially at a time when huge outlays are required to roll out 4G networks. Things have not been helped in that respect by what has happened in India, where the write-offs triggered by the Supreme Court decision in India are bound to make internationalised operators pause to wonder whether the Asia-Pacific region is as secure a place to invest as they thought only a few years ago.

The other problem is that when assets are relatively scarce they tend to be offered, at least initially, at somewhat inflated prices and hence they must be seen as strategically important for an operator based outside the region to take the plunge. Realistically, it is hard to see which among these operators currently needs to follow this route and there is nothing in their recent strategic statements to suggest that it is anywhere near the top of their investment agendas. At the end of the day money certainly counts, and China Mobile, for one, has plenty of cash, but overall it looks very much as though the need to launch data-hungry networks, whether 3G or increasingly 4G, will dominate the agenda in the Asia-Pacific region for some while yet.

NOTES

1. It might be termed the Eurovision Song Contest version of Europe since the investment flows mainly from West to East. The decision by the government of Ukraine in November 2013 to accept Russia as a suitor in preference to the EU – admittedly under huge pressure and in the face of popular unrest – appeared to have stopped this eastward encroachment in its tracks. However, the EU is now pumping money into Ukraine in order to preserve it – minus the Crimea – as part of 'Europe'.
2. Official data for Japan at end 2013 are as follows: total = 136.6 million divided up into NTT DoCoMo (62.2 million), SoftBank (34.8 million) and KDDI (39.6 million).

REFERENCES

Cellular-news (2014), 'China Mobile said to be considering buying a large stake in Vodafone', www.cellular-news.com, 28 January.

Curwen, P. (2009), 'Making the straightforward complicated Indian-style', *info*, **11**(4), 83–5.

Curwen, P. (2010a), 'Empire building in the dock: the case of Vodafone', *info*, **12**(4), 74–6.

Curwen, P. (2010b), 'Making the straightforward complicated Indian-style – reprise', *info*, **12**(6), 139–41.

Curwen, P. (2012a), 'Licence issue and transparency: a cautionary tale', *info*, **14**(1), 70–71.

Curwen, P. (2012b), 'Licence issue and transparency: reprise', *info*, **14**(4), 78–9.

Curwen, P. and J. Whalley (2008), *The Internationalisation of Mobile Telecommunications: Strategic Challenges in a Global Market*, Cheltenham, UK and Northampton, MA, USA: Edward Elgar.

Ronen, S. and O. Shenkar (1985), 'Clustering countries on attitudinal dimensions: a review and synthesis', *Academy of Management Review*, **10**(3), 435–54.

Whalley, J. and P. Curwen (2013), 'Unravelling complex organisational structures among mobile operators', *info*, **15**(4), 3–22.

6. Structural and strategic adjustment among African mobile operators

6.1 INTRODUCTION

Africa is a continent of contrasts, as is demonstrated only too clearly in the sections that follow. However, given such matters as the large number of countries/islands on the continent, their geography, history, culture and access to natural resources, this hardly comes as a surprise. Nevertheless, one thing that has always been a very common feature across the continent is a poor fixed-wire infrastructure, so the advent of mobile communications has been much more significant than in, say, Europe and the USA. Up until the middle of the first decade of the 2000s the total number of mobile subscribers remained fairly small – China and the USA could each muster more subscribers than the whole of Africa in 2005 – although this was clearly not going to last. But potential has to be exploited and, if nothing else, it is hugely expensive to roll out mobile networks across a continent the size of Africa.

This raises a number of related questions. For our purposes, the main ones relate to identifying key differences between countries and the role of both domestic and overseas-based operators. Since the period up to the end of 2007 is amply covered in Curwen and Whalley (2008), what follows will refer to that period only where necessary for the purposes of understanding historical continuity.

6.2 PENETRATION AND OTHER RATIOS

Altogether there are 56 countries and islands in this region. For convenience, these will simply be referred to as countries in what follows, but the fact that the islands can sometimes play an exceptional role is noted where relevant. The precise number of subscribers in each country is a matter of some conjecture for a variety of reasons. One is the issue of activity rates, since it is well known that a significant proportion of subscriber identity modules (SIMs) in circulation is not in regular use. This, in turn, partly reflects the practice of issuing pre-activated SIMs.

The decision to require SIMs to be registered can accordingly have a dramatic effect on subscriber numbers. For example, when the deadline for registration expired in Uganda at the end of August 2013, 1.5 million SIMs remained unregistered, although many unauthorised calls continued to be made (Cellular-news, 2014a). Furthermore, on the stipulated registration date of 27 November 2013, 1.7 million of the total 5.4 million SIMs in Niger had not been registered, causing estimates of subscriber numbers to fall by one-third (Cellular-news, 2013a). Equally, on the stipulated registration date of 31 December 2013, subsequently set back by one month, 2.4 million of the outstanding 10.3 million SIMs in Zambia remained unregistered and were de-activated (Cellular-news, 2014b). In an interesting recent development, the operators in Ghana were instructed to repeat the SIM registration scheme concluded in 2011 on the grounds that 20 per cent of the identity cards used to register SIMs were fakes (TeleGeography, 2014).

No doubt numbers will bounce back somewhat, and these countries may not be particularly typical – SIM registration has typically resulted in 10–15 per cent de-activation and where this occurred before end 2012 the figures in Table 6.1 will already have taken account of this – but it is fair to conclude that a degree of overstatement is endemic in Africa. For this reason, subscriber numbers are given in ranges in Table 6.1.

Of the 56 countries, nine claim to have fewer than 1 million subscribers, five have between 1 and 2 million subscribers, 15 have between 2 and 5 million subscribers and five have between 5 and 10 million subscribers. This totals to 34 countries, or 61 per cent of all countries in the region. Relevant information about these countries is presented in Table 6.1.

Table 6.1 seeks to shed light on any interesting correlations among the number of claimed active subscribers, the number of networks and the population. In theory, the larger the population, the larger the number of subscribers, although in practice disparities in both average income and the shape of the income distribution among African countries would lead one to expect a weak correlation. Equally, both the size of the population and the number of subscribers should have some bearing on the number of sustainable networks.

The first thing that appears extremely clear is that countries with very small populations (fewer than 1 million) have commensurately small numbers of subscribers – true in seven out of eight cases. While it is true that penetration ratios are now comfortably above 100 per cent in most advanced countries, despite the fact that the very young – although they start getting personal devices at a surprisingly young age these days – and the very old are not included, this is far less likely to be the case in an emerging economy. However, the ratio appears to alter somewhat randomly as population size grows. Among countries with populations between 1 and 2 million, two out of six (Botswana and The Gambia) have more than 2 million subscribers, yet this phenomenon applies to none of the seven countries with populations between 2 and 5 million. Equally, four out of five countries with populations between 5 and 10 million – but note Libya below – have fewer than 5 million active subscribers and that is also true of four of the eight countries with the largest populations.

Given the difficulties inherent in determining precise numbers of active subscribers and, in some cases, precise population numbers, it is best not to draw too many conclusions from the above other than that there still appears to be reasonable potential for growth in the larger countries in the sample.

Table 6.1 Countries with fewer than 10 million active[1] subscribers in Africa at end 2012

	<1 million	1–2 million	2–5 million	5–10 million
Countries	9	5	15	5
Networks: 1	5	0	0	0
Networks: 2	2	1	1	0
Networks: 3	2	3	4	3
Networks: 4	0	1	7	0[4]
Networks: 5	0	0	2	2[4]
Networks: 6	0	0	1[2]	0
Population: <1 mn	7	1	0	0
Population: 1–2 mn	1	3	2	0
Population: 2–5 mn	1	1	5	0
Population: 5–10 mn	0	0	4	1
Population: >10 mn	0	0	4[3]	4[5]

Notes:

1. A more accurate description might be 'claimed to be active'.
2. In Somalia only one or two networks tend to be available in any given location, requiring the use of multiple SIMs.
3. Chad (12 mn); Malawi (14 mn); Niger (17 mn); South Sudan (10 mn).
4. One network shut down in Guinea during 2013.
5. Burkina Faso (16 mn); Guinea (11 mn); Madagascar (21 mn); Rwanda (11 mn).

The situation in respect of network numbers is also erratic. Predictably, all countries with a monopoly provider have populations below 1 million, yet two tiny countries have three competing providers. Again predictably, these turn out to be the French islands of La Réunion and Mayotte, and it is indeed the former, with a population of roughly 850 000, that is the exceptional small 'country' with more than 1 million subscribers – in practice, almost 2 million, so the penetration ratio exceeds 200 per cent. In comparison, the island of Madagascar has over 20 million inhabitants but only some 6 million active subscribers, an exceptionally low penetration ratio for a country that is neither extremely poor nor subject to ongoing war or other unrest.

Surprisingly, perhaps, only four countries have duopolies. The commonest number of networks – in 12 out of 34 cases – is three, which on the whole is to be expected, but four countries (Benin, Burundi, Guinea and South Sudan) have five networks and one has six, which appears somewhat excessive but relates to Somalia, where licences have not seemingly been required since 1991 – although some have been issued, their legal status is at best dubious – and the operators coordinate spectrum among themselves. Interestingly, Benin is a highly competitive market, while Burundi has one operator controlling half the market but is generally competitive; Guinea has three relatively large providers with similar market shares; and South Sudan – the most recent country to have been granted independence – appears to be developing a competitive structure.

A final point worth noting, and to which we will return, concerns the interest shown by international operators in the countries discussed so far. Only a handful have attracted the interest of more than two such operators – Burkina Faso, Central African Republic, Gabon, Niger and Rwanda – and Bharti Airtel is to be found in three of these and Etisalat in four, so the other operators are essentially absent. However, it is now time to examine the countries with the largest numbers of active subscribers as these are more likely to attract the attention of international operators.

Twenty-two countries (but predictably no islands) are listed in Table 6.2. Given their relative importance, concentration ratios have been added to the data on the number of networks, population size and active subscribers, the latter two rounded up/down to the nearest million. Once again the initial hypothesis must be that there is a positive link between these three variables, but is this the case in practice?

Table 6.2 Countries with more than 10 million subscribers in Africa at end 2012

	Networks (number)	Population (million)	Active subs (million)	1-firm concentration ratio (%)	2-firm concentration ratio (%)	3-firm concentration ratio (%)
Algeria	3	37	38	46.3	76.6	100
Angola	2	20	12	70.9	100	–
Cameroon	2[1]	20	13	54.3	100	–
Côte d'Ivoire	6	23	18[11]	34.3	68.6	90.3
DR Congo	6	73	19	40.5	76.0	91.0
Egypt	3[2]	82	87	44.9	79.1	100
Ethiopia	1	85	20	100	–	–
Ghana	6	25	25	45.6	66.0	80.8
Kenya	4[3]	42	31	63.8	80.7	90.9
Libya	3	6	17	64.2	99.7	100
Mali	2[4]	16	14	57.7	100	–
Morocco	3	32	38	47.9	78.3	100
Mozambique	3	23	10	60.5	89.2	100
Nigeria	9[5]	173	112	42.3	63.4	83.5
Sénégal	3	13	11	62.3	85.9	100
South Africa	5[6]	52	67	45.5	82.5	97.7
Sudan	3[7]	34	23	53.9	85.7	100
Tanzania	8[8]	45	26	35.0	66.2	89.0
Tunisia	3	11	14	52.0	86.2	100
Uganda	7[9]	33	18	40.1	64.5	81.8
Zambia	3[10]	13	10[12]	41.9	79.8	100
Zimbabwe	3	1	12	62.7	86.1	100

Notes:

1. A third licence was awarded in 2013 and the network will be launched in 2014.
2. A fourth licence was awarded to Telecom Egypt in April 2014.
3. The third-largest network, Essar Telecom (yu), struggled to raise funds for expansion and in March 2014 it agreed to sell its infrastructure to market leader Safaricom and its subscribers to Bharti Airtel. However, this will still leave Airtel half the size of Safaricom (Cellular-news, 2014c) even if it proceeds – the regulator has given its blessing but has imposed tough conditions on the deals and in April 2014 Safaricom decided to withdraw unless the conditions were lifted. Orange is considering the sale of what will potentially become much the smallest of the three remaining networks.
4. A third licence was awarded in 2013.
5. Of which five are GSM and four are CDMA.
6. Majority owner Tata Communications has agreed to sell tiny Neotel to Vodacom in 2014.
7. The Republic of South Sudan achieved independence from the Republic of Sudan in July 2011.
8. There are several other licensed networks yet to be launched.
9. At the beginning of 2013 there were four large networks but Bharti acquired Warid in April 2013. Of the other three, one barely existed. A new licensee entered in March 2014. Orange was put up for sale March 2014.
10. The intention to sell a fourth licence was announced in early 2013.
11. The regulator approximately agrees subscriber numbers for the other five operators, but cites Orange as having nearly 1.5 million more than Orange claims for itself – the figure used here.
12. 2.4 million SIMs were de-activated for failing to meet the registration deadline of 31 December 2013.

It is best to begin by treating Nigeria as something of an anomaly. It has by far the largest population and the largest number of active subscribers, but the penetration ratio is only 65 per cent, which ranks it 15th of the 22 countries in the table. Given the physical size of the country and some tendency to civil unrest, there are bound to be areas where mobile provision is at best scanty. In any event, Nigeria is an oddity in Africa because it has so many networks and a significant proportion of these use CDMA rather than GSM, but it is best not to dwell on the latter phenomenon because CDMA networks account for only 3 million or so subscribers, largely served by Visafone, and this number is on the decline.[1] Once data are switched to a GSM-only basis, Nigeria seems more normal, with 109 million subscribers and five operators and, given that in practice Nitel subsidiary M-Tel is likely to go into liquidation, it can reasonably be regarded as having only four operators. It may be noted that all of this adjustment has virtually no effect on either the penetration ratio or the concentration ratios.

Seven countries appear to have penetration ratios of at least 100 per cent. One, Libya, has a ratio approaching 300 per cent, which may have something to do with the unsettled situation there but is anyway best ignored as an aberration. In Morocco, Tunisia and South Africa the ratio comfortably exceeds 100 per cent, which comes as no great surprise, with Egypt, Algeria and Ghana not far behind. So, in effect, the entire northern strip and the southern tip are behaving much like any relatively advanced economy, while interior countries beset by war and/or famine such as DR Congo and Ethiopia are struggling even to reach a ratio of one-third. The one country with a surprisingly low ratio (60 per cent) is Tanzania, although Kenya's 74 per cent is also perhaps lower than one would have expected.

Nigeria aside, it is surprising to find two other countries with as many as seven networks. It is quite hard to understand why countries with what are moderate populations by international standards should need so many networks. In the case of Tanzania the position is deceptive since there are in practice four large networks and four tiny networks – one of which, Smile Communications, is a recent new entrant – and the latter are actually losing rather than gaining subscribers. Whether they will survive for much longer is accordingly a moot point. As for Uganda, a country known for the use of multiple SIMs to take advantage of lower prices on net (that is, to other subscribers on the same network) compared to off net, there were again four large networks as well as three that were very modest in size at the end of 2012 – and Smile Communications is again a recent new entrant. However, two large networks merged in 2013 and the fifth-largest is for sale currently (as of June 2014), so there is reason to believe that a rational market structure is developing.

But is there any better justification for the six active networks in a further three countries? In Côte d'Ivoire there appears to be a new entrant originally licensed in 2000 – many African countries show evidence of old licences that may or may not have been issued officially and may or may not have led to launches – but the other operators appear to be in reasonable health. DR Congo is a good illustration of over-issuance of licences, since there is evidence of as many as 16 licences having been issued at some time or other, with six currently active. One of these, Africell, appears to have launched as recently as mid-2012, while another, MTN, is barely functional. As for Ghana, it includes Globacom, a new entrant in April 2012 despite being licensed in 2008 (and at present seemingly proving quite popular) and only one operator (Kasapa) that is struggling.

What the above seems to indicate is that the relationships between the number of active subscribers, populations and networks are too random to reveal any useful patterns for analysis, but that is not necessarily true for concentration ratios. As noted, the countries with large numbers of networks tend to be competitive, but not necessarily across the board as it is not customary for even a fourth network to have a market share of any consequence. This is clearly shown in Table 6.2. Needless to say, the sole operator in Ethiopia is state-owned, but in Angola the state has only a minority stake in one of the two operators. Both Cameroon and Mali have recently achieved their aim of establishing more than two networks. In Cameroon, after some false starts due mainly to overpricing the licence, Vietnam's Viettel Group – a somewhat unlikely buyer owned by the armed forces – was awarded a licence in June 2013 that incorporated 3G exclusivity until December 2014. For its part, Mali – where again the state has a minority stake in one operator – issued a licence in February to a consortium including Monaco Télécom.

Libya is arguably a duopoly, since the third licensee, LibyaPhone Mobile, is credited with having subscribers by some sources but not even mentioned in others. However, it is not very relevant as all networks are for the time being state-owned and managed by overseas operators. The intention is to float a stake in the largest operator, Libyana, followed by the issue of a licence to a private operator, so considerable restructuring is underway – at least in principle since Libya is politically very unstable and the penetration ratio is so high that there does not appear to be much room left for a major new network.

These exceptions aside, there are 17 countries left in Table 6.2 with three or more active networks. What is immediately striking is the huge variation in the market shares held by the largest networks. These range from roughly 35 per cent in Côte d'Ivoire and Tanzania to more than 60 per cent in Kenya, Mozambique, Sénégal and Zimbabwe. There is an apparent relationship here between the number of networks and the ratio – many more networks in the former case than in the latter, where there are typically only three. But the relationship is not particularly strong overall – compare, for example, Zambia with Zimbabwe.

The two-firm ratios divide fairly equally – five networks in the 60–70 per cent range, five in the 70–80 per cent range and seven in the 80–90 per cent range. The number in the lowest range is unusually large by comparison with that in Europe, and would appear to indicate a fair number of second-placed operators with modest market shares of below 30 per cent. In practice, this is true in only three cases and (interestingly) is equally true in four of the seven networks to be found in the 80–90 per cent range. The second-placed operator with the lowest market share is found in Kenya, where, predictably, the largest is dominant despite de-registering more than one million subscribers during 2013, yet the situation is quite different in Ghana.

In several cases the second-largest operator is not much smaller than the largest. This is particularly the case in Côte d'Ivoire, where MTN and Orange have been virtually the same size for several years, but also a feature of DR Congo (Bharti Airtel and Vodacom), Tanzania (Bharti Airtel and Vodacom) and Zambia (Bharti Airtel and MTN). The presence of large international operators is worthy of note.

The three-firm ratio is never less than 80 per cent, which is to be expected. In every case, the countries with three-firm ratios below 90 per cent (Ghana, Nigeria, Tanzania and Uganda) have more than four networks, so in effect there is a tail of small, weak operators.

It is interesting that there are six countries where third-placed operators have a market share of at least 20 per cent (Algeria, Côte d'Ivoire, Morocco, Nigeria, Tanzania and Zambia). In three cases there are only three networks, but there is clearly a high level of competition in these markets as there is in Nigeria, where the third-largest has give or take the same market share as the second-largest. However, in other cases (Egypt, Kenya and Mozambique) the third-largest operator is struggling to retain a 10 per cent market share, although it still represents a great many subscribers in Egypt.

However, the situation in Egypt has recently become destabilised by the issue of a licence in April 2014 to Telecom Egypt to act as an MVNO on one of the networks in return for all mobile networks being given the right to re-sell Telecom Egypt's fixed-wire services (Telecom.paper,

2014). Vodafone is threatening to initiate international arbitration proceedings on the grounds that what was being offered was access to the old copper-wire network and not the fibre-optic network needed to convey broadband. In any event, Telecom Egypt is already the minority shareholder in the Vodafone network yet it could conceivably opt to act as an MVNO on a rival network. If so, that might trigger an attempt by Vodafone to buy out its partner, especially since a condition of Telecom Egypt's licence is that it must dispose of its stake in Vodafone Egypt within one year.

Overall, it would appear that, although there are many instances of countries with a relatively large number of networks, the reality is that this disguises the role played by the three or four largest in each case. The larger African countries appear to have taken on board the virtues of private ownership and competition between strong operators as the best means to ensure rapid development of mobile networks in an environment where that strategy has for many years offered a far superior route to national connectivity than investment in fixed-wire networks.

But this leads naturally to a further issue. Presumably most countries would have preferred the private operators to be native to Africa, but that was never a realistic prospect. So who are these operators and where have they invested?

6.3 THE MAIN INVESTORS

Summary data on the main investors are presented in Table 6.3, which identifies operators with a substantive investment in Africa during the eight-year period beginning in 2005. It is important to note that these investments are not necessarily direct but may be made in another operator that actually holds the stake in question. In order to maintain consistency, the indirect holdings for the listed operators are all majority stakes providing operational control – VimpelCom via Orascom (now via Global Telecom Holding), Vivendi via Maroc Télécom and Vodafone via Vodacom. Singapore Telecom would also be included were it not for the fact that all its subscribers are derived from a minority stake in Bharti Airtel (see note 7).

Setting aside VimpelCom, which was a peripheral player at best and which sold half its holdings in 2013 – this had a lesser impact on Orascom, where the Egyptian stake is held independently – 11 other companies are listed in the table. Of these, Maroc Télécom, MTN and Vodacom are the only three based on the continent, and two of these are majority-owned by European operators. This seems rather a small

Table 6.3 Africa mobile operator presence: number of countries and subscribers, end year[1]

	2005	2006	2007	2008	2009	2010	2011	2012	Subs (mn)[2]	Africa subs as % total
MTN	11	16	16	16	16	16	17	17	112.1	77.0
Vodafone[3]	9	9	9	10	10	10	8	8	60.5	16.9
Bharti Airtel	0	0	0	0	0	16[4]	16	17[5]	58.9[9]	24.2
Orange	11	11	14	16	17	19	20	20[6]	58.5	40.3
Vodacom[3]	5	5	5	5	5	5	5	5	41.7	100.0
Etisalat	8	8	9	11	11	11	12	12	29.4	45.4
Maroc Télécom	4	4	4	4	5	5	5	5	24.4	100.0
Millicom	5	7	7	7	7	7	7	7	18.7[10]	48.0[12]
Zain	15	15	15	16	17	2[4]	3	3	14.4	43.6
Vivendi	4	4	6	6	7	7	7	7[7]	14.1	39.3
Orascom	5	5	5	5	5	5	5	5[8]	13.5	17.6
VimpelCom	0	0	0	0	0	0	4	4[8]	6.2[11]	3.9

Notes:

1. Numbers at year-end. It is not always possible to be certain of the year in which holdings were transferred.
2. On 31 December 2012. We refer here not to gross subscriber numbers but to 'proportionate' or 'equity' subscribers: that is, the gross figure is weighted according to the proportion of the equity held by the company in question.
3. There are issues with these totals because in early 2010 there were still technically a further 15 stakes held indirectly via Bharti Airtel even though the stake in Bharti itself had been reduced to a small residual of less than 5 per cent. This is more appropriately treated as a financial investment without control, so the Bharti holdings have been excluded. Vodafone also owns a majority stake in Vodacom, which is accordingly counted. It should be noted that Network Partnership Agreements are not included within the Vodafone total as no shareholding is involved.
4. Zain was formerly The Mobile Telecommunications Co. In May 2005, it acquired 85 per cent of Celtel, a pan-African operator with 14 networks. In 2010, it sold 15 networks in Africa (essentially Celtel) to Bharti Airtel.

5. Bharti bought Warid Uganda in April 2013 but was already present in the country.
6. Formerly France Télécom. Orange Uganda was put up for sale in March 2014 and Orange Kenya in April 2014.
7. Partially derived from its (slight) majority stake in Maroc Télécom.
8. Two holdings were sold in 2013 yielding roughly 2 million proportionate subscribers to Orascom and 1 million to VimpelCom.
9. Singapore Telecom has a 32.3 per cent stake in Bharti Airtel and hence indirectly has 19.0 million subscribers in Africa, which accounts for 19.1 per cent of its total subscribers.
10. 18.2 million at end 2013.
11. Entirely derived from its (slight) majority stake in Orascom Telecom Holding (now Global Telecom Holding).
12. 44.0 per cent at end 2013.

number, but one has to bear in mind the difficulties faced by an African company in raising the requisite finance to acquire new assets while simultaneously investing heavily in existing networks.

There are some notable absentees among the internationalised operators listed, such as T-Mobile, Telefónica and Telenor. It can be argued – see Chapter 10 – that all operators are to a considerable extent regionalised, so some are bound to be absent from each continent, and there are limited links between the missing operators and Africa, although Telefónica was once present in Morocco and also via its stake in Portugal Telecom, which itself ceased to be an important investor in Africa when it agreed to sell its stake in Morocco in 2009. Furthermore, Germany once had some colonial presence and that is, of course, a relevant feature as one would expect an ongoing presence from operators based in England and France, which historically had extensive colonial holdings and whose languages are in widespread use today. In fact, not many countries outside Africa are represented in Table 6.3 since Millicom has no home base as an operator and, outside Western Europe and the Middle East, there is only India's Bharti Airtel.

There are three dimensions to Table 6.3 – the number of countries, the number of proportionate subscribers and the relationship between the latter and the total number for a given operator. If all three yield a relatively high figure, then this clearly identifies an Africa-centric operator. On this basis, MTN is the only operator that qualifies. Orange and Bharti Airtel qualify on the first two criteria but not on the third, while Maroc Télécom and Vodacom operate exclusively in Africa but are light on both networks and proportionate subscribers. It is a notable feature of Vodacom that it has operated the same number of networks during the entire period and hence has significantly lagged behind MTN, its main rival in Africa, although it has expressed interest in Angola, Malawi, Nigeria, Sierra Leone and Zambia in recent years.

Overall, given that five years have passed since Curwen and Whalley (2008: 86) provided end-year data for 2007, the rather modest growth in subscriber numbers is quite striking for a continent in desperate need of mobile connectivity. Bharti Airtel is an exception, although its appearance simply mirrors the disappearance of Zain, and Etisalat has made considerable strides. But the larger operators in 2007 have progressed less than might have been expected.

6.4 LOCATION OF NETWORKS

At this point it is useful to identify where the main operators in Africa had their networks during the period 2007 to 2012 inclusive, which is shown in Table 6.4. As has been pointed out, the acquisition of licences or purchase of stakes is governed by chance to a considerable extent, but one would intuitively assume that the large internationalised operators would home in wherever possible on the largest or fastest-growing markets. The downside, of course, is that such behaviour will tend to make these markets much more competitive.

Table 6.4 is illuminating in this respect. It does not contain precisely the same sample set as Table 6.3 because Maroc Télécom is effectively represented by Vivendi, VimpelCom is effectively represented by Orascom (which retained its holding in Egypt independently), and Zain has ceased to be an important player on the continent. However, Vodafone and Vodacom are both listed because each is important to the narrative in its own right.

Only a modest number of countries currently (end 2013) attract the interest of more than two of the listed operators now that VimpelCom has exited the Central African Republic – Burkina Faso (3), Côte d'Ivoire (3), DR Congo (5), Egypt (4), Gabon (3), Ghana (4), Kenya (3), Niger (3), Nigeria (3), Rwanda (3), South Africa (3 – but only if counting both Vodacom and Vodafone separately), Tanzania (5) and Uganda (3). The most notable feature is that Bharti Airtel is to be found in ten of these 13 (but not in South Africa), while Etisalat, MTN and Orange, which also have investments in more than ten countries, can each muster holdings in no more than six. This number is not exceeded by any other operator, although it is notable that Vodafone is present in six cases out of a total of eight investments in Africa.

Many of the significant African markets are included in the above, although it is of some interest that Algeria, Morocco and Tunisia are absent and only Egypt is present out of all the countries spanning the northern shore of Africa. This may partly reflect the fact that all these countries had three operators in 2012 and 2013, so there is anyway limited room for participation unless companies have stakes in the same network. This is the case in Egypt, where Orascom has a small residual stake in the Orange network, but nowhere else. Indeed, shared participation in a network by any two of the sample operators in Table 6.3 does not exist in Africa anywhere other than in Egypt.

In summary, it would appear that there is no longer much leeway in most of the important markets in Africa for the larger operators to

Table 6.4 Location of networks, main African operators: 31 December 2012

Country	Networks number	Bharti	Etisalat	Millicom	MTN	Orange	Orascom[1]	Vivendi[2]	Vodacom[3]	Vodafone[3]
Algeria	3	–	–	–	–	–	Y	–	–	–
Angola	2	–	–	–	–	–	–	–	–	–
Benin	5	–	Y	–	Y	–	–	–	–	–
Botswana	3	–	–	–	Y	Y	–	–	–	–
Burkina Faso	3	Y	Y	–	–	–	–	Y	–	–
Burundi	5	–	–	–	–	–	Y[1]	–	–	–
Cameroon	2[4]	–	–	–	Y	Y	–	–	–	–
Cape Verde Is	2	–	–	–	–	–	–	–	–	–
Cen. Af. Rep.	4	–	Y	–	–	Y	Y[1]	–	–	–
Chad	3	Y	–	Y	–	–	–	–	–	–
Comoros Isles	1	–	–	–	–	–	–	–	–	–
Congo-Brazz.	4	Y	–	–	Y	–	–	–	–	–
Côte d'Ivoire	6	–	Y	–	Y	Y	–	–	–	–
Djibouti	1	–	–	–	–	–	–	–	–	–
DR Congo	6	Y	–	Y	–	Y	–	–	Y	Y
Egypt	3[5]	–	Y	–	–	Y	Y	–	–	Y
Equat. Guinea	3	–	–	–	–	Y	–	–	–	–
Eritrea	1	–	–	–	–	–	–	–	–	–
Ethiopia	1	–	–	–	–	–	–	–	–	–
Gabon	4	Y	Y	–	–	–	–	Y	–	–
Gambia	5	–	–	–	–	–	–	–	–	–
Ghana	6	Y	–	Y	Y	–	–	–	–	Y
Guinea-Bissau	3	–	–	–	Y	Y	–	–	–	–
Guinea	5	–	–	–	Y	Y	–	–	–	–
Kenya	4	Y	–	–	–	Y[9]	–	–	–	Y
La Réunion	3	–	–	–	–	Y	–	Y	–	–
Lesotho	2	–	–	–	–	–	–	–	Y	Y
Liberia	4	–	–	–	Y	–	–	–	–	–

Libya	3	–	–	–	–	–	–	–	–	–
Madagascar	3	Y	–	–	–	Y	–	–	–	–
Malawi	2	Y	–	–	–	–	–	–	–	–
Mali	2[4]	–	–	–	–	Y	–	Y	–	–
Mauritania	3	–	–	–	–	–	–	Y	–	–
Mauritius	3	–	–	Y	–	Y	–	–	–	–
Mayotte	3	–	–	–	–	Y	–	Y	–	–
Morocco	3	–	–	–	–	Y	–	Y	–	–
Mozambique	3	–	–	–	–	–	–	–	Y	Y
Namibia	4	–	–	–	–	–	–	–	–	–
Niger	4	Y	Y	–	–	Y	–	–	–	–
Nigeria	9[6]	Y	Y	–	Y	–	–	–	–	–
Rwanda	3	Y	–	Y	Y	–	–	–	–	–
São Tomé	1	–	–	–	–	–	–	–	–	–
Sénégal	3	–	–	Y	–	Y	–	–	–	–
Seychelles	2	Y	–	–	–	–	–	–	–	–
Sierra Leone	4	Y	–	–	–	–	–	–	–	–
Somalia	6	–	–	–	–	–	–	–	–	–
South Africa	5	–	–	–	Y	–	–	–	Y	Y
South Sudan[7]	5	–	Y	–	Y	–	–	–	–	–
Sudan[7]	3	–	Y	–	Y	–	–	–	–	–
Swaziland	1	–	–	–	Y	–	–	–	–	–
Tanzania	7	Y[8]	Y	Y	–	–	–	–	Y	Y
Togo	3	–	Y	–	–	–	–	–	–	–
Tunisia	3	–	–	–	–	Y	–	–	–	–
Uganda	8	Y	–	–	Y	Y[9]	–	–	–	–
Zambia	3	Y	–	–	Y	–	–	–	–	–
Zimbabwe	3	–	–	–	–	–	Y	–	–	–
Total		17	12	7	17	20	5	7	5	8

Notes:

1. Orascom is now known as Global Telecom Holding and is majority-owned by VimpelCom, but the latter did not acquire its stake in Egypt. The holdings in Burundi and CAR were sold in 2013. Telecel Zimbabwe was alleged not to have paid its licence renewal fee in January 2014.
2. Other than in the islands of La Réunion and Mayotte, subscribers come via Maroc Télécom, its 53 per cent owned subsidiary that was provisionally sold to Etisalat in November 2013 with a view to completion in 2014.
3. Vodafone includes the networks owned by majority-owned subsidiary Vodacom, which is also listed separately. Technically, Vodafone owns a very small residual stake in Bharti Airtel. However, this produces such a small average number of equity subscribers per network that it is best treated as an investment. As there are 12 non-overlapping countries, other sources could, in principle, cite Vodafone's total as 22.
4. A third licence was awarded to a non-listed operator in 2013.
5. A fourth licence was awarded to a non-listed operator in 2014.
6. Of which five are GSM and four are CDMA. The entries in the table are all in GSM networks.
7. The Republic of South Sudan achieved independence from the Republic of Sudan in July 2011.
8. The government is negotiating to buy this minority stake but the two parties are far apart on valuation.
9. Orange Uganda is to be sold in 2014, and possibly also Orange Kenya. They are likely to be bought by other operators listed in the table.

establish an independent presence and they do not indulge in network sharing. This would suggest that there is no longer much scope for major restructuring in these markets, but no reason to expect that competition will not be fierce, a matter to which we will return after examining the strategies of individual operators.

6.5 OPERATOR STRATEGIES

This section examines the strategies of some of the individual operators listed in Table 6.4 in so far as this can be divined by examination of the data outlined above – the companies themselves rarely respond to speculation in the media. It may be noted, first, that there appear superficially to be both Africa-based and foreign-based investors but the reality does not quite bear this out. There are two large Africa-based operators relying on the continent for at least the majority of their proportionate subscribers, but Vodacom has only five investments and is majority-owned by Vodafone. Equally, MTN is a public company majority-owned by international financial institutions.

None of the other operators relies on Africa for as much as half of their proportionate subscribers, so they cannot really be regarded as Africa-centric. However, this is about to alter in one respect; see below.

In November 2013, France's Vivendi agreed (subject to regulatory approvals) to sell to Etisalat its 53 per cent stake in Maroc Télécom. While it operates a major network in France (SFR), where it fairly recently bought out its minority partner, Vodafone, Vivendi has never really been a telecommunications company as such even thought did take a controlling stake in Maroc Télécom, its only overseas mobile investment – La Réunion and Mayotte technically count as French overseas operations. Obviously, Maroc Télécom gave Vivendi an entry into francophone Africa but it was never likely to be plain sailing. The other main shareholder is the government of Morocco (with 30 per cent), and Maroc Télécom is no longer growing – Vivendi obtained roughly 11 million proportionate subscribers from its stake for each of the three years 2009 to 2012 inclusive – nor generating healthy margins (Smith, 2013). It may be noted that Vivendi is also negotiating to sell its stake in SFR.

Furthermore, the asset transfer will result in Etisalat having two separate network holdings in both Burkina Faso – where it will be the majority shareholder in both the largest and third-largest operators – and Gabon – where it will be the largest shareholder in the second-largest and third-largest networks. This seems likely to cause problems, although so far nothing explicit has been mentioned. Overall, therefore, Etisalat will gain three new African countries and roughly 25 million proportionate subscribers, bringing the total in 2014 to perhaps 55 million, which will represent 60–65 per cent of its overall subscribers.

It is worth noting that Etisalat had only two holdings in Africa at the end of 2004, at which point it agreed to purchase a 50 per cent stake in Atlantique Télécom. The stake was bought in April 2005 and raised progressively until it reached 100 per cent in 2010. Etisalat failed to acquire a stake in Tunisia in 2006 and withdrew from the bidding for a 3G licence in Morocco, but was the successful majority participant in the Nile Telecom consortium that acquired the third licence in Egypt (which was followed by a launch in 2007). Subsequently, entries were effected into Nigeria and Sudan in 2008 and, after Sudan was split up, into South Sudan in 2011. It is worthy of note that by the end of 2012 only three countries – Egypt, Nigeria and Côte d'Ivoire – accounted for roughly 22 million of its 29 million proportionate subscribers in Africa, so its portfolio there is not particularly well balanced.

On the whole, Etisalat's foreign incursions have not shown acceptable returns, partly because it overpaid initially for assets such as the Egyptian licence. It can be argued that the new acquisition is more sensibly structured, with its emphasis on a portfolio linked to Arabic-speaking countries despite the potential problems already alluded to. But a further issue is that Etisalat is acquiring an operator with its base in a market with a penetration ratio of over 100 per cent and two strong competitors, one part-owned by Orange and one part-owned by Zain.

As noted in Curwen and Whalley (2008: 93–4), MTN entered the mid-2000s with a heavy dependence on its home market. It then underwent a turbulent period that saw a series of unsuccessful attempts to expand into new countries but eventually settled on the acquisition of Investcom in 2006. However, Investcom brought with it only one real prize in Africa, namely the largest (by far) network in Ghana. Hence MTN remained reliant on South Africa and increasingly on Nigeria, where it had acquired a near-complete stake in the largest network – but see also Section 6.6 below. For the past several years, the Investcom portfolio has continued to provide one-quarter of MTN's proportionate subscribers, while the home market has continued to decline in importance. Indeed, it currently provides far fewer proportionate subscribers than Nigeria and, interestingly, the near-half stake in an Iranian network is fast catching up. In total, Ghana, Iran, Nigeria and South Africa produce two-thirds of MTN's proportionate subscribers.

Clearly, MTN must continue to work hard to avoid its market share in these four markets from being eroded by competition, but there has long been a cloud of a different kind on the horizon in Iran, where Turkcell is currently suing MTN for $4.2 billion in damages relating to the manner of the original acquisition of the licence (Maylie, 2013). Sanctions have also prevented MTN from repatriating its share of the network's profits.

Understandably, therefore, MTN has been casting around for opportunities to expand its portfolio in Africa as well as outside the continent on occasion – where India, Iraq and Myanmar have attracted its attention to no good effect. There was even a rumoured attempt at a merger with Bharti Airtel in 2009. Within Africa, it has shown interest in networks in DR Congo, Kenya and Zimbabwe and in licences in Angola and Ethiopia, not to mention the Vivendi stake in Maroc Télécom, but so far to no avail. Very recently, MTN has allegedly entered talks with a view to acquiring its small loss-making domestic rival Telkom, but this will have no material effect on its situation in South Africa. On the other side of the coin, MTN has some issues in Iran, Syria and Sudan because US investors do not like being linked with companies that have operations in those countries. Overall, therefore, MTN seems to be stuck in something of a rut, realising that its portfolio of assets is not all that it should be but unable to do much about it for the time being.

As for domestic rival Vodacom, it is also seeking to merge with an even smaller domestic rival than Telkom, in this case Neotel, which is majority-owned by Tata Communications. Neotel's small mobile network, which has only 150 000 subscribers, is not really the prize, however, but rather its fixed-wire network – note the comparison with the recent behaviour of Vodacom's majority owner Vodafone. In any event, Vodacom may not be allowed to transfer Neotel's 800 MHz licence even it succeeds with the takeover. Nevertheless, the crucial point is that Vodacom has remained locked into its five-country portfolio for a good many years. As is noted in Curwen and Whalley (2008: 94–5), once the restriction on investment above the equator imposed by majority owner Vodafone was lifted, Vodacom sprang into action, expressing interest in Algeria, Angola, Ghana, Nigeria and taking a stake in Libya Africa Portfolio (LAP Green), but for one reason or another they all came to nothing. Subsequently, Malawi and Sierra Leone have been mentioned and the strategy appears to have been restricted to countries that are politically stable with densely populated cities and room for growth. Unfortunately, that sounds like a sensible investment recipe for any operator seeking to expand in Africa, so Vodacom is hardly likely to have the floor to itself if assets or licences are on offer.

One can argue that Vodacom is not really a free agent with Vodafone looking over its shoulder. Vodacom is not particularly important for Vodafone, as it contributed only 7.5 per cent of its proportionate subscribers at the end of 2012 (even without adjusting down to take account of Vodacom's relatively high inactivity rates), whereas Egypt contributed 6.1 per cent, Kenya 2.3 per cent and Ghana 1 per cent. Superficially, it might seem sensible for Vodafone to use Vodacom as the agency through which to expand its portfolio in Africa, especially as Vodafone stated in 2010 that it would be open to selling its stake in Egypt in order to concentrate on sub-Saharan Africa, but if that is indeed the case then Vodafone is clearly not pressurising its subsidiary to act. Obviously, Vodafone has recently had its hands full with developments in the USA, but even though it is now in receipt of a sufficiently large payment to underwrite investments anywhere in the world that it wishes, it does not seem likely that Africa will be uppermost in its thoughts – in some ways a surprising conclusion to draw about the operator with more proportionate subscribers in Africa than any operator other than MTN.

Orange is also something of an anomaly. Parent France Télécom stated some years ago that it wanted to concentrate on markets where it had majority control over the relevant operator and where it was one of the largest networks. However, the reality in Africa does not altogether match the rhetoric. For example, it does not have direct stakes in several cases, operating via a minority stake in Sénégal's Sonatel; in other cases such as Equatorial Guinea and Mauritius it holds a minority stake; in yet other cases such as Guinea-Bissau and Uganda it has few subscribers; and in Tunisia it has yet to do much real catching up with old-established licensees.

So, while Orange had the most networks – a matter not entirely unrelated to France's colonial past in Africa – and accounted for nearly 60 million proportionate subscribers at the end of 2012, this looks much less impressive if one is divided by the other, yielding an average of 3 million proportionate subscribers per network. As Table 6.3 shows, this is by no means unique, but it is quite surprising that Orange has done nothing until very recently to rationalise its African portfolio – indeed, the number of networks has done nothing but grow. In practice, Egypt alone accounted for 28 million proportionate subscribers in 2012 and, if Cameroon, Côte d'Ivoire and Morocco are added in, roughly 43 million of the 58 million were accounted for by just four of the 20 networks. In the past, there was a suggestion that Orange was intent on emulating Telefónica's role in Latin America, but it has yet to achieve anything like the same percentage in a single overseas region in respect of proportionate subscribers, let alone revenues, which are generally modest in the many African countries with low ARPUs.

So just how are things likely to proceed? Orange has stated at various times in recent years that it was interested in Tanzania but not South

Africa or Nigeria – too anglophone, perhaps; that it wanted to focus on emerging markets in Africa; and that it was interested in acquiring a group with operations in Africa or alternatively individual licences. In early 2011, the stated aim was to double revenue in emerging markets by 2015, partly through acquisitions in Africa, and that despite problems in North Africa it was not the degree of risk that concerned the company but rather the regulatory framework and the potential for growth. In early 2012, Orange threatened to exit emerging markets where it was not a top-two operator, but by early 2013 the focus was back on expansion with an expression of interest in entering Burkina Faso, Gabon and Mauritania. The Vivendi stake in Maroc Télécom was also the subject of some interest, but in the event Orange was unwilling to take on Etisalat. In October 2013, Orange expressed an interest in setting up an MVNO operation in South Africa despite its previous claims not to be interested in that country, and it was even alleged that Orange would like to acquire Vodacom's network there, although Vodafone's opinion on the matter was not recorded.

At the end of 2013, Orange acknowledged that it would like to dispose of its Ugandan network, which was trading poorly, and that the Kenyan network was under review. Some kind of partnership is possible in either case if a sale is not forthcoming. Hence, taken overall, there appear to be some rather mixed messages emanating from Orange, but it is certainly enthused (compared, for example, with Vodafone) to be playing a role of consequence in Africa and seems likely to expand into new countries if circumstances permit unless it is constrained by its existing burden of debt.

Given Millicom's static holdings, the fact that the percentage of proportionate subscribers in Africa rose from 25 per cent at the end of 2006 to 43 per cent at the end of 2009 demonstrated that it was rebalancing its asset portfolio towards Africa, and its departure from the Asia-Pacific region reinforced that conclusion. However, it is worth noting that its reason for cutting back in that region was the excessive number of licences in certain countries, which is equally true of many African countries.

6.6 THE BHARTI/ZAIN AFFAIR AND ITS FALLOUT

At this point it is of interest to explore the events of the late 2000s when the most significant restructuring of mobile portfolios in Africa took place. This primarily involved Zain and Bharti Airtel, but several others among the companies listed in Table 6.3 became involved and matters became extremely complicated, as outlined below.

Zain, based in Kuwait, was previously known as The Mobile Telecommunications Co (MTC). It was 24.6 per cent owned by the Kuwait Investment Authority (KIA) with the State of Kuwait holding a direct plus indirect stake of 34.5 per cent. In March 2005, Celtel International, the former MSI Cellular based in the Netherlands with 14 networks across Africa, agreed to be taken over. The purchase of the initial 85 per cent stake was completed in May 2005. In April 2006, MTC agreed to buy 65 per cent of Nigeria's Vmobile with an option to purchase the rest. MTC announced a so-called 3×3×3 strategy – a nine-year three-phase strategy moving it from regional to international to global status. In January 2007, it announced a target of 70 million subscribers by 2011.

In July 2007, a spokesman declared that 'Africa's going to be the next major growth area for us' (Lennighan, 2007). Celtel agreed to acquire 65 per cent of Westel Ghana in July 2007, so by the end of 2008 Zain held 16 stakes across Africa, as shown in Table 6.3. While it was true that the assets being acquired had some geographic logic, it was hard to believe that Zain possessed the requisite trained personnel to operate its acquisitions. Furthermore, even though it was cash-rich, having long enjoyed a domestic monopoly and being supported by state agencies, the drain on its resources was unremitting.

The first clear signs that Zain had bitten off more than it could chew emerged in July 2009. At this point, the KIA announced that it would be willing to sell its entire stake at the right price and it was rumoured that Etisalat was enquiring about the availability of a 51 per cent stake. However, in the event it was Vivendi Universal that proved to be holding talks about taking a stake of between 51 per cent and 65 per cent in Zain's African operations. At the time, Vivendi held six stakes across Africa but these did not generate large numbers of subscribers, so there was a case to be made for scaling up its operations in the region. Roughly 32 million proportionate subscribers were available in total and these were valued by Zain at between $10 billion and $11 billion. However, Vivendi/Maroc Télécom had little experience outside francophone Africa, the synergies were limited and Vivendi would probably have been obliged to make an unpopular rights offer in order to proceed without damaging its credit rating, so its subsequent loss of interest was understandable (Parker, 2009).

A deal was then brokered by the Kharafi Group whereby an Indian/Malaysian consortium agreed to buy 46 per cent of Zain for $13.7 billion, but the consortium was unable to demonstrate that it could raise the cash and the deal fizzled out.

But in many ways MTN appeared to be a more attractive target than Zain. Both held assets in the Middle East and across Africa, and both held large, controlling stakes in almost every network. However, MTN was considerably larger, which made it somewhat more attractive, although, by the same token, it also suggested that prospective buyers needed to be thinking in terms of taking a small majority stake rather than making an outright takeover.

Bharti Airtel made its move on 4 May 2008. It initially made an indicative bid for a 51 per cent stake, which was subsequently raised so as to be equivalent to valuing the whole of MTN at $45 billion, but MTN demanded to be valued at $50 billion and that the MTN CEO be placed in charge of Bharti-MTN. MTN also wanted the Bharti-MTN HQ to be located in Johannesburg, and it would evidently be necessary for Bharti to seek a secondary listing in South Africa before execution of the takeover.

Bharti sought the support of its minority owner Singapore Telecom as it was struggling to finance a full takeover, but finding enough cash was not the most serious obstacle to the deal. Although a provisional structure with Bharti in control was agreed, it was rejected by the MTN board, which – supported by strong nationalist sentiments expressed by everyone from government ministers downwards – argued in favour of Bharti becoming a subsidiary of MTN while allowing Bharti Airtel's parent, Bharti Enterprises, to hold a majority stake in what would now become MTN-Bharti. Unfortunately, this was treated as too damaging to Indian pride because it would prevent Bharti from claiming to be a global telecoms player, and the deal promptly fell apart (*New York Times*, 2008).

Despite the difficulties faced during 2008, Bharti had never given up on its desire to become a force on the global mobile stage, and at the end of the day its only expertise lay in emerging markets so there was little choice but to keep its acquisitive gaze turned towards Africa. And with no other suitable prospects surfacing during the period through to April 2009, there was little choice but to reprise its previous attempt to acquire MTN. On this occasion, however, the plan was that Bharti would initially take a 49 per cent stake in MTN while MTN shareholders would in return take a 36 per cent stake in Bharti. A full merger would, it was hoped, follow as soon as was practicable.

In the event, after months of tortuous negotiations, the South African government effectively decided to scupper the deal. The Communications Ministry announced that it would only support a dual listing in India and South Africa and that it would not sanction any future takeover of MTN by Bharti. However, such a listing was incompatible with existing Indian laws and regulations. With governments in both countries now up in arms, MTN shareholders complaining about an inadequate share price premium and Bharti left with the prospect of permanent minority ownership and lack of control of MTN, negotiations were duly terminated at the end of September.

These events explain why Zain's African assets became the new focus of attention for Bharti Airtel in October 2009. The network in Sudan was not included in the sale and nor was Wana (formerly Maroc Connect) in Morocco, so what Bharti would realistically be getting for its money was roughly 30 million proportionate subscribers. The value placed on these subscribers by Bharti in February 2010 was $10.7 billion in cash and debt.

But what really interested analysts were the true reasons for Zain's willingness to sell. Rumours began to circulate towards the end of 2009 to the effect that Nasser Al Kharafi, head of the Kharafi Group, was in financial difficulties due to the 'credit crunch'. It was, Al Kharafi explained, a matter of simple strategy – Zain wanted to concentrate on the Middle East while Bharti wanted to enter Africa (Wigglesworth, 2010). But why, in turn, was Bharti willing to pay so high a price? According to the available figures covering the nine-month period January–September 2009, Zain Africa accounted for 38 per cent of Zain's revenues but made a net loss of $112 million, with seven networks in the red. The obvious answer was that Bharti expected to be able to put its experience of selling profitably in a market with a very low ARPU to good use in markets where ARPU was considerably higher. But this presumed that India and Africa were readily substitutable, and by no means all analysts agreed with that proposition. However, at the end of the day, the reality was that Bharti was facing increasing competition in its home market and desperately needed to expand overseas.

The deal was formally accepted by both parties on 23 March 2010 (Leahy and Wigglesworth, 2010). However, regulatory permissions would need to be sought in more than ten countries and the governments of Gabon and Congo-Brazzaville initially raised objections, while in Nigeria there was an ongoing dispute over ownership of the network being sold. As a result, the true size of post-takeover Bharti had yet to become clear.

Negotiations proved to be tortuous, but in June 2010 Bharti was finally able to announce officially that it had acquired the Zain assets, while noting that litigation was still ongoing in Nigeria. Bharti began the re-branding of Zain networks in mid-October – it may be noted that in Kenya, for example, Kencell (2000) begat Celtel Kenya (2004), which begat Zain Kenya (2008), which begat Bharti Airtel Kenya (2010) – and now uses the 'airtel' brand throughout its networks. By the end of 2010, Bharti had increased its stake to 97 per cent via an open offer.

But where did that leave MTN? Clearly, it would need to look at smaller, less desirable, collections of assets if it wanted to respond to Bharti. The only candidate that offered any possibilities was Orascom, even though at the time it was feuding with Orange over the ownership of the jointly held network in Egypt, thereby rendering that network unappealing. Elsewhere in Africa, Orascom – controlled by Weather Investments II – was, first, the 97 per cent owner of a profitable network in Algeria with over 14 million subscribers, although a feud with the Algerian tax authorities had left Orascom feeling unwanted; second, it owned a half-share in a modest network in Tunisia; and, third, it was accruing a negligible number of subscribers in four other countries. Not surprisingly, rumours of interest by MTN initially referred only to Algeria before expanding to include the six networks outside Egypt.

At the end of April, trading in Orascom was suspended pending an announcement about the six stakes, now valued at roughly $6 billion. It was thought that Orascom would end up with some kind of stake in MTN, but not one that would offend South African sensibilities. However, the status of MTN as the eternal bridesmaid was seemingly enhanced by the announcement in mid-June that it was no longer in negotiations to buy any networks from Orascom (Telecom.paper, 2010). As a consequence, Orascom appeared to have no potential suitors for its African assets – at least until it emerged in September 2010 that the board of Russian mobile operator VimpelCom, controlled by Russia's Altimo (the Alfa Group) and Norway's Telenor, had approved the acquisition of a majority stake in Orascom. On the face of it, this was a surprising strategic move given that VimpelCom's existing assets were located in the CIS and Asia, but many of these could be said to be emerging markets with low penetration and high growth potential not that dissimilar to those held by Orascom.

VimpelCom indicated that it would be willing to sell the Algerian network to the government, but talked in terms of a price tag of $7.8 billion, which far exceeded what the government was willing to pay. Despite the uncertainty this created, VimpelCom agreed to buy 51.7 per cent of Orascom (plus Wind in Italy), although the transaction excluded Orascom's minority stake in Egypt (Middleton, 2010). Further uncertainty was created by the poor quality of most of the Orascom assets outside Algeria.

An agreement was struck between VimpelCom and Orascom in November to the effect that Orascom Tunisia would be sold to Qatar

Telecom for \$1.2 billion since it could neither be controlled nor consolidated by VimpelCom, but in December Telenor reiterated its opposition to any deal on the grounds that VimpelCom was overpaying and that, in acquiring Wind, it was not adhering to its strategic objective of expanding in emerging markets (Parker, 2010). Despite this, in late December, the board voted six to three (Telenor's votes) to proceed and a subsequent shareholders' meeting approved the merger.

6.7 CONCLUSIONS

The above sections have reviewed a cross-section of evidence relating to the restructuring of the mobile sector in Africa, in terms of both the individual countries and the main operating groups to be found there. Given its overall size, and the large number of countries and islands involved, some variability was only to be expected. One slight oddity is that some countries appear to be intent on issuing new licences when the general trend worldwide, and especially in more advanced countries, is towards consolidation because it is problematic to make a financial success of an operation that is not among the three largest in a market. This is borne out by the high three-firm concentration ratios evident in Table 6.2.

Another factor of interest is that only a handful of internationalised operators now play a meaningful role in Africa – MTN, Vodafone/Vodacom, Orange, Bharti Airtel and Etisalat. Five years ago the situation was more fluid, but the fallout from the Bharti/Zain affair described above, which, if nothing else, exposed how complicated restructuring can be in practice, combined with a tendency for operators with weak African portfolios (Orascom/VimpelCom), financial difficulties (Zain) or irons in other fires (Vivendi) to exit, has taken its toll. It is of further interest that this residual group contains only two Africa-based companies, MTN and Vodacom, one of which is controlled by a European operator.

There appear to be only two companies that remain noticeably expansion-minded – Etisalat, where Africa is its most important sphere of operations, and Orange, where Europe is more important overall but which seems intent on building up an empire in francophone Africa. It accordingly comes as no great surprise that it is Etisalat that has agreed to acquire the majority stake in Maroc Télécom being off-loaded by Vivendi. However, although this represents a fairly significant event if only in respect of the exit of Vivendi from mainland Africa, it is dwarfed by the Bharti/Zain affair and there is no obvious fallout to come.

In noting the generally conservative attitude to expansion within Africa, it is important to bear in mind that companies operating in Africa are constrained by the view taken by the financial markets of the countries in which they have networks. For example, Moody's published its assessment of MTN in December 2013, which might appear from what has been discussed above to be one of Africa's more solid companies with a thriving business (Cellular-news, 2013b). However, Moody's expressed considerable concern about the political, social, regulatory and economic environments of key countries in which MTN had networks, singling out South Africa and Nigeria (as well as Iran). Hence, despite making positive comments about MTN's credit profile, subscriber growth in most markets and prudent management of its liquidity, the overall assessment was ultimately influenced by the fact that there were problems in these three big markets. These problems will not disappear in the foreseeable future, which will necessarily affect MTN's strategic behaviour, especially in relation to its ability to finance expansion.

It may also be noted that the transfer of a substantial portfolio of African assets from one non-African investor to another is of interest for a number of reasons. The first, clearly, is that, even allowing for the above, it is easier to acquire such a portfolio – ultimately only financial muscle is a necessary requirement – than to operate it successfully. Zain had the money and it was able to build brand visibility, but it ultimately lacked the managerial skills to succeed, especially when confronted with the mounting depreciation, amortisation and interest costs that were the consequence of its investment spree. It may be argued that this reflected a desire to start big rather than engage in an initial learning curve, but this was compounded by its desire to run things from its headquarters in Kuwait. Africa is not merely an immense continent, but one that contains a wide variety of cultures, and a presence on the ground is needed if these are to be catered for successfully.

But if Zain failed, why should Bharti (which ultimately inherited Zain's African holdings), Etisalat or MTN do any better? In the case of Bharti, the argument has always been that it would be able to transfer its experience in the low-ARPU Indian market to (not quite so) low-ARPU markets in Africa. However, it is somewhat easier to deal with a single regulator and government – even in India – than to deal with highly diverse regulators and governments in a number of African countries, as the responses in Gabon and Congo-Brazzaville initially demonstrated. Even MTN, itself based in Africa, is not necessarily seen as in tune with the concerns of less developed and more volatile countries to the north.

So, finally, is a period of consolidation to be expected for the foreseeable future? On the whole, the evidence points in that direction. In the past, a number of highly desirable prizes such as the third licence in Egypt have understandably attracted the interest of a significant number of internationalised operators. However, even though there are still a few duopolies and some apparently underdeveloped markets, there are usually good reasons for prospective new entrants to proceed with considerable caution, most notably political and/or other sources of instability and well-entrenched incumbents. Furthermore, there is likely to be a considerable gap between what the government thinks a licence or a stake in a company being privatised is worth and the valuation placed upon it by prospective purchasers.

One final factor to take into account is the behaviour of the relative minnows on the continent. While these are too small to attract the attention of the international operators, they may be accumulated by a prospective new entrant or, more likely, be merged with other smaller players, thereby helping to consolidate oversupplied markets. It is known, for example, that Sudatel Telecom Group (Expresso) is looking to sell its networks in six countries including Ghana (TeleGeography, 2013a), as is BinTel (Azur) in three countries (TeleGeography, 2013b).

NOTE

1. In November 2012, investment firm Capcom stated that it intended to merge its MTS and Multi-Links operations with Starcomms. The merged entity would end up with 20 MHz of contiguous spectrum in the 1900 MHz band, sufficient to roll out LTE. However, the merger is yet to take place.

REFERENCES

Cellular-news (2013a), 'Third of mobile phones cut off in Niger', www.cellular-news.com, 28 November.

Cellular-news (2013b), 'Moody's affirms MTN debt rating, changes outlook to stable', www.cellular-news.com, 17 December.

Cellular-news (2014a), 'Unregistered SIM cards still able to make calls in Uganda', www.cellular-news.com, 16 January.

Cellular-news (2014b), 'Zambia cuts off 2.4 million SIM cards', www.cellular-news.com, 11 February.

Cellular-news (2014c), 'Kenya's Yu to close network – sell off infrastructure and customer base', www.cellular-news.com, 3 March.

Curwen, P. and J. Whalley (2008), *The Internationalisation of Mobile Telecommunications: Strategic Challenges in a Global Market*, Cheltenham, UK and Northampton, MA, USA: Edward Elgar.
Leahy, J. and R. Wigglesworth (2010), 'Zain board agrees $10.7bn Bharti deal', www.ft.com, 24 March.
Lennighan, M. (2007), 'MTC invests in Lebanon, Africa', www.totaltele.com, 2 July.
Maylie, D. (2013), 'Turkcell files lawsuit against MTN in South African court', www.totaltele.com, 27 November.
Middleton, J. (2010), 'VimpelCom, Weather merge to create new telco giant', www.telecoms.com, 12 October.
New York Times (2008), '$50 billion Bharti-MTN deal falls apart', http://dealbook.blogs.nytimes.com, 24 May.
Parker, A. (2009), 'Vivendi halts Zain talks on Africa assets deal', www.ft.com, 20 July.
Parker, A. (2010), 'VimpelCom's $6.6bn deal hangs in the balance', www.ft.com, 22 December.
Smith, M. (2013), 'Vivendi to sell Maroc Tel stake to Etisalat for $5.7 billion', www.reuters.com, 5 November.
Telecom.paper (2010), 'Orascom ends asset sale talks with MTN', www.telecompaper.com, 10 June.
Telecom.paper (2014), 'Telecom Egypt acquires mobile licence for EGP 2.5 bln', www.telecompaper.com, 3 April.
TeleGeography (2013a), 'Sudatel plots mass exodus; African subsidiaries up for grabs', www.telegeography.com, 22 March.
TeleGeography (2013b), 'Out of Africa? BinTel seeks buyers for Azur units', www.telegeography.com, 22 March.
TeleGeography (2014), 'Ghanaian cellcos ordered to re-run SIM registration', www.telegeography.com, 17 March.
Wigglesworth, R. (2010), 'Zain deal motives under scrutiny', www.ft.com, 18 February.

7. Structural and strategic adjustment among European mobile operators

7.1 INTRODUCTION

One issue that must be immediately addressed is the manner in which Europe is being treated for the purposes of this chapter. In Curwen and Whalley (2008), 51 countries are listed under the heading of Europe. However, the number to be considered in this chapter has been raised to 60. Kosovo has been included even though it has not been universally recognised as an independent country, whereas the inclusion of Russia reflects the emergence of the national operators at the expense of the regional networks. The other seven – Armenia, Azerbaijan, Kazakhstan, Kyrgyzstan, Tajikistan, Turkmenistan and Uzbekistan – could in principle be listed under the Middle East (which is not discussed independently in this book) and/or the Asia-Pacific region. However, their inclusion in this chapter makes it easier to understand how certain operators such as Telenor and TeliaSonera view their operational footprints.

It has been clear for some time that Europe as a region has been steadily encroaching on territory to the east of its original Western European boundaries. The first major sea change occurred in May 2004 when ten countries joined the European Union, effectively bringing the former Eastern Europe into the fold, and the EU has now expanded to encompass 28 countries, with the addition of Bulgaria and Romania in 2007 and Croatia in July 2013. The underlying presumption is that this number will continue to grow, although recent events in Turkey and Ukraine are indicative of the major issues associated with prospective member states. Nevertheless, it is fair to say that what might be termed Eurasia – incorporating the countries formed after the break-up of the USSR – is currently viewed as the preserve of operators based in Europe (including Russia) rather than of those in the Asia-Pacific region.

In the main sections that follow, the ownership of mobile communication licences across Europe, according to the wide definition of the region that has been chosen, is clarified because some operators use their own names as brands wherever possible whereas others trade under a variety of names, and the operational footprints of those that are most active internationally will be discussed in detail. As will become evident, Europe has become the main region outside the USA where operators have been forced to address whether their footprints, which tended historically to grow fairly haphazardly as and when opportunities arose, make good sense in the present-day environment for mobile communications.

7.2 MARKET STATUS

It is helpful to begin by examining the market positions of the main operators present in Europe, as depicted in Table 7.1. In order to make the table more manageable, the countries where none of these operators is present are excluded. The table also omits América Móvil because its home base is not in Europe and it has fewer proportionate subscribers in Europe than the 11 operators listed, although it will have an important role to play in the analysis of recent developments in Europe.

Table 7.1 identifies the situation as of end 2012 but changes during 2013 are included as footnotes – these will be discussed in more detail below. The table must be treated with caution at this stage as it identifies the market position but does not illustrate either the overall size of the market or the relative size of networks, both of which are more conveniently left for Table 7.2. It is of some interest to compare the table with Table 5.2 in Curwen and Whalley (2008). It may be noted, first, that Telecom Italia appears in neither case. Second, KPN, OTE and TDC appear in Table 5.2 but not in Table 7.1 – although OTE is present indirectly via the stake held by Deutsche Telekom and KPN will form part of the narrative when discussing América Móvil – and are replaced by Hutchison Whampoa, Mobile TeleSystems (MTS) and VimpelCom. The latter two operators are Russian and have no direct role to play in the EU, but VimpelCom, in particular, is important in relation to TeliaSonera, one of its major shareholders. In any event, their footprints do not make them suitable for discussion within the context of the Asia-Pacific region.

Given that both KPN and OTE form part of the narrative below, a few words about the third departing party, TDC, are in order. In January 2006, the Nordic Telephone Company acquired 88.2 per cent of the shares of TDC via a tender. In November 2006, the CEO stated that only Denmark constituted a core business and hence everything else was up for grabs at the right price. In practice, 'everything' did not amount to all that much in terms of Europe, but it had no difficulty in disposing of its assets in Austria (a minority stake sold in June 2007), Germany (sold to debitel in July 2007) and Latvia and Lithuania (sold to Mid Europa

Table 7.1 Market position, 31 December 2012

Country	Deutsche Telekom	France Télécom	Hutchison Whampoa	MTS	Tele2	Telekom Austria	Telefónica	Telenor[1]	Telia Sonera	VimpelCom	Vodafone
Albania	2	–	–	–	–	–	–	2	–	–	1
Armenia	–	3	–	1	–	–	–	V	–	2	–
Austria	2	[3[3]]	[4[3]]	–	–	1	–	–	–	–	–
Azerbaijan	–	–	–	–	–	–	–	–	1	–	–
Belarus	–	–	–	1	–	2	–	–	3	–	–
Belgium	–	2	–	–	–	–	–	–	–	–	–
Bosnia/Herzeg.	3	–	–	–	–	–	–	–	–	–	–
Bulgaria	[2[2]]	–	–	–	–	1	–	[2[2]]	–	–	–
Croatia	2	–	–	–	3	1	–	–	–	–	–
Czech Republic	1	–	–	–	–	–	2	–	–	–	3
Denmark	–	–	4	–	–	–	–	2	3	–	–
Estonia	–	–	–	–	3	–	–	–	1	–	–
Finland	–	–	–	–	–	–	–	–	2	–	–
France	–	1	–	–	–	–	–	–	–	–	–
Georgia	–	–	–	–	–	–	–	V	1	3	–
Germany	1	–	–	–	mvno	–	4	–	mvno	–	2
Greece	1	–	–	–	–	–	–	–	–	–	2
Hungary	1	–	–	–	–	–	–	2	–	–	3
Ireland	–	–	4	–	–	–	2	–	–	–	1
Italy	–	–	4	–	–	–	1	V	–	3	2
Kazakhstan	–	–	–	–	3	–	–	V	1	2	–
Kyrgyzstan	–	–	–	–	–	–	–	V	–	2	–

Latvia	–	–	–	–	2	–	–	–	1	–	–
Liechtenstein	–	2	–	–	–	3	–	–	–	–	–
Lithuania	–	–	–	–	1	–	–	–	2	–	–
Luxembourg	–	3	–	–	–	–	–	–	–	–	–
Macedonia	1	–	–	–	–	2	–	–	–	–	–
Malta	–	–	–	–	–	–	–	–	–	–	1
Moldova	–	1	–	–	–	–	–	–	2	–	–
Montenegro	2	–	–	–	–	–	–	1	–	–	–
Netherlands	3	–	–	–	4	–	–	–	–	–	2
Norway	–	–	–	–	3	–	–	1	2	–	–
Poland	1	2	–	–	–	–	–	–	–	–	–
Portugal	–	[3[4]]	–	–	–	–	[[5]]	–	–	–	2
Romania	3	1	–	–	–	–	–	–	–	–	2
Russia	–	–	–	1	[4[6]]	–	–	V	3	2	–
Serbia	–	–	–	–	–	3	–	2	–	–	–
Slovakia	2	1	–	–	–	–	3	–	–	–	–
Slovenia	–	–	–	–	–	2	–	–	–	–	–
Spain	–	3	–	–	–	–	1	–	4	–	2
Sweden	–	–	4	–	2	–	–	3	1	–	–
Tajikistan	–	–	–	–	–	–	–	V	1	3	–
Turkey	–	–	–	–	–	–	–	–	1	–	2
Turkmenistan	–	–	–	1	–	–	–	–	–	–	–

Country	Deutsche Telekom	France Télécom	Hutchison Whampoa	MTS	Tele2	Telekom Austria	Telefónica	Telenor[1]	Telia Sonera	VimpelCom	Vodafone
UK	1	1	4	–	–	–	2	–	–	–	3
Ukraine	–	–	–	2	–	–	–	V	3	1	–
Uzbekistan	–	–	–	[7]	–	–	–	V	2	1	–

Notes:

1. V indicates via stake in VimpelCom.
2. OTE's network in Bulgaria (in which Deutsche Telekom held a stake) was sold to Telenor in August 2013.
3. Hutchison bought Orange in 2013, thereby becoming the third-largest operator (of three).
4. Sonae exercised its option to buy out Orange in August 2013.
5. Present via a 2 per cent stake (reduced to 0.5 per cent in 2013Q3) in Portugal Telecom which owns the largest network.
6. Sold in 2013.
7. Licence revoked during 2012.

Partners in January 2007). By end 2007, it had 7 million proportionate subscribers in Denmark, Poland and Switzerland, but the 19.6 per cent stake in Poland was itself sold off in December 2008. In 2010, TDC Schweiz (trading as Sunrise) was bought by CVC Capital Partners, effectively leaving TDC as solely a domestic operator. This is unquestionably the most extensive dismantling of a European operator's international portfolio yet witnessed.

The most noteworthy feature of Table 7.1 is arguably that there is only one apparent example of two operators having stakes in the same network. However, even this is somewhat illusory since whereas France Télécom (Orange) and Deutsche Telekom (T-Mobile) do indeed share infrastructure in the UK and use the EE (formerly Everything Everywhere) brand that they jointly own, they also continue to use their original Orange and T-Mobile brands to attract customers and they maintain separate Internet domains.

A second feature is that the listed operators do not have stakes in one another, with the somewhat peripheral exception of Telenor's minority stake in VimpelCom. However, there are a number of other such stakes that are not apparent in the table, such as América Móvil's minority stakes in KPN and Telekom Austria, Deutsche Telekom's minority stake in OTE and Telefónica's minority stakes in Telecom Italia and Portugal Telecom (now tiny). By and large, such stakes would seem to be of uncertain value, given a clear preference for majority control wherever possible, as evidenced by the likes of Vodafone, Telekom Austria and Hutchison Whampoa. However, this does not necessarily mean that minority stakes cannot confer some degree of control over strategy – for example, this is understood to be the case for Telefónica in relation to Telecom Italia and for Deutsche Telekom in relation to OTE.

A third feature is the link between an operator and its market position in its home country. It is to be expected that a domestically based operator will be in first position, and indeed this is generally the case, but it does not necessarily hold true. For example, Vodafone is the only UK operator to be based in the UK but it holds only third place in the market. Equally, both MTS and VimpelCom cannot by definition both hold first place in Russia, nor TeliaSonera and Tele2 in Sweden.

Looking across the list of operator names in Table 7.1, there is little sense of new entry (Curwen and Whalley, 2012a). Indeed, only Hutchison Whampoa – based in Hong Kong and trading as '3' in Europe – seems out of place in that it emerged via the acquisition of 3G rather than 2G licences. It is noteworthy that it occupies fourth position in each of its six markets although it has recently moved up into third (but still last) place in Austria by buying up the France Télécom network. Hutchison basically sought to compete by acquiring 3G licences and using roaming agreements to access GSM subscribers (Curwen and Whalley, 2012b), which effectively required it to use relatively low prices to attract subscribers – a strategy that would inevitably be loss-making for (a good) many years. What is remarkable – and very hard to explain – is that its market share in its six European markets varies only from 8 per cent (Italy) to 11 per cent (Denmark) with three of the other four at 10 per cent. No other operator would dream of trying to cope with the financial consequences of such a pattern, but they do not have the deep pockets of Hutchison Whampoa on which to fall back.

7.3 MARKET CONCENTRATION

A summary picture of mobile market concentration in Europe is presented in Table 7.2.

Table 7.2 Mobile market concentration in Europe, 31 December 2012

Total subscriber range	Total countries	Number of countries 1 network	Number of countries 2 networks	Number of countries 3 networks	Number of countries 4 networks
0 to 100 000	9	5	2	1	1
100 000 to 1 mn	4	0	0	4	0
1 mn to 5 mn	13	0	3	9	1
5 mn to 10 mn	10	0	0	5[2]	5[2]
10 mn to 25 mn	13	0	0	10[3]	3
25 mn to 50 mn	2[1]	0	0	0	2
50 mn to 75 mn	5	0	0	1	4[4]
75mn to 100 mn	2	0	0	0	2
>100 mn	2	0	0	0[5]	2[5]
Total	60	5	5	30	20

Notes:

1. Both having roughly 26 million subscribers.
2. The purchase of the Telefónica network in Ireland by Hutchison ('3'), if approved, would affect these totals.
3. One of these has licensed a fourth operator in 2013, while another licensed a fourth operator only for its licence to be revoked.
4. Poland strictly has six operators but two are so small as not to affect the analysis.
5. The purchase of the KPN network in Germany, if approved, would affect these totals.

A number of conclusions can initially be drawn:

- The only monopolies are to be found among the countries with fewer than 100 000 subscribers.
- Only ten countries (nine excluding Kosovo) have fewer than three networks. They are all small.
- However, three is the predominant number, comprising 50 per cent of the overall sample, but 57 per cent of networks with more than 100 000 subscribers and 53 per cent of all networks with more than 1 million subscribers.
- Countries with four networks (including Poland) are to be found across a wide range of subscribers, but include all bar one of the networks with more than 25 million subscribers. The exception is Turkey, which is arguably Asian if defined in terms of its territory.

The boundaries between subscriber ranges are inevitably somewhat arbitrary, and it must be acknowledged that data on subscribers are always subject to a degree of disagreement, although that tends to be relatively modest in Europe compared, say, to Africa. The main issue is whether to rely on the operators' own figures or those issued by regulators. However, Table 7.3, which sets out concentration ratios – the proportion of the total market controlled by a given number of operators – is concerned with relative rather than absolute subscriber numbers, and the ratios can be treated as accurate. Concentration ratios do not vary much on an annual basis, so those shown in the table for end 2012 can be assumed to hold one year later – subject to the consequences of M&A activity discussed elsewhere.

The first point to note is that at the end of 2007 (Curwen and Whalley, 2008: table 5.2) – excluding those countries with fewer than three networks – the largest operator very commonly had a market share of between 40 per cent and 50 per cent. The exceptions were the UK (below 30 per cent) and Germany, Greece, Montenegro and Poland (between 30 per cent and 40 per cent). The second point is that only in Germany and the UK (plus tiny Liechtenstein) did the fourth-largest operator control more than 10 per cent of the market. Furthermore, in roughly half of the sample, the largest two operators controlled at least 80 per cent of the market – a reasonable definition of a duopoly from a real-world perspective – and only in Liechtenstein and Poland (narrowly) and the UK (by a considerable margin) was the ratio below 70 per cent. Where three or more operators had been licensed, a considerable gap often existed between the number of subscribers controlled by the second-largest operator and the number controlled by the third-largest operator.

Table 7.3 Mobile market concentration in Europe,[1] 31 December 2012

Country	GSM networks[2]	Total subscribers (000s)	% share largest operator	% share largest 2 operators	% share largest 3 operators
Albania	4	5 200	36.7	71.3	99.4
Andorra	1	65	100.0	–	–
Armenia	3	3 800	62.9	84.2	100.0
Austria	4[4]	13 000	41.2	72.7	89.9
Azerbaijan	3	10 100	43.8	73.2	100.0
Belarus	3	11 000	46.9	90.5	100.0
Belgium	3	12 400	41.7	76.1	100.0
Bosnia/Herzeg.	3	3 400	43.2	84.9	100.0
Bulgaria	3[5]	12 500	44.2	79.9	100.0
Croatia	3	5 200	46.4	84.9	100.0
Cyprus (South)	2[6]	1 150	72.6	100.0	–
Czech Republic	3	13 900	39.3	73.7	100.0
Denmark	4	7 700	44.0	70.3	89.0
Estonia	3	1 900	43.8	73.1	100.0
Faroe Isles	2	60	69.0	100.0	–
Finland	3	9 500	40.6	74.7	100.0
France	4[7]	67 100[18]	41.9	76.2	92.8
Georgia	3[8]	4 600	44.4	78.9	100.0
Germany	4[9]	107 900	33.6	62.4	82.3
Gibraltar	1[10]	30	100.0	–	–

Greece	3	16 500	46.9	80.3	100.0
Greenland	1	60	100.0	–	–
Guernsey	3	70	67.6	84.5	100.0
Hungary	3[11]	11 000	45.9	77.5	100.0
Iceland	3	400	38.2	71.9	100.0
Ireland	4[12]	5 300	41.0	70.4	91.0
Isle of Man	2	95	93.7	100.0	–
Italy	4	86 900[19]	37.0	67.2	92.0
Jersey	3	200	35.4	68.0	100.0
Kazakhstan	4	26 200	49.9	82.7	94.9
Kosovo[3]	2	3 000	85.4	100.0	–
Kyrgyzstan	4	5 200	48.6	95.7	98.6
Latvia	3	2 600	41.7	82.3	100.0
Liechtenstein	4	35	42.2	67.7	84.8
Lithuania	3	4 900	38.8	76.7	100.0
Luxembourg	3	800	52.6	86.6	100.0
Macedonia	3	2 400	50.2	76.4	100.0
Malta	3	500	52.1	86.9	100.0
Moldova	4	3 500	55.3	89.5	99.0
Monaco	1	30	100.0	–	–
Montenegro	3	1 000	40.3	74.5	100.0
Netherlands	3	18 600	48.5	74.5	100.0
Norway	3	6 100	54.2	81.5	100.0
Poland	6[13]	53 500	29.5	57.2	82.8
Portugal	3	17 300	43.2	79.4	100.0

Country	GSM networks[2]	Total subscribers (000s)	% share largest operator	% share largest 2 operators	% share largest 3 operators
Romania	4	26 200	39.2	69.7	94.1
Russia	4[14]	212 300	33.4	62.9	89.5
San Marino	1	n/a	100.0	–	–
Serbia	3	10 500	51.8	82.5	100.0
Slovakia	3	6 500	44.1	79.7	100.0
Slovenia	3	2 200	57.3	87.7	100.0
Spain	4	52 900[20]	39.2	67.5	93.1
Sweden	4	14 200	46.6	73.2	89.9
Switzerland	3	10 000	61.8	83.3	100.0
Tajikistan	5[15]	6 200	44.4	73.0	89.8
Turkey	3	67 300	52.2	80.1	100.0
Turkmenistan	2	2 030	56.3	100.0	–
UK	4	80 600	36.7	67.2	88.9
Ukraine	4[16]	54 800	46.7	83.4	98.2
Uzbekistan	4[17]	19 900	48.8	97.2	99.8

Notes:

1. Broadly defined.
2. The table excludes CDMA450/800 networks. Only operational networks are included.
3. Kosovo is not universally recognised as an independent state.
4. In January 2013, Hutchison (fourth-largest) took over Orange (third-largest).
5. A fourth licence was awarded in January 2013.
6. In January 2014, a third 2G/3G licence was issued to PrimeTel. There are two further operators in North Cyprus, controlled by Turkey, which shared a further 0.6 million subscribers in 2012.

7. The fourth operator, Iliad (Free Mobile), entered the market in 2012. Third-placed Bouygues Télécom is intent on buying second-placed SFR and, if successful, will sell its base stations to Iliad.
8. Aquafon and A-Mobile operate in Abkhazia and MegaFon in South Ossetia. Subscriber numbers are modest.
9. KPN (third-largest) received a takeover bid from Telefónica (fourth-largest) in 2013. The European Commission is investigating.
10. In September 2013, a subsidiary of Eazi Telecom, Shine Mobile, was launched.
11. A fourth licence was awarded in January 2012 to MPVI Mobil but annulled in September. However, the government appears to be determined to introduce a state-owned new entrant.
12. Hutchison (fourth-largest) made an agreed bid for Telefónica (second-largest) in 2013. It awaits regulatory clearance.
13. However, CenterNET and Mobyland have fewer than 1 million subscribers between them.
14. As well as the four nationwide operators, there are a number of regional networks.
15. This may be 4, as subscriber data are reported for two networks that appear to have the same primary shareholder.
16. Ukraine also has some CDMA networks of some size – for example, Intertelecom had 1.4 million subscribers at the end of 2013.
17. The largest network, MTS, had its licence revoked in August 2012, causing most of its subscribers to switch to other networks. It was declared insolvent in 2013 and its assets were passed to state-owned Uzbektelecom, whose subsidiary Uzmobile was granted mobile licences on favourable terms in March 2014. Obviously, these events had a marked effect on concentration ratios.
18. Based on operators' own claims. According to the regulator, the total was 70.5 million. Iliad claimed to have been the fastest-growing operator in 2013 and to have gained a 12 per cent market share.
19. According to the regulator, subscriber numbers fell from roughly 93 million to 91.7 million at end 2013. The market share of Wind (third-largest) was given as 23.3 per cent at end 2012 as against 24.8 per cent in the table.
20. Representing a reduction of 5 per cent during 2012.

The really unusual case was the UK, where there were four almost equal-sized operators together with the new entrant 3 UK. In the majority of cases, the most recent operator to launch was the one with the fewest subscribers.

The list of countries in Table 7.3 is larger than that used in Curwen and Whalley (2008), so comparisons must be drawn with care. In practice, however, only two of the ten additional countries listed in Table 7.3 show any unusual features, namely Armenia, where the largest operator has a 63 per cent market share, and Russia, where the equivalent figure is a mere 33.4 per cent. If we restrict ourselves to a comparison using only the countries listed in 2008, again omitting duopolies, the following conclusions can be drawn.

In the first place – as was to be expected given a five-year period under examination – the clear majority of countries still have an operator with a market share lying between 40 per cent and 50 per cent. However, it would be misleading to infer that the hallmark of the period was stability.

Those countries of any size where the largest operator had a market share of less than 40 per cent at the end of 2012 were Albania, Czech Republic, Germany, Italy, Lithuania, Poland, Romania, Spain and the UK. This list is longer than that for end 2007, even though Greece and Montenegro have disappeared. In practice, the upward shift in Montenegro was only marginal, raising it to just over 40 per cent, whereas that for Greece was 7 per cent. It is unclear for now what to make of Greece. In essence, the overall market has collapsed since 2009, when it peaked at roughly 20 million, due to the financial crisis in parts of the EU. The reduction in subscriber numbers at OTE and Vodafone between end 2009 and end 2012 was less than 20 per cent, but it was roughly 30 per cent at Wind, which went into administration in early 2010. Whether the relationship between these three operators is set to stabilise is a moot point because their relative market shares have been very volatile on an annual basis, and it is evident that the financial situation in Greece itself remains far from stable.

In the case of Germany, concentration ratios have fallen since end 2007 irrespective of the number of operators involved. This is primarily because the number of subscribers at the two smaller networks has grown, albeit rather slowly, over the past few years whereas the number at the two larger networks has fluctuated considerably on an annual basis and is no higher than at the end of 2009. T-Mobile may still have twice as many customers as Telefónica, but both Telefónica and E-Plus (KPN) have shown themselves to be extremely competitive. The big issue here is that Telefónica has provisionally acquired E-Plus – this is discussed below – and the regulator's decision on whether to permit this to proceed will clearly have major repercussions for concentration ratios in Germany (and possibly elsewhere) in future years.

It is possible that France will not need to be included in the 'elsewhere' category because in March 2014 Bouygues Télécom made a bid worth $20 billion for SFR – subsequently raised twice with increasing emphasis on the cash element of the bid – which Vivendi is anxious to dispose of having previously bought out its partner Vodafone. Bouygues was not the only bidder and its rivals for SFR were not other mobile operators, but the implication of a successful bid by Bouygues would have been to create a market leader with a market share of around 50 per cent and a third-placed operator with roughly a 10 per cent market share – a potential situation that was most unlikely to commend itself to the regulatory bodies in France, as Bouygues knew perfectly well. Partly in order to forestall this, Bouygues arranged to sell its network of 15 000 base stations to rival network Iliad for €1.8 billion as well as to dispose of some spectrum. In response, the industry minister declared that there were virtues in this arrangement since it would greatly strengthen the third player, whereas the takeover of SFR by Altice Group subsidiary Numéricable – the obvious alternative – would leave the market in a state of destructive competition.

However, it was the latter's offer that was provisionally accepted, only for it to run into difficulties not in mainland France but in La Réunion and Mayotte off the coast of Africa, where Altice Group is the majority owner of Outremer, a rival of SFR subsidiary SRR. In addition, Bouygues raised the cash element of its offer to €15.5 billion, but Altice sealed the deal by raising its own bid to €13.5 billion (plus a possible further €750 million at a later date) together with 20 per cent in the merged entity (*The Economist*, 2014).

It therefore appears to be the case that France will retain its four incumbents for the time being – with Altice holding a 60 per cent stake in the new merged entity – irrespective of the outcome in Germany. However, the most interesting aspect is that it is indicative of a preference for fixed–mobile convergence over consolidation and that the decision runs contrary to the wishes of the French establishment, which has been used to getting its own way on matters of industrial structure. It also means that the market remains unstable because both Bouygues and Iliad are weak in relation to Orange and Altice-SFR. This could lead to an attempt to merge with the larger Iliad, effectively executing a takeover, which is unlikely to be well received in many quarters (TeleGeography, 2014c), although the economy minister went on record in April 2014 to reiterate the desire of the government to reduce the number of incumbents to three. For the moment, however, the two parties place wholly incompatible views on the value of the Bouygues network.

The position in the UK is different in so far as T-Mobile and Orange have operated jointly as EE since July 2010. At the end of 2007, the UK was unusual in boasting four networks of a fairly similar size, but EE, combining two of these, has obviously become the market leader with a market share up from 25 per cent to 36.7 per cent at end 2012. The reason this is no larger is due partly to a gradual loss of customers to all other networks and partly to the relative success of Hutchison Whampoa (3 UK), which continues to prop up the market but has noticeably closed the distance between itself and its larger competitors.

As for Poland, this was the most equally divided of all European markets at the end of 2007, with three incumbents sharing almost all the customers. But since then a new entrant, Netia, has performed very creditably – two further new entrants have yet to make much of an impact – growing to become more than half the size, in terms of subscribers, of the largest network. Interestingly, although the original three operators do not differ all that much in size, the market leader is no longer Polkomtel, which has slipped to third place. T-Mobile (as it now trades) surged into top spot during 2012, and this can be attributed in good part to the final settlement of the dispute over the ownership of PTC between Deutsche Telekom, Vivendi and Elektrim, which enabled the first, as sole owner, to pursue a more coherent strategy.

The other countries where the largest operator had a below 40 per cent market share at the end of 2012 have a variety of stories to tell. In Albania, for example, the share fell from over 50 per cent at the end of 2007 as a result of slow growth in recent years by the original duopoly, Vodafone and OTE, combined with an outstanding performance by new entrant Eagle Mobile, which has almost caught them up – OTE subsidiary AMC is actually shrinking, due in part, no doubt, to problems at OTE itself.

As for the Czech Republic and Lithuania – both of which have been stagnant markets for many years – the percentage has (predictably) fallen back only slightly from just above 40 per cent at the end of 2007. In Italy, the reduction was just over 3 per cent, in line with the market as a whole, which is not significant, but the 5 per cent and 6 per cent reductions in Romania and Spain respectively are more worthy of note. The Romanian market has shown some growth overall, but market leader Orange has struggled to maintain existing numbers. However, this was an improvement on second-placed Vodafone, which has performed particularly badly since end 2010, shedding nearly 2 million subscribers. As a result, the two-firm ratio also fell – but more severely – from 83.4 per cent at end 2007 to 69.7 per cent at end 2012. A further stagnant performance in recent years at third-placed OTE was accordingly enough to increase its market share and to ensure that new entrant RSC&RDS did not make significant gains at the expense of all the incumbents.

Spain is somewhat different in that there was some, albeit modest, growth from end 2007 to end 2011, but 2012 witnessed a severe fall in overall numbers of roughly 5 per cent due to the financial crisis. Interestingly, this was not an across-the-board phenomenon. Market leader Telefónica was badly hit, shedding several million subscribers, while second-placed Vodafone also suffered significant losses. However, the smaller operators continued to gain customers and this clearly contributed to the 10 per cent reduction in the two-firm ratio between end 2007 and end 2012.

Some significant reductions in the one-firm ratio also occurred in countries where the ratio nevertheless remained above the 40 per cent mark. The main examples are Macedonia, Norway, Serbia and Slovenia. Macedonia is easily explained by the entry of a third operator – Telekom Austria, trading as 'Vip' – in 2007. This is one of the most successful instances of new entry in that Telekom Austria is now bigger than OTE and the market has three credible networks with the market leader – its main (indirect) shareholder is T-Mobile – suffering a steady loss of subscribers. This pattern is more or less replicated in the case of Norway, in that what was essentially a duopoly at the end of 2007 now has a viable third network operated by Tele2. However, one feature that is different is that both market leader Telenor and second-placed TeliaSonera have been marking time for several years while Tele2 showed a sudden spurt in subscriber numbers during 2011, the year in which it took over full ownership of the network.

Slovenia again shares certain features in that a new entrant in 2007, Tusmobil, has gathered a modest market share, while market leader Telekom Slovenije has been shedding small numbers of subscribers over a period of years. These factors have had the effect of reducing the one-firm ratio from over 70 per cent at end 2007 to 57.3 per cent at end 2012. The position in Serbia is that market leader Telekom Srbija has also been shedding small numbers of subscribers in recent years, whereas its two main competitors have been growing, thereby reducing both one-firm and two-firm concentration ratios by almost 10 per cent.

The other side of the coin would naturally constitute an increase in the one-firm ratio in certain countries during the same period. The situation in Greece, where this phenomenon occurred, has already been noted, but it is also noteworthy that there are no other comparable examples. Indeed, the one generalisation that can safely be made is that this period was notable for the downward trend in concentration ratios. On the face of it, that appears to contradict the argument presented in this text that consolidation would be desirable in a number of markets, with three as the ideal number of networks in all bar the largest countries. However, Table 7.3 reveals an interesting pattern. As things stood at the end of 2012, only 21 of the 60 countries still had four or more networks. In six cases the three-firm ratio was above 95 per cent, in 12 cases it was above 90 per cent and in 18 cases it was above 85 per cent. Only in Liechtenstein, Germany and Poland was it below 85 per cent and all are unusual cases in some respects – for example, in the case of Germany an attempt to merge two networks is under way at the time of writing (June 2014).

Hence, in 51 cases overall, either there was no fourth network or it existed but with a less than 10 per cent market share, which is unlikely to prove profitable in a competitive market requiring heavy investment in new technology. In other words, the trend towards a reduction in the one-firm ratio and (less frequently) the two-firm ratio mostly reflects competitive pressure exerted by the second-largest and (primarily) the third-largest operators rather than by struggling fourth-placed networks. The only operator that could arguably be seen as exerting pressure from a fourth-placed position across a number of markets is Hutchison Whampoa, yet even here, with a parent able to absorb enormous losses over a period of years, the subsidiary's market share in every case still hovers only around the 10 per cent mark.

It can reasonably be argued that, without the presence of a Hutchison Whampoa, a market is likely to settle into a cosy relationship among three large networks. However, if a fourth-placed operator is loss-making, then it cannot be forced to stay in business just to keep the regulator happy and there is bound to be ongoing pressure for M&A activity that will result in a sub-scale network losing its independence. If a three-firm market is the norm, then that is a fact of life that regulators should address in devising policies to ensure that competition is maintained.

7.4 EXPANSION AND CONSOLIDATION

So far, the emphasis of this chapter has been on countries and their networks without worrying much about network ownership as such. Hence the time has come to examine this topic because what happens in one country cannot entirely be isolated from what happens in another if the same operator is involved in both cases. For convenience, the operators are presented in pairs in tabular form, although no particular significance should be placed on the precise pairings chosen or the precise sequence of pairings.

Vodafone

The history and organisational structure of Vodafone Group is laid out in detail in Chapter 3 and hence does not warrant detailed repetition at this point. However, there is a need to reflect upon key recent events within the somewhat different context of this chapter.

The first point, as is obvious from a glance at Table 7.4, is that Vodafone is the major player in the European mobile sector. As such, it might have been expected to remain a predator, yet, interestingly, that has not been the case over the past five years. Indeed, Vodafone relinquished stakes in two major European countries in 2011. The clue as to why this has happened lies in the fact that every stake listed in Table 7.4 is a majority. However, in France, Vivendi Universal held a controlling 55.8 per cent stake in SFR, which it consistently refused to sell to Vodafone, so, eventually, under pressure from shareholders, Vodafone decided to accept a good price (€7.95 billion) for a non-controlled asset in an increasingly competitive market. This meant a loss of 10 million proportionate subscribers, but this was not too serious a loss for Vodafone given its overall numbers in the region. Interestingly, Vivendi is currently negotiating to sell SFR – but not with Vodafone.

The disposal of the SFR stake in mid-2011 was promptly followed by the sale of its only other minority stake in Europe – 24.4 per cent of Poland's Polkomtel – to Spartan Capital Holdings, an investment vehicle controlled by Zygmunt Solorz-Zak. The sale realised a modest €920 million but meant a loss of only 3.5 million proportionate subscribers.

It is of interest to reflect on the fact that adjusting for the loss of these 12.6 million proportionate subscribers would have left Vodafone with 161.2 million such subscribers at the end of 2010, yet only 139.6 million are shown at the end of 2012. In other words, proportionate subscriber numbers were falling elsewhere in Europe during this two-year period, and in practice this was primarily caused by reductions of roughly 6 million in Italy, 3 million in Germany and 2 million in Spain and Romania (although there was also a 2 million improvement in Turkey). Somewhat ironically, neither SFR nor Polkomtel suffered losses.

In August 2011, with revenue in virtual freefall in Greece, Vodafone announced that it had initiated talks with a view to a merger with Wind Hellas that would create a network larger than that of the existing market leader OTE (branded as Cosmote) in which Deutsche Telekom had recently increased its stake. The obvious downside was that it would leave the sector as a duopoly and the European Commission immediately expressed its reservations. However, when the deal was finally called off in February 2012 after Vodafone had written off £450 million of its investment in Greece, the main reason appeared to be Vodafone's concerns about how the deal would be financed if Greece were to be forced out of the eurozone. Subsequently, the two operators agreed to share infrastructure for 2G and 3G – but not LTE – service provision primarily in rural areas.

Obviously, Vodafone could do little to prevent the general downturn in European markets in response to the financial crisis in the eurozone, but in any event it had more important matters on its plate. The key issue on the minds of everyone at Vodafone during 2012 and 2013 concerned the impending transfer of the stake in Verizon Wireless, which was completed in February 2014. Before this point Vodafone had been viewed as a predator but now, in one stroke, it became a potential prey, especially as its acquisition would no longer be subject to competition issues in the USA.

As was immediately obvious, there were several good reasons for AT&T to make a move for Vodafone, in the process turning itself into the most internationalised operator in the world. In the first place, it was technically available, in the sense that neither the UK government, nor the US government nor Vodafone's shareholders were likely to raise objections – which had certainly not been the case when AT&T tried to take over Telefónica and the response to a bid for either T-Mobile or Orange would be expected to be just as negative. Second, AT&T evidently believed that it could raise the necessary finance. Third, Vodafone was profitable and had consolidated tax losses waiting to be put to good use. Fourth, the outlook in Europe was, it was hoped, improving. Finally, AT&T had already established its own LTE network in the USA, so had experience that would be useful when applied to Vodafone's European networks.

On the minus side of the equation, AT&T had no experience of operating in Vodafone's overseas markets, although it would have the option to divest some of these at a later date. Second, the tax losses would not be available to a company headquartered outside Europe. Third, as a primarily mobile operator, AT&T would probably feel uncomfortable with Vodafone's recent forays into the fixed-wire market (see below) but would be unable to unravel any completed deals. Fourth, the loss of yet another iconic British company would not go unopposed and there would be a good deal of regulatory uncertainty in Europe in general and in many of Vodafone's individual markets – the deal would have to be authorised in every market in which Vodafone operated.

Buying Vodafone would be a very far cry from taking a small non-controlling stake in América Móvil.

In the event, AT&T was forced by the financial regulators to clarify its intentions in January 2014 and in effect stated that it had decided not to proceed this time, but not necessarily for all time (Thomas, 2014). This statement was followed almost immediately by a report that China Mobile, in which Vodafone had until recently held a small stake, was interested in acquiring anything from 5 per cent to 20 per cent of Vodafone (Cellular-news, 2014a). Although this would not constitute a controlling stake as such, it would potentially be the largest individual stake in Vodafone and, given the buyer, would create something of a crisis among regulators in European countries, possibly including the UK, where Vodafone operates.

Any attempt to make a move on Vodafone will take a good many months to come to fruition, so Vodafone must carry on regardless in the interim. A key element in its short-term strategy is the continuation of its move into fixed-wire services. Traditionally, Vodafone saw itself as a mobile-only operator that happened to have acquired some limited fixed-wire assets via M&A activity. However, the growth of high-speed data transfers meant that there was much increased interest in packages that would, for example, enable a customer to create a Wi-Fi network in the home via a fixed-wire link while also using a smartphone with a mobile contract where there was no Wi-Fi signal. Such a fixed-wire link could be provided via an arrangement with another operator, but it would be more profitable in the longer term if the link were owned by the party providing the mobile service.

Vodafone made a major commitment to this new strategy when it acquired the fixed-wire assets of Tele2 in Spain and Italy in October 2007. However, the roughly $1 billion cost of this transaction was fairly modest, and it subsequently appeared to be problematic to get the financial markets to show support for the strategy. What has made a real difference recently is the huge amount of cash forthcoming from the disposal of the Verizon Wireless stake. Vodafone has committed £7 billion of this by March 2016 to Project Spring, which is designed to improve the delivery of high-speed data, and is moving quickly to strike deals across Europe (Gabriel, 2013).

Naturally, where there is a great deal of cash available, potential targets appear out of the woodwork even if, as in this case, Vodafone itself had not specifically mentioned them. However, some targets were indeed real rather than imaginary. For example, Vodafone had had previous dealings with Cable & Wireless in that it had bought its modest mobile business in 1999. Subsequently, Cable & Wireless rose in value and in 2010 it was split into two parts – Cable & Wireless Communications serving the Caribbean, and Cable & Wireless Worldwide (CWW) providing services to, among others, multinationals in the UK. Due to a combination of poor management and difficult trading conditions, the share price of CWW fell from 90p to 15p at the end of 2011, at which point Vodafone began to show an interest that drove up the value of CWW to 38p per share or just over £1 billion ($1.5 billion) in total. The deal was agreed in June 2012 in the face of competition from Tata Communications (Thomas, 2012).

Vodafone had suffered badly in Southern Europe as a result of the financial crisis in the eurozone, but far from treating that as a reason for withdrawal, it chose instead to invest heavily in the expectation of a recovery. Thus, for example, in January 2014, Vodafone entered negotiations to buy Spanish cable operator Ono. At the time, Ono's private equity owners were anyway considering an IPO valuing Ono in the €6–7 billion range, including debt, a substantial sum but well within Vodafone's capabilities. Not only did Ono have 1.9 million residential and 100 000 small business customers at the time, but it was also a successful mobile MVNO. Unlike Telefónica (its MVNO network partner) and Vodafone itself, which had both shed subscribers during 2013, Ono had significantly expanded its subscriber base by 600 000 to over 1 million. In mid-February, Vodafone made a formal offer worth €6.9 billion, but this was considered to be inadequate and Ono decided to proceed towards an IPO in April. The offer was raised to €7 billion and again to €7.2 billion before finally being accepted in mid-March subject to regulatory approvals (Telecom.paper, 2014b).

Also in January, Vodafone acquired a 6.5 per cent stake in Greece's ForthNet, although this did not come with any degree of control, and it is allegedly attempting to increase its 18.5 per cent stake in Hellas Online. Meanwhile, Italy's Fastweb, a broadband and pay-TV operator with a modest mobile MVNO operation, which is owned by Swisscom, is on the list of potential targets because, the purchase of Tele2 Italy notwithstanding, Vodafone's fixed-wire network in Italy remains limited in scope and there are no cable networks. The cost is estimated at no less than €3 billion but in any event Swisscom is showing no interest in a sale at the time of writing (June 2014).

In early 2013, Vodafone approached Kabel Deutschland with a view to making a takeover bid worth roughly €10 billion including debt. Although Vodafone had inherited some fixed-wire assets when it took over Mannesmann more than ten years previously, it had struggled in a highly competitive market also supplied by the likes of Kabel and Liberty Global. At the time, Kabel had 8.5 million connected households in 13 of

Table 7.4 Vodafone and Deutsche Telekom: European stakes, 31 December 2012

Vodafone			Deutsche Telekom		
Country	Stake %	Proportionate subscribers	Country	Stake %	Proportionate subscribers
Albania	99.7	2 014 000	Albania	39.0	700 000
Czech Rep.	100.0	3 375 000	Austria	100.0	4 104 000
Germany	100.0	33 890 000	Bosnia	19.6	100 000
Greece	99.9	4 400 000	Bulgaria	40.0[1]	1 803 000
Hungary	100.0	2 631 000	Croatia	51.0	1 186 000
Ireland	100.0	2 201 000	Czech Rep.	60.8[2]	3 342 000
Italy	76.9	17 369 000	Germany	100.0	36 586 000
Malta	100.0	310 000	Greece	40.0	3 095 000
Netherlands	99.9	5 200 000	Hungary	59.2	2 864 000
Portugal	100.0	6 267 000	Macedonia	30.2	364 000
Romania	100.0	8 122 000	Montenegro	45.4	162 000
Spain	100.0	15 344 000	Netherlands	100.0	4 720 000
Turkey	100.0	18 928 000	Poland	100.0	16 040 000
UK	100.0	19 544 000	Romania	34.5	2 196 000
			Slovakia	51.0[3]	1 179 000
			UK	50.0	14 109 000
Total Europe		139 595 000	Total Europe		92 050 000

Notes:

1. Held via OTE and sold to Telenor in August 2013.
2. Raised to 100 per cent in February 2014.
3. To be raised to 100 per cent during 2014.

the 16 states in Germany, but that also proved attractive to US-based Liberty Global, which had recently taken over Virgin Media in the UK. In June, Liberty Global offered slightly more than Vodafone, but Vodafone responded by offering €87 per share as against the €85 allegedly offered by Liberty (in addition to taking on the outstanding debt) and its offer was accepted by 76 per cent of Kabel's shareholders (BBC, 2013).

The much-mooted possibility of a tie-up with BSkyB in the UK looks to be less feasible. Valued at roughly $22 billion, BSkyB's main shareholder is Rupert Murdoch, who was prevented from acquiring the residual shares in 2011 and is (2014) mired in a phone-hacking scandal. However, he has ambitions to expand rather than contract his European empire.

On a more modest note, but symbolic of Vodafone's new-found emphasis on fixed-wire assets, it appears to have been the only bidder in February 2014 to show an interest in a joint venture with the Irish Electricity Supply Board, to be formed with the intention of stringing telecommunications cables along the electricity network.

Deutsche Telekom

If the number of networks and the number of proportionate subscribers are taken together, then Deutsche Telekom – or T-Mobile, as it is mostly known – is arguably the most important operator in Europe. Unlike Vodafone, however, it has full ownership of only a handful of its networks. Part of the reason for this concerns its relationship with OTE. Although OTE is not a particularly significant contributor to T-Mobile's overall subscriber numbers, and has suffered badly since the financial crisis hit the EU, it is of interest to explore briefly how the two operators became linked.

Rumours that Deutsche Telekom intended to take a stake in OTE of Greece, owner of 98.6 per cent of mobile operator Cosmote, surfaced in December 2007. The government had attempted to find a strategic partner for OTE earlier in the year, but had responded to the lack of interested parties by placing a 10.7 per cent stake on the stock exchange, thereby lowering its direct holding to 28 per cent (together with a further 5 per cent held indirectly). Domestic investment group MIG held a further 20 per cent, and it was this stake that Deutsche Telekom initially agreed to buy in March 2008 (Parker and Hope, 2008).

The purchase was dependent on the Greek government's ceding of control over OTE to Deutsche Telekom, which intended to build up its stake to approximately 30 per cent, thereby making it the largest shareholder with rights to appoint the CEO and the majority of the board.

In April, the government proposed that both it and Deutsche Telekom should end up with 25 per cent plus 1 share stakes, with the former nominating the chairman and the latter the CEO of OTE. This arrangement was widely seen as a preliminary step that would eventually lead to Deutsche Telekom taking over OTE, in part because the government proposal included an option for Deutsche Telekom to acquire an additional 15 per cent stake at a later date. The deal was agreed in May, with Deutsche Telekom stating its intention to buy 20 per cent from MIG, 3 per cent from the government and 2 per cent on the open market. Regulatory clearance by the European Commission came in October, subject to the disposal of Cosmofon in Macedonia where Deutsche Telekom was already present.

The rationale for an eventual takeover was clear. In 2007, Deutsche Telekom was wholly dependent on Europe, broadly defined, other than in respect of its substantial US network. Furthermore, its European holdings were very much biased towards the former Eastern European states. OTE, meanwhile, was heavily oriented towards the Balkans. Nevertheless, there was a problem in Macedonia, where both operated and OTE had a wholly owned network – to which the solution was for OTE to sell its network in March 2009. There were only two other overlaps, in Bosnia and Montenegro, where their respective networks competed (although OTE only had a small stake in both cases indirectly via Telekom Srbija and not many proportionate subscribers), so Deutsche Telekom would not only be able to fill in some of the blanks in its European coverage but would be able to rationalise the overlapping holdings.

An option to sell a further 5 per cent to T-Mobile was agreed in June 2009 and the shares were transferred in August. At this point OTE still operated in Albania, Bosnia, Bulgaria, Greece, Montenegro, Romania and Serbia, but contributed only 6.7 million proportionate subscribers to T-Mobile. To boost this, Deutsche Telekom bought a further 10 per cent stake in OTE in June 2011. However, OTE was beginning to suffer the consequences of the financial meltdown in the eurozone and in 2011 it sought to sell out of Serbia, with its stake in Telekom Srbija being acquired by the state in 2012. Since OTE, like Deutsche Telekom, operated in Bosnia and Montenegro, albeit indirectly via Telekom Srbija, OTE also exited from those countries (O'Brien, 2011). In April 2012, the stakes in Albania, Bulgaria and Romania were offered for sale and it was the wholly owned Bulgarian network, with nearly 5 million subscribers, that was the first to be transferred (to Telenor) in August 2013. That meant that OTE was contributing to T-Mobile at that point in time 15 million gross subscribers and a modest 6 million proportionate subscribers, plus a presence in Albania, Greece and Romania.

In the circumstances, it might seem reasonable to query whether Deutsche Telekom any longer regarded its investment in OTE as a wise move. If so, the answer is surprisingly positive since, in January 2014, Deutsche Telekom approached the government with a view to acquiring its residual 10 per cent stake and claimed that OTE was an investment of strategic importance with long-term prospects, adding that it had transformed itself into a high-performance company with healthy financial figures and low borrowing (TeleGeography, 2014a). It accordingly pledged to invest a further €1.2 billion to add to the €2 billion invested previously.

It is also worth noting that opportunities may slip away without an operator having any control over events. For example, Mobitel was the mobile subsidiary of Telekom Slovenije, which was 62.5 per cent owned by the state, with the other shares held by public and private institutions and the general public. In July 2007, the government offered for sale a 49.13 per cent stake in Telekom Slovenije, hoping to complete the transaction by the year-end. One party that expressed interest in the stake in October was Croatian incumbent T-Hrvatski Telekom, owner of mobile operator T-Mobile Hrvatska, in which Deutsche Telekom held a bare majority stake. Curiously, another potential bidder was another operator in which Deutsche Telekom held a majority (59.2 per cent) stake, Magyar Telekom, and in practice only the latter submitted a bid in January 2008, which was rejected. In early March, the remaining bids were also rejected. A collapse in the Telekom Slovenije share price held back the government from proceeding during 2008 and nothing much has changed since then, although it now appears that the stake may shortly be re-offered at a much-reduced price. As a result, Slovenia remains one of the missing links in Deutsche Telekom's coverage in this corner of Europe.

It can be argued that Deutsche Telekom has something of a history of being involved in complicated transactions in Europe. For example, the position in Poland may currently look stable, whereas the story behind is anything but. A protracted and increasingly acrimonious dispute between Deutsche Telekom and Vivendi Universal regarding who would take control of PTC, which had been rumbling on for several years, appeared to have ended in late 2006 when the courts decided in Deutsche Telekom's favour (Curwen and Whalley, 2008: 127). Deutsche Telekom paid €600 million to acquire a 48 per cent stake in PTC but, despite this, Vivendi Universal continued to dispute the various court rulings and it was not until December 2010 that an agreement was reached whereby

Deutsche Telekom paid €1.4 billion to Vivendi and Elektrim, a transaction that was concluded in January 2011, to settle the case and take control of PTC (Deutsche Telekom, 2010: 48).

The most recent complication has arisen in the context of the UK. After much speculation that highlighted the limited profitability of the company and the highly competitive nature of the market, Deutsche Telekom agreed in November 2009 to merge T-Mobile UK with the UK operations of Orange (Deutsche Telekom, 2009: 54). The merger formally occurred on 1 April 2010, with the resultant equally owned company being known as Everything Everywhere (subsequently renamed EE). In the 2010 annual report, the company was given a 'fair value' of €7.3 billion compared to its asset value of €5.5 billion. However, this gain was wiped out by accumulated losses of €2.2 billion.

After the merger, it was alleged that the owners of EE were looking to leave the UK mobile market. Orange suggested that once the synergies underpinning the merger began to be realised, an IPO would be a possibility, a suggestion echoed by Deutsche Telekom. It was rumoured that both parties would reduce their stake by half through the IPO, which, in practice, was put on hold in January 2014. Alternatively, it was suggested that Orange might buy out T-Mobile UK, although this was thought to be too 'financially challenging' for the French operator. It has also been suggested that EE could be sold to a consortium led by a former CEO or to AT&T. Needless to say, any future restructuring of EE will attract the eagle eye of the regulator, Ofcom.

The fact that at the end of 2012 T-Mobile had 93 million proportionate subscribers in Europe covering 16 countries sounds more impressive than the implied average if the three large networks in Germany, Poland and the UK are removed, which is just under 2 million. On the other hand, the networks themselves – only two of which are wholly owned – are mostly well placed in their respective markets – four are market leaders and all three in third place have market shares above 20 per cent.

However, like Vodafone, Deutsche Telekom is well aware that the era of the mobile-only presence in European countries is drawing to a close. Hence, during 2013, it was forced to address the issue as to whether to exit such countries, which in its case included the Czech Republic and Poland, or to bolster its presence there by acquiring fixed-wire and broadband assets. In practice, it responded by offering to buy GTS Central Europe (less its Slovakian network and hence four countries in all) for €546 million ($730 million) in November, stating that a core part of its strategy was to strengthen its position with business customers (Henning, 2013) – it may be noted that it increased its stake in T-Mobile Czech Republic to 100 per cent at the same time at a cost of €800 million. This purchase was almost certainly overpriced, although it will never be possible to tell what would have happened in its absence. What is rather curious, however, is that it simultaneously agreed to sell its fixed-wire broadband activities in the Netherlands to CDS Group (Telecom.paper, 2013). In February 2014, the new CEO stated that 'The long-term goal is to create an integrated pan-European network of mobile and fixed networks based on all-IP' (Wood, 2014a).

It may finally be noted that, its purchase in the Czech Republic notwithstanding, Deutsche Telekom specifically ruled out the purchase of any further minority stakes in mobile networks. Despite this, it promptly announced at the end of February 2014 that it was considering the purchase of the state's 48.8 per cent stake in Slovakia's Slovenské Telekomunikácie, in relation to which it had the right of first refusal, at some point during 2014.

France Télécom/Orange

The names France Télécom and Orange have long tended to be used interchangeably when referring to the mobile activities of the French incumbent, but it was the latter that became dominant during the late 2000s. As a result, the shareholders voted to make Orange the formal name of the operator as from 1 July 2013 – see the Orange press release of 28 May. However, as the discussion that follows is largely concerned with the prior period, the former parent's name will be used for events during that period to avoid any confusion.

It is fair to say that France Télécom is not the first name that springs to mind in the context of mergers between European operators, if only because of the inevitable network overlaps in a number of countries. Nevertheless, as a potential acquisition target, TeliaSonera appeared to have its attractions in that it would provide exposure to emerging markets in 'Eurasia', significantly expand European coverage – the only overlaps are in Moldova and Spain – without generating antitrust issues and provide scale for the ramping up of mobile Internet activities. Hence it came as no great surprise when France Télécom revealed its interest in acquiring TeliaSonera in April 2008, simultaneously approaching banks for a €10 billion loan facility to finance a takeover.

The response was almost entirely negative. For example, the Swedish government let it be known that it was concerned about the security implications of a sale to a foreign operator – especially, perhaps, one that was 27.5 per cent owned by the state. As for the individual shareholders, they had bought shares at SKr85 each when the operator was first privatised in 2000, and were less than thrilled at the prospect of a compulsory buyout at anything near the existing price of SKr54, even though, after adjusting for reinvested dividends, the break-even price was roughly SKr60 (Hall and Hughes, 2008).

France Télécom nevertheless went ahead, tabling a provisional offer in early June worth €26.5 billion ($41.6 billion), comprising cash (at SKr56.22 per share for up to 52 per cent of the deal's total value) and an equity exchange. This was immediately rejected as too low. The financial markets expressed their opinion in their usual manner, promptly forcing down the bidder's share price by 8 per cent and hence the total value of the offer. The France Télécom CFO warned that any further fall in the company's share price could induce it to walk away, yet the Swedish government was holding out for an offer well above SKr60 per share – a level that would undoubtedly generate precisely such a fall.

At the end of June, France Télécom decided not to proceed as it was not prepared to raise its provisional offer. Its share price immediately rose by 7.2 per cent – reflecting the markets' relief that the company was no longer pursuing a bid lacking industrial logic – while that of TeliaSonera fell by 13.5 per cent to SKr43.1. However, it was widely agreed that France Télécom had suffered such a severe dent to its credibility that it would be most unlikely to chase after any other European incumbents for some time – and that would include Telenor, which had been mentioned in passing as an alternative to TeliaSonera.

Subsequently, France Télécom/Orange has not tried to do anything remotely as dramatic, despite the fact that Europe has long provided the majority of its proportionate subscribers, so one might have expected that it would try continuously to bolster its presence in the region. In practice, the opposite is increasingly the case because, like its main rivals, Orange is anxious to dispose of non-core sub-scale networks. As shown in Table 7.1, Orange is the third-placed operator in a number of countries, but these exclude Switzerland, which it finally exited in March 2012. The original agreement to merge with TDC Schweiz – France Télécom had only 1.5 million proportionate subscribers at the time – was signed in April 2010, with France Télécom due to take a 75 per cent stake in the merged entity. However, the regulator refused permission in 2011, despite the dominant position held by Swisscom (see Table 7.3) and the deal was dropped. France Télécom then put the network up for sale and it was bought by Matterhorn Mobile (Apex Partners) in March 2012.

France Télécom was also successful in its desire to exit Austria, where its 35 per cent minority stake in a network with no growth provided fewer than 800 000 proportionate subscribers. The buyer was the even smaller network operated by Hutchison Whampoa, with the deal completed in early 2013. The situation in Portugal was very similar, except that the stake in Sonae.com amounted only to 20 per cent, and the main shareholder, Sonae, exercised its option to buy out Orange in August 2013.

Orange is alleged to have agreed to buy the KPN network in Belgium, where Orange already has a majority-owned network, but so far there is no evidence to support this, although the KPN network is certainly for sale at the right price. Given that France Télécom did not succeed in acquiring a stake in PTK Kosovo in July 2012, nor to acquire Yoigo, which was withdrawn from sale in October 2012, the only acquisition success consists of a licence won in Armenia – a peripheral country in European terms – in August 2008, which led to a launch in November 2009 but modest numbers of subscribers thereafter. Hence, overall, it can be stated with certainty that, in relation to Europe, Orange is concerned primarily with ensuring that retention of its remaining European networks can be justified.

The situation in relation to EE in the UK has already been discussed in the context of T-Mobile and does not bear repeating. However, Orange currently faces much the same issues as the other major European operators in that it is not properly geared up to provide quad-play services in many countries where it is present as a mobile operator. One such is Spain, where, as noted, Ono has been acquired by Vodafone. Hence, in February 2014, Orange was alleged to be considering a bid for either Jazztel – the share price of which had recently risen sharply, which would serve as something of a discouragement – or Yoigo (the mobile subsidiary of TeliaSonera). Given that Orange held third place, with Yoigo lagging well behind in fourth place in the market, this would not appear to present a significant regulatory issue and TeliaSonera is certainly willing to sell (at the right price). However, it would do little to improve Orange's quad-play capabilities.

In January 2014, Moody's downgraded the credit rating of Orange to Baa1 from A3. It stated that, although Orange had taken steps to deal with its ongoing problems via cost reductions, asset sales, improved marketing and less generous payouts to shareholders, it would continue to feel downward pressure on revenues and cash flows due to competition, regulation and a difficult economic environment. As indicated, this may lead to further disposals in Europe, although there is no evidence currently that this is under serious consideration.

Telefónica

The recent history of Telefónica in terms of its European assets is unusually complicated. This is particularly the case in relation to its

Table 7.5 France Télécom and Telefónica: European stakes, 31 December 2012

France Télécom			Telefónica		
Country	Stake %	Proportionate subscribers	Country	Stake %	Proportionate subscribers
Armenia	100.0	595 000	Czech Rep.[2]	70.8	3 528 000
Austria	35.0	782 000	Germany	76.8	14 822 000
Belgium	52.9	2 258 000	Ireland[3]	100.0	1 542 000
France	100.0	28 100 000	Italy	10.3	3 314 000
Liechtenstein	100.0	8 000	Portugal	2.0[4]	152 000
Luxembourg	52.9	56 000	Slovakia[2]	70.8	940 000
Moldova	94.3	847 000	Spain	100.0	20 531 000
Poland	49.8	7 418 000	UK	100.0	22 864 000
Portugal[1]	20.0	713 000			
Romania	96.8	9 922 000			
Slovakia	100.0	2 985 000			
Spain	99.9	12 727 000			
UK	50.0	14 109 000			
Total Europe		81 520 000	Total Europe		67 693 000

Notes:

1. Sonae exercised its option to buy out Orange in August 2013.
2. 65.9 per cent of these networks were provisionally sold in November 2013 to the PPF Group, which, now that the deal has been approved by the European Commission (in January 2014), is required to make an offer to buy the minority stakes. However, Telefónica intends to retain a 4.9 per cent stake previously held as unissued Treasury shares.
3. Under offer.
4. Reduced to 0.5 per cent in 2013Q3.

indirect stake in Telecom Italia, but this is left for discussion in a later section as it has ramifications beyond those relating specifically to Telefónica.

It has been apparent since 2010, and arguably even earlier, that Telefónica had developed into a somewhat unruly organisation with, at the end of 2009, roughly the same number of proportionate subscribers in Latin America as elsewhere in the world, including China. Although Europe was clearly its second-most important region, it contained only three countries, each majority or wholly controlled, that produced significant numbers of proportionate subscribers, as shown in Table 7.5. Clearly, therefore, there was a strong case for rationalisation via the disposal of the non-core networks. This was underlined by the need to pay down €7–8 billion per year of its massive debt mountain in order to protect its debt rating.

It should also be borne in mind, however, that Telefónica has long been torn between the desire to build an international empire and the need to be cost-efficient. Hence it negotiated unsuccessfully for a licence in Georgia in 2009 and expressed an interest in acquiring a 40 per cent stake in Telekom Srbija in June 2010, although it eventually decided not to make a formal offer. Further interest was noted in Polkomtel in February 2011 and in Yoigo in February 2013. However, it is notable that none of the potential acquisitions came to fruition, whereas the potential disposals were mostly pursued successfully. This is no great surprise, given that it committed itself in February 2010 to a policy of organic growth rather than expansion, and a general antipathy in the financial markets towards any further outlays on acquisitions.

Aside from Italy, seven other European assets are shown in Table 7.5. These exclude the Isle of Man, which was sold to HgCapital and CPS Partners in June 2010, but as it represented fewer than 100 000 subscribers it is of no real significance. The first of the seven is the Czech Republic, which is broadly covered by footnote 2. It is not altogether clear why Telefónica is holding on to the previously unissued Treasury shares and these may well be sold off in the near future. As the network in Slovakia is held as a subsidiary of the network in the Czech Republic, the same consideration applies there also.

Telefónica's strategy in Germany is, frankly, rather odd. Until 2011, it was the sole owner of the smallest of the four incumbents – not a particularly comfortable position but, with 18 million subscribers at the end of 2011, it was operating on a perfectly adequate scale. Then, in June 2012, it made an offer for E-Plus, a network of broadly similar size owned by KPN, only to suffer rejection on the grounds of undervaluation despite the fact that KPN was highly anxious to discourage the attentions of América Móvil, which had begun to acquire a stake in KPN. After due consideration, Telefónica evidently decided that, if it could not scale up, it might as well reduce its overall commitment by selling off part of its stake in its own network – while maintaining majority control – via an IPO. This took place in October 2012, raising €1.45 billion, and served to reduce Telefónica's stake to 76.8 per cent – equivalent, at the end of 2012, to a loss of 4.5 million proportionate subscribers.

However, the case for merging the third- and fourth-largest networks and thereby becoming the market leader still made economic sense. As a result, in July 2013, Telefónica made an offer to KPN for E-Plus worth €8.1 billion – although in practice it comprised €5 in cash plus a 17.6 per cent stake in the enlarged Telefónica Deutschland (Van Daalen, 2013a). The shares given to KPN would be newly issued, so Telefónica's own stake would fall from 76.8 per cent to roughly 65 per cent. If it were approved by regulators – a fairly big 'if' – the deal would create the market leader in Germany, with over 40 million subscribers. In effect, this would be equivalent to a market share of roughly 37 per cent, with the remaining two networks not all that far behind. This bid took place in a complicated context – discussed below – which also involved an attempt by América Móvil, already a significant shareholder, to take over the whole of KPN, and it was in order to secure the approval of the former that the bid was sweetened in October to provide KPN with 20.5 per cent of post-merger Telefónica Deutschland.

As things stand currently (2014), the deal is being scrutinised by the European Commission. Although the market would be reduced to three players by the deal, with all three of roughly equivalent size and owned by major operators, which is arguably an ideal structure for a mobile market in Europe, the fact remains that this is by far the largest market and regulators will be well aware that if a three-firm structure is deemed to be appropriate for Germany, then there can be little justification for a market with more than three firms elsewhere in Europe.

The initial response of the Commission was published in February 2014. This claimed that the average price of a mobile plan would rise by roughly one-third for many pre-paid customers but only by roughly 10 per cent for those with contracts. This was shrugged off by Telefónica as simply the traditional first salvo in merger negotiations with the Commission. However, in April, Telefónica predictably offered, *inter alia*, to sell spectrum to a new operator – equivalent to roughly 40 MHz in bands above 1 GHz across a fair spread of urban areas – and provide it with national roaming. Provision would also be made for up to three MVNOs. The offer would be open until the year-end. Initial feedback suggests that the Commission will demand more far-reaching concessions if the deal is to get the go-ahead.

The third country in the table is Ireland. In June 2013, Hutchison Whampoa made an offer of up to €850 million ($1.15 billion) for Telefónica's network – a somewhat unusual move in that Telefónica was the second-largest and Hutchison the smallest of the four operators. Not surprisingly, even given that a country the size of Ireland has no obvious need for four networks, the European Commission intervened in November to set up a full inquiry into the takeover with a view to presenting its report in March 2014 – a date that was not met. Presumably, the Commission's opinion about competition in Ireland will have to be compatible with that on competition in Germany, although it may be argued that in the former case (unlike in Germany) the third-placed network will have a much smaller market share (at roughly 20 per cent) than its two larger competitors.

Meanwhile, Hutchison tried to influence that decision by announcing at the end of March that it was willing to assist with the setting up of a new MVNO – rumoured to be the largest cable operator UPC – which would be bequeathed 3 Ireland's youth-oriented MVNO branded as '48' with at least 50 000 subscribers. This new operator would then be given some of the combined Hutchison/Telefónica spectrum, permitting it to evolve into a nationwide network via roaming rights with Hutchison (TeleGeography, 2014b).

The stake in Portugal Telecom dates back to a cross-shareholding agreement that resulted in Telefónica accumulating 10 per cent of its Iberian neighbour by mid-2010. The network in Portugal – the only one owned in Europe by Portugal Telecom – had roughly 7 million subscribers at the time and was clearly non-core, so Telefónica initially disposed of 8 per cent in June 2010 at the time when it was about to acquire Portugal Telecom's stake in Vivo Brazil, although it was alleged in some quarters that Telefónica retained control. In December 2012, Telefónica was alleged to have sold off the residual 2 per cent, although it was still in evidence in the 2012 annual report. In May 2013, the stake was definitely reduced – apparently to 0.5 per cent, since this is what appears on the latest Telefónica website bulletin dated September 2013. No doubt this will also be disposed of in the near future.

One unusual feature of Telefónica's recent strategy is that, while other operators have been buying up fixed-wire assets, it has been doing the reverse. In March 2013, it sold its fixed-wire broadband business in the UK to BSkyB and subsequently signed two deals for LTE backhaul with BT and cable operator Virgin Media.

In February 2014, Telefónica stated that it had had a better-than-expected year in calendar 2013, partly because it had not been necessary to repeat the write-offs of 2012. Significantly, it had been able to cut its net debt to €45 billion, an improvement of more than €2 billion on the target for the year. As a result, it has set a new target of €43 billion for 2013, which in principle means that there will be funds available for investment although nothing specific was mentioned in its statement. However, it was claimed that Telefónica was close to completing a deal to acquire media group Prisa's controlling 56 per cent stake in loss-making pay-TV platform Digital+ – Telefónica already held a 22 per cent stake.

Telenor

Telenor is one of the largest operators on a worldwide basis, which, given that it is the incumbent in a small European country located on Europe's northern fringe, is somewhat surprising. However, what is immediately striking when viewing its European assets in Table 7.6 is that it is very unlike any other operator in Western Europe or the EU (of which Norway is not a member) in that its networks lie mostly to the east and include several parts of the former USSR. Given that the majority of its proportionate subscribers are located in the Asia-Pacific region, it is evident that, even compared to TeliaSonera, its eyes are firmly fixed on the eastern horizon.

The main reason for this can be summed up in one word – VimpelCom. However, it is sensible to begin with matters not related to this operator. As shown in Table 7.1, this means the networks in Albania and Montenegro, which are insignificant, together with Denmark, Hungary, Norway, Serbia and Sweden. It may be noted that, despite Telenor's full ownership, each of these latter five yielded only between 2 million and 4 million proportionate subscribers at the end of 2012. Obviously, this had something to do with the modest overall size of each market, plus the fact that Telenor was the largest network in two of the smallest markets – Montenegro and Norway – but the third-largest in the much bigger Swedish market.

It may also be noted that none of these markets is growing significantly, so it may readily be concluded that, VimpelCom aside, Telenor is indeed a minor player on the European scene. It has periodically made some efforts to improve on this, but with limited success. It was allegedly interested in a stake in Poland's Netia P4 in December 2008 and in acquiring Wind independently of VimpelCom in September 2010. It made some progress towards the purchase of Poland's Polkomtel in

February 2011 but failed to complete any deal in July. A licence in Romania also attracted its interest in March 2012, as did Tele2 Croatia in February 2013, but the only real success was the acquisition of the Bulgarian network shed by OTE in August 2013. With the better part of 5 million subscribers, this makes it the largest European network independently held by Telenor, although it is nevertheless not the market leader.

Tracing the fluctuations in both the equity and voting share percentages of VimpelCom held by Telenor is a complicated matter. However, the salient feature is that the percentage of the equity held during the past several years has fluctuated within the 30 per cent to 40 per cent range. This means that it is a large minority shareholder, but in this case it is not the largest shareholder, which is Russia's Altimo, and hence Telenor does not have operational control over VimpelCom. Altimo appears to have taken control of more than one-half of the equity in 2013 and hence now appears to be very much in the driving seat, but its relationship with Telenor has always been somewhat antagonistic and it is difficult to say how recent events will alter this.

So far, Telenor has been willing to accept this constant bickering over the strategic objectives for VimpelCom as the price to be paid for maintaining access to VimpelCom's networks, by no means all of which lie within Europe. And as noted above, it is via VimpelCom that Telenor acquires the majority of its own proportionate subscribers within Europe. The full data for VimpelCom appear in Table 7.8, and most of the equivalent stakes that appear in Table 7.6 are 35.7 per cent, which was Telenor's equity interest at the end of 2012 – it appears to have fallen to 33 per cent in April 2013.

The development of VimpelCom's European holdings is discussed below, but a few comments are in order at this point. First, it may be noted that Telenor was present in Ukraine in 2009 as majority owner of a network, Kyivstar (in which Altimo was also involved), providing over 12 million proportionate subscribers. In 2010, this was merged with a small network owned by VimpelCom such that Telenor's interest in Ukraine was now derived indirectly via its stake in VimpelCom. This network has barely grown for several years. It must be assumed that the present troubles in the country will affect subscriptions, but precisely how is impossible to predict.

Second, Russia (in particular) and Italy, both inherited indirectly via VimpelCom, are the biggest sources of proportionate subscribers for Telenor in Europe. The Italian link is only the third-largest network and, like its larger rivals, is no longer growing in a saturated and highly competitive market. It may be argued that this is generally the case

Table 7.6 Telenor and TeliaSonera: European stakes, 31 December 2012

Telenor			TeliaSonera		
Country[1]	Stake %	Proportionate subscribers	Country	Stake %	Proportionate subscribers
Albania	2.9	52 000	Azerbaijan	38.1	1 683 000
Armenia	32.1	257 000	Belarus	30.4	319 000
Denmark	100.0	2 015 000	Denmark	100.0	1 462 000
Georgia	18.2	176 000	Estonia	100.0	868 000
Hungary	100.0	3 489 000	Finland	100.0	3 249 000
Italy	35.7	7 693 000	Georgia	74.3	1 541 000
Kazakhstan	26.8	2 300 000	Germany	38.0	200 000
Kyrgyzstan	35.7	886 000	Kazakhstan	61.9	8 333 000
Montenegro	100.0	400 000	Latvia	60.3	645 000
Norway	100.0	3 175 000	Lithuania	100.0	1 953 000
Russia	35.7	20 031 000	Moldova	74.3	929 000
Serbia	100.0	3 207 000	Norway	100.0	1 641 000
Sweden	100.0	2 385 000	Russia	25.2	15 797 000
Tajikistan	35.7	323 000	Spain	76.6	2 840 000
Ukraine	35.7	9 267 000	Sweden	100.0	6 587 000
Uzbekistan	35.7	3 741 000	Tajikistan[2]	–[2]	2 135 000
			Turkey[3]	38.0	13 334 000
			Ukraine	20.8	1 690 000
			Uzbekistan	94.0	8 907 000
Total Europe		59 397 000	Total Europe		74 113 000

Notes:

1. Bulgaria added in 2013.
2. Both via a 60 per cent stake in a direct subsidiary and indirectly via Russia's MegaFon.
3. Turkcell also has a subsidiary in North Cyprus which is not counted as independent.

everywhere in the EU, so Telenor's portfolio of assets may prove to be a wise choice. Certainly, despite its fluctuating stake in VimpelCom, its overall performance in terms of proportionate subscribers has been relatively impressive, rising from 100 million at the end of 2009 to 177 million at the end of 2012.

TeliaSonera

As shown in Table 7.1, TeliaSonera has the broadest range of assets in Europe (widely defined). It is, predictably, much the largest network in Sweden but, surprisingly perhaps, not as large as Radiolinja in Finland. Overall, it has a stake in the largest operator in an impressive eight countries. In calculating its stakes, it must be borne in mind that several are held by Fintur Holdings, its majority-owned joint enterprise with Turkcell in which it also has a minority stake.

Thirteen of its networks are majority-owned (directly and indirectly), which is also fairly impressive, but only six of these are market leaders, so the overall picture is somewhat mixed. Furthermore, the two really big networks are in Russia and Turkey, where TeliaSonera is very much the non-controlling shareholder.[1] Hence, when its proportionate subscribers are reviewed in Table 7.6, it becomes clear that no single network is especially important. Indeed, the issue could be seen from the opposite perspective as an excess of insignificant networks, given that 11 yield fewer than 2 million proportionate subscribers.

Naturally, that could be seen as the price to be paid for operating primarily in countries outside the large markets to the West, and, if Spain is the exception, then it is not a particularly desirable one as Yoigo is a poorly placed fourth in the market and has been considered for many years as a prime candidate for being swallowed up by one of its larger rivals – it was offered for sale but subsequently withdrawn in both 2012 and 2013.

In other respects, TeliaSonera has been structurally very consistent in Europe for quite a long time. It was noted above that France Télécom revealed an interest in acquiring TeliaSonera in April 2008. The response was almost entirely negative, but France Télécom nevertheless went ahead, tabling a provisional offer in early June worth €26.5 billion ($41.6 billion), which was immediately rejected as too low. At the end of June, as noted, France Télécom decided not to proceed as it was not prepared to raise its provisional offer.

Apparently safe from predation, TeliaSonera has made little effort to seek new pastures in Europe. In February 2011, it expressed an interest in Poland's Polkomtel but dropped out in June. In March 2012, it expressed an interest in Hi3G Denmark 'at the right price', but evidently the price was not right. More curiously, despite being the only bidder for 49 per cent of MegaCom in Kyrgyzstan in November 2011, its bid was not accepted and it took no further action.

In April 2012, it was claimed that henceforth its strategy would be to sell non-core assets while increasing control over core assets. While it had made some progress in the latter respect in previous years, this appears to have stalled of late and, given that it has failed to sell Yoigo, it has not made any progress in the former respect. Indeed, given the paucity of proportionate subscribers in so many markets, the distinction between core and non-core might appear to be fairly fine, but it is easy to understand why the cluster of networks in the Baltic region is not going to be broken up.

América Móvil/KPN/Telekom Austria

América Móvil is at first sight a somewhat unlikely operator to appear in a discussion about Europe since it traditionally restricted itself to the USA and Latin America. However, it has recently decided – whether because of, or despite, difficult trading conditions in the region is hard to say – to acquire stakes in relatively small European operators, initially KPN and Telekom Austria.

Aside from its small MVNO operations in several countries, KPN has long been associated with only three wholly owned networks of any size, in Belgium, Germany and its home base the Netherlands. The former two comprised the third-largest networks in the country, which is commonly associated with a struggle to become profitable. It can reasonably be argued that the incumbent from such a small country cannot expect to build up and maintain an international portfolio of assets, but this has evidently not deterred the likes of Telekom Austria and Telenor. Nevertheless, KPN was always more likely to end up as prey rather than as predator.

Indeed, rumours of potential bids were rife. In April 2008, for example, it was alleged that Telefónica was interested in acquiring KPN, which it denied. Given the financial cloud that at the time hung over third-placed operators, it was hard to see why a much larger operator would want to spend scarce resources acquiring KPN at a large premium to its ruling market value. However, the fact that Telefónica was struggling in fourth place in Germany, and hence would presumably have welcomed a merger with KPN's network, provided some kind of rationale. It is also of some significance that KPN did not have the state as a shareholder, so (in principle) anyone could acquire it without creating a political storm.

In the event, KPN soldiered on alone but, with minimal subscriber growth and a heavy burden of debt, no other European operator was likely to find it an attractive target. The fact that the predator eventually emerged in 2012 in the guise of a non-European operator, América Móvil, was therefore both logical and surprising at the same time. The initial offer, for 27.7 per cent of the equity at €8 per share, was made in May 2012. Despite (frankly implausible) claims by KPN that it was being undervalued, the offer was successful and América Móvil ultimately ended up with a 29.8 per cent stake, just below the 30 per cent cut-off point that would have necessitated a takeover bid for the whole of KPN (Steinglass, 2012). However, the percentage of the voting rights acquired represented only half that of the equity holding.

At the time, KPN attempted to fend off América Móvil by offering for sale its networks in Belgium and Germany, but could not attract any interest at an acceptable price. As a result, it had to scrap its dividend and make a €3 billion rights issue in early 2013. In July, an offer from Telefónica to buy E-Plus in Germany was tabled (see above). América Móvil was initially opposed to the deal but subsequently acquiesced after extracting improved terms from Telefónica. However, América Móvil still preferred to keep KPN intact, so it launched a takeover bid for the whole of KPN, valuing the residual 70.2 per cent of the equity it sought to acquire at €2.40 per share or €7.2 billion in total (Van Daalen, 2013b).

At this point, in August, the Stichting Preferente Aandelen KPN – an independent foundation set up to protect the interests of KPN's shareholders – exercised its right to acquire roughly half of all the voting shares in KPN, thereby effectively preventing América Móvil from exercising control over KPN's strategy. In October, América Móvil withdrew its offer, indicating that it was not prepared to raise the offer to the level (allegedly €3 per share) that would have overcome any opposition.

In February 2014, after revealing patchy results for calendar 2013, KPN made it known that, whereas its relationship with América Móvil had deteriorated at the end of 2013, harmony had been restored. América Móvil promptly reduced its stake to 27.1 per cent in an unspecified manner and for unspecified reasons. In March 2014, Moody's downgraded KPN's debt to Baa3, noting that it would be relying upon the sale of E-Plus to finance its future operations. Quite where América Móvil goes from here is uncertain.

Table 7.7 América Móvil and Telekom Austria: European stakes, 31 December 2012

América Móvil			Telekom Austria		
Country	Stake %	Proportionate subscribers	Country	Stake %	Proportionate subscribers
Austria	23.7[1]	1 275 000	Austria	100.0	5 380 000
Belarus	23.7	1 138 000	Belarus	100.0	4 800 000
Belgium	29.8	1 019 000	Bulgaria	100.0	5 574 000
Bulgaria	23.7	1 321 000	Croatia	100.0	1 921 000
Croatia	23.7	455 000	Liechtenstein	100.0	6 000
France	29.8	134 000	Macedonia	100.0	632 000
Germany	29.8	6 973 000	Serbia	100.0	1 860 000
Liechtenstein	23.7	1 000	Slovenia	100.0	663 000
Macedonia	23.7	150 000			
Netherlands	29.8	2 688 000			
Serbia	23.7	441 000			
Slovenia	23.7	157 000			
Spain	29.8	60 000			
Switzerland	29.8	15 000			
Total Europe		15 827 000	Total Europe		20 836 000

Note:

1. In January 2014, various stakes owned by the Carlos Slim family, 26.8 per cent in all, were brought together under the control of European subsidiary Telecom Carso.

This means that América Móvil now holds a large minority but non-controlling stake in an operator that is potentially losing one of its main assets to Telefónica and is valued at far less per share than América Móvil paid for its stake. It is difficult to see this as a success story. But has it fared any better in relation to Telekom Austria? One positive distinction is that it has been made welcome as a shareholder. In early 2012, ÖIAG, the state holding company, owned 28.4 per cent of Telekom Austria and was committed to retaining a blocking minority of 25 per cent. The remainder was privately held, with stakeholders including a consortium led by Ronny Pecik, which had bought a 20.1 per cent stake in October 2011. In June 2012, América Móvil initially acquired a 5 per cent stake and sought regulatory approval to buy the consortium's shares at what was believed to be a price of €9.50 per share – uncertain because the deal was private.

ÖIAG concluded that América Móvil would satisfy the objectives that had been laid down for an investor from outside the EU to hold more than a 25 per cent stake. In October, América Móvil stated that it now held 22.8 per cent, which had risen to 23.7 per cent by the year-end. In March 2014, América Móvil bought a further 3.1 per cent, thereby accumulating a stake large enough to form a blocking minority but not so large as to necessitate a takeover bid for the whole of Telekom Austria.

In March 2014, ÖIAG gave its backing for Telekom Austria's management to negotiate a shareholders' syndicate agreement with América Móvil, provided the latter promised to retain the HQ in Vienna and the government retained a blocking minority of 25 per cent plus one share. The respective shareholdings would be combined, followed by an offer for the minorities and a capital increase. América Móvil would be expected to end up with a stake in excess of 30 per cent. In April, América Móvil formally signed up for the deal and launched a bid worth €7.15 per share for the minority holdings (Wood, 2014b).

Precisely what was acquired can be seen in the list of Telekom Austria's assets in Table 7.7. At the end of 2012, Telekom Austria was present in seven countries but, most unusually, all were wholly owned subsidiaries. Among the seven, as shown in Table 7.1, three were market leaders, although the network in Belarus was not far behind MTS and well ahead of the two other operators. As can be seen, Austria, Belarus and Bulgaria provided between them 76 per cent of total proportionate subscribers, so Telekom Austria could not be said to be well balanced.

It should be appreciated that this was not what was intended, since during the period beginning in 2005 Telekom Austria appeared to be highly ambitious. However, its attempts to expand met with mixed results. For example:

- It failed to win 51.1 per cent of Telekom Montenegro in March 2005.
- It failed to win a licence in Slovakia in July 2006.
- It failed to acquire Mobi 63 of Serbia in July 2006.
- It failed to win 65 per cent of Telekom Srpske in Bosnia in 2006.
- It failed to win the second GSM licence in Kosovo in January 2007.
- It expressed an interest in a stake in OTE in 2007 but withdrew.

Hence, although Telekom Austria was successful in entering Belarus and Serbia, it must be concluded that its stated objective, which was to become the leading player in Southeastern Europe and to continue expanding in Central and Eastern Europe despite the failures of mid-2006, was not achieved to its satisfaction.

In the circumstances, it must be questioned, as in the case of KPN, whether América Móvil has made a sensible move in seeking to enter Europe via a stake in Telekom Austria. The one fairly certain thing is that, even if the investment pays off, it will not be in the immediate future. In November 2013, Telekom Austria reported that its operations in Central and Eastern Europe were suffering in the context of highly competitive markets, a challenging macroeconomic climate and especially a tough regulatory regime – the loss of roaming fees when Croatia entered the EU was especially damaging. Furthermore, the purchase of new spectrum for LTE, especially in Austria, had been more costly than expected. The Telekom Austria CEO made the point that the difference between the four networks in the EU and the four outside it could be summarised very easily – the latter were more profitable.

However, the CEO of América Móvil went on record in February 2014 to the effect that his company would remain long-term investors in Telekom Austria and that he regarded the latter's networks as valuable and promising for the future. The state-owned holding company, ÖIAG, which is the largest stakeholder in Telekom Austria, has recently expressed its willingness for América Móvil to replace it in that capacity, although taking a stake above 30 per cent would trigger a compulsory takeover bid. Another possibility that has surfaced recently is that América Móvil and ÖIAG will set up a shareholders' syndicate agreement, although that, in turn, will trigger a compulsory bid for minorities.

It is superficially hard to view Telekom Austria as a major force to be reckoned with in Europe and, realistically, its 20 million or so proportionate subscribers and lack of presence in core European markets does not suggest that it is likely to become one in any hurry. But if América Móvil is to use it as its primary vehicle for expansion within Europe, as seems probable, Telekom Austria may come to play a much bigger role than its current status indicates.

Mobile TeleSystems

Both Mobile TeleSystems (MTS) and VimpelCom are to a degree peripheral in terms of Europe as traditionally defined because neither Western Europe nor the EU is central to their activities. However, both are substantial operators with European connections, and Telenor is a major shareholder in VimpelCom, which has its HQ in Amsterdam, suggesting that it sees itself more in terms of Europe than of Asia.

The oddest thing about MTS is perhaps that, being based in Russia and hence in principle well acquainted with the somewhat dubious trading practices of the former USSR and subsequently the CIS, it should find itself in some difficulties in the latter – for example, the licence of its network in Turkmenistan, Barash Communication Technologies, was cancelled in December 2010. Service had to be halted as it was claimed that the licence had expired even though MTS claimed that it ran until 2012. As a result, no subscribers were recorded during 2011. The network was reopened in August 2012 on the basis of a 2G/3G three-year licence covering up to 2.4 million users, a number it has yet to reach.

Furthermore, despite having acquired a majority stake in an operator in Kyrgyzstan, or so it thought, this was annexed and never returned, despite legal action that extended well into the period under consideration. In practice, this was of little consequence in terms of foregone proportionate subscribers, but what happened in Uzbekistan is an entirely different matter. At the end of 2011, there were three fairly large networks in Uzbekistan, of which the largest, Uzdunrobita, was owned by MTS and another by TeliaSonera. What exactly happened in 2012 is a matter of dispute, although the most popular version has it that MTS was subjected to some kind of blackmail operation by members of the government who wished to acquire control (TeleGeography, 2013). The outcome was that MTS had its licence suspended in August 2012 and was declared insolvent in January 2013. Although it is appealing for compensation before an international tribunal, it seems unlikely that it will actually receive any, irrespective of the ultimate verdict.

This has been quite costly, as Uzdunrobita would have expected to have grown its subscriber base to roughly 10 million and MTS has only two other networks still operating on a larger scale. One, predictably, is in Russia and accounts for over 70 per cent of all proportionate subscribers. The other, also fairly predictably, is in Ukraine and is growing slowly due to fierce competition especially from the largest

Table 7.8 Mobile TeleSystems and VimpelCom: European stakes, 31 December 2012

Mobile TeleSystems			VimpelCom		
Country	Stake %	Proportionate subscribers	Country	Stake %	Proportionate subscribers
Armenia	80.0	1 904 000	Armenia	90.0	72 000
Belarus	49.0	2 563 000	Georgia	51.0	494 000
Russia	100.0	70 930 000	Italy	100.0	21 550 000
Turkmenistan	100.0	1 140 000	Kazakhstan	75.0	6 442 000
Ukraine	100.0	20 120 000	Kyrgyzstan	100.0	2 482 000
			Russia	100.0	56 110 000
			Tajikistan	80.0	906 000
			Ukraine	100.0	25 960 000
			Uzbekistan	100.0	10 194 000
Total Europe		96 657 000	Total Europe		124 858 000

network, owned by VimpelCom. It could accordingly be argued that MTS needs to expand into new countries, although the CIS may have lost some of its attractions of late. In practice, it expressed an interest in entering both Kazakhstan and Azerbaijan in 2009, although no progress was made, but most of its so far doomed attempts to internationalise have covered a rather ill-assorted bunch of countries, other than in the sense that they have all been fairly proximate in geographical terms, and it is hard not to conclude that it is either unwilling or unable to come up with the requisite finances to pursue its objectives – this was, after all, a company that once boasted of its desire to become one of the world's top ten by market value by 2010.

MTS initially expressed an interest in the third licence in Syria in 2008 but failed to be short-listed in November 2010. More importantly, it was alleged to have offered to buy Telekom Austria in July of that year, a move that (if it did indeed take place) was probably not viewed with much enthusiasm by the Austrian government. More recently, in July 2013, it allegedly expressed an interest in acquiring Telefónica CZ, but this was rapidly snapped up by a local bidder. Where, if anywhere, Mobile TeleSystems goes from here is accordingly difficult to forecast.

VimpelCom

As for MTS's rival, VimpelCom, some of the recent history has already been covered in the previous discussion of Telenor's activities. However, there have been other complications that need to be unravelled. In essence, before the end of 2010, VimpelCom was a very peripheral player on the European mobile scene. Its networks were entirely located in Russia and the CIS countries, and until the merger with Telenor in Ukraine in 2010 discussed above, Russia was very much the predominant source of proportionate subscribers in Europe.

However, in April 2011, a merger took place with Wind Telecom such that VimpelCom acquired, *inter alia*, the whole of a substantial network in Italy branded as Wind Telecom, although it was nevertheless only the third-largest in terms of subscribers. Italy, along with Russia, Ukraine and Kazakhstan, are the designated priority markets for VimpelCom in Europe, which is hardly surprising given that they provided roughly 110 million of the 125 million proportionate subscribers in Europe at the end of 2012.

It is apparent that Wind Italy is intended to be something of a one-off in that VimpelCom itself has not expressed any further interest in investing in the EU. It expressed some temporary interest in Telekom

Srbija in December 2010, but the other countries that have periodically emerged as of potential interest are the likes of Belarus and Moldova.

However, a complication is that in April 2011 VimpelCom also acquired a 51.7 per cent stake in the then Orascom Telecom Holding – renamed as Global Telecom Holding in September 2013. In April 2013, an offer to acquire the rest of Orascom was tabled, but this appears to have been done by Altimo rather than VimpelCom and was not successful. Although Orascom had no subsidiaries in Europe, it did have some history of interest in acquiring European assets in the run-up to its takeover by VimpelCom and, in October 2011, it expressed an interest in acquiring Orange's networks in Austria and Switzerland, although nothing came of this. As several years have passed since then, it must be assumed that VimpelCom does not intend to use Global Telecom as a vehicle for acquiring further assets in Europe.

This belief is reinforced by the fact that VimpelCom has recently found itself in a situation not unknown among its rivals, namely having to do some kind of penance for what is perceived as overinvestment in its networks leading to excessive levels of debt leading to a virtual wipe-out of dividends. Whether this penance will take the form of asset disposals is yet to be determined.

Hutchison Whampoa

Hutchison Whampoa has already been covered in the above narrative, so there is no need to repeat that material in detail. However, it is useful to remind ourselves of the recent activities that have involved consolidation in some form or another.

In the first place, France Télécom sought to exit Austria, where its 35 per cent minority stake in a network with no growth provided fewer than 800 000 proportionate subscribers. The buyer was the even smaller network operated by Hutchison, with the deal completed in early 2013. However, in order to secure agreement to the takeover from the European Commission, Hutchison was obliged to divest some spectrum, base stations and an MVNO while offering privileged access to the enlarged network to independent MVNOs.

Second, in June 2013, Hutchison made an offer of up to €850 million ($1.15 billion) for Telefónica's network in Ireland – a somewhat unusual move in that Telefónica was the second-largest and Hutchison the smallest of the four operators. Not surprisingly, even given the fact that a country the size of Ireland has no obvious need for four networks, the European Commission intervened in November to set up a full inquiry into the takeover with a view to presenting its report in May 2014.

Hutchison tried to improve its prospects by offering to dispose of some of its spectrum and to maintain its existing infrastructure sharing agreement with Eircom.

Third, in July 2013, it was announced that a proposed merger between the Italian networks of Telecom Italia and Hutchison H3G was no longer on the cards due to a dispute over the valuation placed on the country's (much the) smallest network. This was essentially a repeat of a process initiated in 2010, but it proved impossible to secure agreement on the appropriate value to be placed on H3G's accumulated losses if used to offset taxation. It may be added that once again the proposed merger would have left a 'big four' country with only three networks and hence stimulated a strong regulatory response both within Italy and the European Commission.

In early 2014, Hutchison decided to try again, but this time it engaged in talks with VimpelCom with a view to merging their respective Italian networks (3 Italia and Wind Telecomunicazioni). The latest version of this tie-up envisages the much smaller 3 Italia effectively taking over its rival, with VimpelCom retaining a small equity interest. The merger would create a third network similar in size to Telecom Italia and Vodafone, but this is unlikely to be popular with the regulatory authorities despite the superficial improvement in the relative strength of the remaining networks, which was not a feature of the proposed merger with Telecom Italia. It has been suggested that, post-merger, there would be a disposal of Wind's fixed-wire business Infostrada. If so, both Telecom Italia and Vodafone would doubtless wish to acquire them.

It is easy to appreciate Hutchison's dilemma. The parent company has lost large amounts of money pursuing its dream of becoming a force in the mobile sector, so the only real options are to bulk up or get out. In the former context, Hutchison's unsuccessful attempt to acquire Ireland's Eircom in April 2012 may be added to the above, as well as the somewhat perfunctory interest expressed in acquiring stakes in Polkomtel and Telekom Srbija in 2010. As for disposals, the fundamental issue is, understandably, that Hutchison wants to exit at an acceptable price but that is difficult to determine when a potential buyer may be primarily interested in accumulated losses that can be set against taxes that have yet to be quantified. After all, it is not as though it is providing many subscribers – especially in Denmark, which is effectively for sale but deemed to be overpriced – even in Italy since it is the smallest surviving network in every case.

In February 2014, Hutchison made positive noises about the six European countries that form the 3 Group, indicating that subscriber numbers had improved during 2013 other than in the UK, but when a

Table 7.9 Hutchison Whampoa and Tele2: European stakes, 31 December 2012

Hutchison Whampoa			Tele2		
Country	Stake %	Proportionate subscribers	Country	Stake %	Proportionate subscribers
Austria	100.0	1 293 000	Croatia	93.0	722 000
Denmark	60.0	499 000	Estonia	100.0	513 000
Ireland	100.0	819 000	Germany	100.0	50 000
Italy	97.4	6 745 000	Kazakhstan	51.0	1 644 000
Sweden	60.0	883 000	Latvia	100.0	1 043 000
UK	100.0	8 950 000	Lithuania	100.0	1 903 000
			Netherlands	99.7	478 000
			Norway	100.0	1 129 000
			Russia	100.0	22 529 000
			Sweden	87.3	3 296 000
Total Europe		19 189 000	Total Europe		33 307 000

company is talking about reaching a milestone of positive operating cash flow (defined as EBITDA minus capital expenditure) after so many years of operation, it is evident that it is still a long way from making anything resembling a true profit. That being so, it may be surmised that Hutchison will continue to see itself cast in the role of a disruptive force and hence it is likely to remain a regular feature of future discussions of consolidation in Europe. Nevertheless, there will evidently be no coming together of Hutchison and Telefónica in the UK, which was ruled out in April 2014 on the grounds that 3 UK would be successful as an independent network.

Tele2

Tele2 is definitely an odd man out in terms of Europe, where it has all of its operations. This is partly because of its emphasis on countries to the east, although that is common to most Nordic-based operators, and partly because at the end of 2012 it owned only one stake of any significance, which it then proceeded to sell. In April 2013, it sold out in Russia to VTB Group, leaving itself with eight networks and an MVNO (in Germany) with a total of little more than 10 million proportionate subscribers.

It can be argued that Tele2 has been shrinking steadily since it disposed of its MVNO in Austria in early 2008, followed by its stakes in Liechtenstein and Luxembourg in July and Switzerland in September. France followed in 2009, although it acquired 51 per cent of Neo in Kazakhstan in March 2010 and the outstanding 50 per cent of Mobile Norway in September 2011, so it was not entirely a one-way process. In March 2013, it stated that it would henceforth be concentrating on the Netherlands, Norway, Sweden and Russia, but in the last case it evidently meant that it was concentrating on its disposal. Furthermore, having failed to win new spectrum in the 2013 auction in Norway, Tele2 has indicated that it may either sell the network (possibly to Hutchison Whampoa) or form a joint venture with Access Industries (Ice.net).

In its current state, Tele2 is market leader only in Lithuania, while in Kazakhstan, where it was already seriously lagging the two larger networks, it ran into difficulties at the end of 2013. Just where Tele2 goes from here is a moot point.

Telecom Italia

One operator that has yet to be discussed independently is Telecom Italia. This may come as something of a surprise, given that this is the incumbent in one of the ‘big four’ European economies. However, this was also the case in Curwen and Whalley (2008) – in other words, Telecom Italia has been a bit player on the European mobile scene for a good many years. Indeed, after selling stakes in the Czech Republic in March 2005, Greece in July 2005 and Turkey in July 2006, Telecom Italy no longer had a European stake other than in its home market.

In its weakened state at that time, Telecom Italia appeared vulnerable to predation, although it seemed very unlikely that the Italian government would permit a takeover bid for the entire company given that it was also responsible for the fixed-wire backbone. But with AT&T and América Móvil expressing preliminary interest in making some kind of offer, there was felt to be an urgent need for an intra-European solution.

In fact, Telecom Italia had been taken over earlier. The first occasion on which Telecom Italia was taken over involved a typical Italian solution whereby potential foreign predators such as Deutsche Telekom were warned off by the government while a more or less acceptable domestic buyer – in the form of Olivetti – was allowed to proceed (Curwen, 1999). Subsequently, Olivetti sold on control over Telecom Italia to another Italian company, Pirelli (Curwen, 2003). However, Pirelli soon tired of the problems associated with Telecom Italia’s poor trading performance and effectively offered to sell its controlling stake, exercised via Olimpia, part of a complex holding structure, in December 2006.

Telefónica allegedly responded with an offer to buy a 30 per cent stake in Olimpia at roughly €3 per share, but appeared to back away when the government expressed its habitual opposition to foreign predators. However, it re-emerged as part of the Telco consortium, which successfully bid to acquire Olimpia at the end of April. Telefónica’s contribution would entitle it to receive 42.3 per cent of the equity of Telecom Italia owned by the various parties in the consortium once they had bought Olimpia. This was claimed to be 23.6 per cent at the time, so Telefónica’s indirect stake appeared to be roughly 10 per cent, but it subsequently became clear that Telco would own 23.6 per cent of the ‘voting rights’ but only 16.3 per cent of the ‘equity’ because Telco would exclusively own ordinary shares in Telecom Italia with attached voting rights that were not available to holders of the preferred shares. Hence Telefónica’s indirect equity stake would amount only to 6.9 per cent.

The remainder of the consortium equity – 57.7 per cent, equivalent to 13.6 per cent of the voting rights and 9.4 per cent of the equity in Telecom Italia – would be held by Italian companies (Generali, 28.1 per cent; Mediobanca, 10.6 per cent; Intesa Sanpaolo, 10.6 per cent; Benetton, 8.4 per cent), so technically the Italian identity of Telecom Italia was maintained.

In late March 2008, Telco raised its voting rights to 24.5 per cent. Naturally, this was interpreted in some quarters as presaging an attempt by Telco, or even by Telefónica acting alone, to take majority control of Telecom Italia, but these rumours were firmly squashed in June when the Telefónica chairman stated that 'we have no ambitions for the time being for [acquiring] anything else'. Subsequently, the Telecom Italia CEO stated that he ruled out 'hostile deals' and that it would not be possible for anyone to take over Telecom Italia if the country 'did not agree'.

Unfortunately, Telecom Italia continued to perform badly, resulting in a progressive collapse of the share price, and the members of Telco were forced to write off a large part of their investments. By 2013, their thoughts had turned towards dissolving the shareholder pact governing the ownership of Telco, which could take place either during 15–30 June 2014 or during 1–15 February 2015.

Fairly obviously, Telecom Italia had to be recapitalised if it was not to drown under the weight of its debts. Its debt rating stood at Baa3, the lowest investment grade, which in turn meant that raising money in the open market would be expensive because of the risk element. In the medium term, therefore, the only obvious solution appeared to be the disposal of its Brazilian subsidiary, TIM Participaçoes, which was worth roughly $11 billion, and if that was set in hand then Telefónica stood prepared to cover short-term funding difficulties.

However, this proposal was not as straightforward as it appeared to be because Telefónica not only controlled its own network, Vivo, in Brazil, but also had an indirect stake in TIM Participaçoes through Telco. As of end 2012, Telefónica owned 46.2 per cent of Telco, which in turn owned 22.4 per cent of Telecom Italia. Hence Telefónica's indirect stake was 10.35 per cent. The problem arose in September 2013, when Telefónica sought to raise its stake in Telco to 66 per cent, in that this triggered an adverse response from the Brazilian regulator. As is only too common, media reports have significantly misrepresented the various figures involved. The first point to bear in mind is that in 2009 Telefónica owned half of a joint venture with Portugal Telecom, which in turn owned 59.4 per cent of Vivo. In July 2010, Telefónica agreed to buy out its partner, subsequently buying some of the minorities in March 2011 and raising its stake to 62.3 per cent. In March 2011, Vivo was merged with Telesp and Telefónica emerged with a 73.8 per cent overall stake in Vivo.

Meanwhile, Telecom Italia owned 69.8 per cent of TIM Participaçoes. As can be determined from the above, Telefónica at that point accordingly owned a doubly indirect stake of roughly 7 per cent in TIM Participaçoes. If Telefónica were to raise its stake in Telco to 66 per cent, this indirect stake would itself rise to over 10 per cent and this appeared to be a step too far for the Brazilian regulator, especially given the possibility that Telefónica might end up owning Telco in its entirety.

The regulator's response was to demand that Telefónica either restore its situation in 2010 when it owned 29.7 per cent of Vivo and a smaller stake in Telco, or, if that was impractical, to find a way of exiting from TIM Participaçoes (Edgerton and Galvao, 2013). Given that Telefónica intended to take control of Telco, which in turn effectively controlled Telecom Italia, it certainly appeared to be feasible in due course for Telefónica to force Telecom Italia to sell its Brazilian subsidiary. As a result, intense speculation broke out in the Italian press that Telefónica was looking to set up a joint venture with rivals América Móvil and Oi with a view to taking over TIM Participaçoes and later breaking it up. However, this was utterly refuted by Telefónica in January 2014 (Telecom.paper, 2014a), adding that it was considering legal action against the regulator for imposing an unreasonable solution. For its part, Telecom Italia continues to insist that TIM Participaçoes is a strategic asset.

In order to succeed in becoming the main shareholder in Telecom Italia in its own right, Telefónica also appeared to need regulatory approval in Argentina. However, this issue has been resolved since Telecom Italia agreed in November 2013 to sell its stake in the country for $960 million to Fintech Group (Reuters, 2013) – a first tranche was sold in December. This still leaves the need to pacify nationalistic forces in Italy, given that the pretence that Telecom Italia is still under Italian control could no longer be sustained. To achieve that will require the separation of the fixed-wire network and its transfer into Italian ownership.

One further factor of interest in what is turning into a fascinating power struggle is the behaviour of BlackRock, the world's biggest money manager based in the USA. BlackRock owns a 3.9 per cent stake in Telefónica and in December 2013 it raised its stake in Telecom Italia to 10.1 per cent (but with only 7.8 per cent of the voting rights). Precisely what motivated BlackRock's move remains to be seen. It was presumably not attracted by Telecom Italia's trading prospects as its debt had been reduced to junk status by S&P in November 2013.

Finally, there is a potential spoiler in the shape of Naguib Sawiris, owner of the rump of Orascom not sold to VimpelCom and a minority shareholder in Telecom Italia. In December 2012, Sawiris offered to invest €3 billion in Telecom Italia, but this was rejected. Nevertheless, when he again indicated his interest to the government in January 2014, with the proviso that this was contingent on Telco selling its stake, he received a much more enthusiastic welcome.

In March 2014, Telecom Italia published its interim accounts for 2013. These revealed a third straight annual loss – on this occasion of €674 million. However, asset disposals meant that net debt had fallen to €26.8 billion, just within its target. For the first time ever as a listed company, no dividend was declared on the ordinary shares.

7.5 CONCLUSIONS

Europe, widely defined, is currently beset by a host of problems that are forcing operators heavily dependent on the region to reassess their strategies. Some of these problems are broader than Europe itself in so far as they relate to economic and political developments on a worldwide scale, but at the same time the problems within the eurozone and more widely the whole of the EU – not to mention the current turmoil in countries such as Turkey and Ukraine that aspire to become member states – are of great significance to the likes of Vodafone, Orange, T-Mobile and others.

Perhaps surprisingly, the theme that crops up most frequently is over-regulation. The argument is that the move towards quad-play is something to embrace for the most part and can be dealt with to a greater or lesser extent by the acquisition of fixed-wire rather than mobile assets. As for increased competition, this is simply a fact of life and underpins the need to improve efficiency and, if appropriate, to reconfigure the portfolio of assets.

Not that living with competition is a straightforward strategy if that competition becomes vicious, and there are indications that the only real strategy perceived to be open to operators with relatively small market shares is to be disruptive. This was and largely remains the Hutchison Whampoa philosophy, and it has not exactly been profitable. But if the overall market is in recession, needs must. A recent interesting illustration occurred in France when, in February 2014, Bouygues Télécom, languishing in third place in the market, chose to cut the price of its triple-play service in larger cities from €30 a month to €20 a month, together with a provision to cancel with immediate effect. Arguably even more interesting was Bouygues' subsequent bid for Vivendi's French subsidiary SFR, which, had it been successful, would certainly have caused a major disruption of the market since it would have created a new market leader and left two operators in a major European country with a 90 per cent share of the mobile market. The mooted merger with Iliad would have far less effect.

However, over-regulation is treated more seriously. The view taken by the operators, predictably enough, is that the purpose of regulation is to make things run more efficiently and to support the operations of European operators in the face of competition from operators based elsewhere. What regulators should not be doing is so to depress the returns on running networks that the financial health of operators becomes threatened, yet that, operators claim, is the ultimate outcome of the current regulatory climate within Europe and especially the EU. It is certainly the case that margins have been forced down by reductions in, for example, roaming charges and it has been argued above that the attitude towards the consolidation of markets often looks perverse, so there is definitely a case to be answered.

The one outstanding feature that comes across from the above discussion is that mobile operators in Europe are no longer much interested in acquiring one another – if they are interested in expensive purchases, then they are much more likely to take the form of fixed-wire networks. As a result, Europe has become a potential target for predatory and well-funded operators from a variety of other regions – for example, América Móvil (Latin America), AT&T (USA) and China Mobile (Asia-Pacific). So far only modest inroads have been made, but if AT&T makes an attempt to acquire Vodafone in a complete reversal of the historic trend for US-based operators, the fallout could be far-reaching.

However, AT&T appears to be back-tracking in relation to Vodafone specifically and to Europe in general. The reason is twofold. In the first place, what AT&T had to offer was the experience of launching a large network based on LTE at a time when Europe was lagging by some margin. However, each passing month reduces AT&T's advantage in this respect as the licences are issued to other operators and the experience of launches becomes more widespread. Furthermore, the takeover bid by Comcast for Time Warner Cable in the USA is forcing AT&T to re-evaluate the speed at which it needs to roll out its technically advanced fixed-wire and mobile networks – more speed means more costly (Moneybeat, 2014).

It may also be the case that European targets could turn out to be more expensive than current share prices indicate. SFR in France, for example, witnessed a significant increase in its market value once Altice Group and Bouygues Télécom had both entered initial takeover bids. Interestingly, even though it was unsuccessful, Bouygues is backed by the deep pockets of GIC (Singapore's sovereign-wealth fund) and the Ontario Teachers' Pension Plan Board, so if these institutions were willing to back Bouygues, then there must be some willingness outside Europe to pursue targets in Europe on an opportunistic basis.

NOTE

1. TeliaSonera is the largest direct shareholder in Turkcell but, for the moment, control over Turkcell is in the hands of a smaller shareholder, Çukurova. Disputes over the ownership of Turkcell have been raging for years, but are in principle close to a resolution, which may result in Altimo acquiring control (Cellular-news, 2014b).

REFERENCES

BBC (2013), 'Vodafone agrees 7.7bn-euro deal to buy Kabel Deutschland', www.bbc.co.uk, 24 June.

Cellular-news (2014a), 'China Mobile said to be considering buying a large stake in Vodafone', www.cellular-news.com, 28 January.

Cellular-news (2014b), 'Çukurova again given more time to resolve Turkcell shareholder dispute', www.cellular-news.com, 13 February.

Curwen, P. (1999), 'Has Telecom Italia broken the mould of European capitalism?', *info*, **1**(4), 343–54.

Curwen, P. (2003), 'Telecom Italia takeovers: Chinese boxing clever', *European Business Review*, **15**, 38–47.

Curwen, P. and J. Whalley (2008), *The Internationalisation of Mobile Telecommunications: Strategic Challenges in a Global Market*, Cheltenham, UK and Northampton, MA, USA: Edward Elgar.

Curwen, P. and J. Whalley (2012a), 'Incumbency and market share within European mobile telecommunications networks', *Telecommunications Policy*, **36**(3), 222–36.

Curwen, P. and J. Whalley (2012b), 'Third time lucky? An exploration of Hutchison Whampoa's involvement in the mobile telecommunications industry', *info*, **14**(2), 17–41.

Deutsche Telekom (2009), *Toward the new Deutsche Telekom. The 2009 financial year*, Bonn, Germany.

Deutsche Telekom (2010), *Life is for sharing. The 2010 financial year*, Bonn, Germany.

The Economist (2014), 'Numericable gains SFR. What now for Bouygues?', www.economist.com, 6 April.

Edgerton, A. and A. Galvao (2013), 'Telefonica told by Brazil to engineer Tim sale or undo deals', www.bloomberg.com, 5 December.

Gabriel, C. (2013), 'Vodafone banks on massive capex to reverse sales', www.rethink-wireless.com, 13 November.

Hall, B. and C. Hughes (2008), 'F Telecom arranging financing for Nordic buy', www.ft.com, 29 May.

Henning, E. (2013), 'Deutsche Telekom buys GTS, strengthens Eastern European operations', www.totaltele.com, 11 November.

Moneybeat (2014), 'AT&T says window is closing in Europe', http://blogs.wsj.com, 7 March.

O'Brien, K. (2011), 'Greek Telephone operator sells Serbian stake', http://dealbook.nytimes.com, 30 December.

Parker, A. and K. Hope (2008), 'Deutsche Telekom in €2.5bn Greek buy', www.ft.com, 17 March.
Reuters (2013), 'Telefonica says no plan to take control of Telecom Italia', www.reuters.com, 14 November.
Steinglass, M. (2012), 'América Móvil launches offer for KPN', www.ft.com, 30 May.
Telecom.paper (2013), 'T-Mobile sells broadband activities to CanalDigitaal owner', www.telecom.paper.com, 19 November.
Telecom.paper (2014a), 'Telefonica "not part of" any deal to sell off TIM Brasil', www.telecom.paper.com, 6 January.
Telecom.paper (2014b), 'Vodafone to acquire Ono for EUR 7.3 bln', www.telecom.paper.com, 17 March.
TeleGeography (2013), 'Uzbek authorities offer to purchase assets of bankrupt MTS subsidiary', www.telegeography.com, 15 August.
TeleGeography (2014a), 'DT renews interest in state's 10% OTE stake', www.telegeography.com, 31 January.
TeleGeography (2014b), '3 Ireland concessions in O2 deal could create a new operator', www.telegeography.com, 24 March.
TeleGeography (2014c), 'Iliad in discussions to acquire Bouygues Telecom', www.telegeography.com, 8 April.
Thomas, D. (2012), 'CWW shareholders vote for Vodafone takeover', www.ft.com, 18 June.
Thomas, D. (2014), 'AT&T rules out bid for Vodafone', www.ft.com, 27 January.
Van Daalen, R. (2013a), 'Telefónica to buy KPN's German unit in $10.7 billion deal', http://pnline.wsj.com, 23 July.
Van Daalen, R. (2013b), 'Carlos Slim's America Movil says it may pull KPN offer', http://pnline.wsj.com, 30 August.
Wood, N. (2014a), 'Deutsche Telekom to plough €1.2bn into Greece ops', http://pnline.wsj.com, 4 February.
Wood, N. (2014b), 'America Movil brokers deal for Telekom Austria takeover', www.totaltele.com, 24 April.

8. Structural and strategic adjustment among mobile operators in Latin America

8.1 INTRODUCTION

Latin America is generally held to comprise the countries lying south of the US border together with the Falkland Islands. However, this is not a clear-cut matter because of the existence of the 25 or so groups of islands in the Caribbean. Among these are to be found Puerto Rico and the US Virgin Islands, which are more sensibly treated as part of the USA for analytical purposes – and that is how they are treated in this book. That said, the analysis in this chapter concentrates on the countries on the mainland because the islands generally have too few subscribers to warrant consideration. The one obvious exception is the Dominican Republic, which had roughly 10 million subscribers at the end of 2012, but it is also necessary to draw Jamaica into the overall picture because of events linking it to other countries under discussion.

As usual, events up until the end of 2007 are dealt with in considerable detail in Curwen and Whalley (2008) and do not warrant lengthy repetition. Nevertheless, it is important to bear in mind that the modern history of mobile provision in Latin America is inextricably linked with the decision by US-based operators such as the then BellSouth to turn aside from activities outside the USA in order to concentrate on the home market – events covered in Chapter 9.

8.2 BACKGROUND

As discussed in Curwen and Whalley (2008: 148–9), a great deal of structural reorganisation took place with regard to the provision of mobile services in Latin America during 2003/04. By the end of 2004, three dominant operators had seemingly emerged that between them controlled nearly 60 per cent of the subscribers in the 20 main Latin American markets – América Móvil, Telefónica and Telecom Italia (which was in the process of reintegrating its mobile subsidiary TIM). However, even among these three, the disparity in terms of proportionate subscribers was already very marked – 57 million, 32 million and 9 million respectively. Their main rival was Millicom, with 3 million proportionate subscribers, while other operators with an even more modest presence included NII Holdings and Verizon Communications. However, so far as the islands were concerned, the two operators with much the most widespread presence were Cable and Wireless (LIME) and Digicel.

During 2005, América Móvil entered Chile, Paraguay and Peru, and hence was present in 13 countries in the region, but Telefónica's merger with BellSouth meant that it was now present in 13 (plus Puerto Rico). Given that Telecom Italia was present in only six, and had far fewer subscribers (mostly in Brazil), it can reasonably be argued that by the end of 2005 the region was dominated by a 'big two' rather than a 'big three'. The other central point of interest was that América Móvil had only a single presence outside the region in the form of its US network (Tracfone), which essentially provided a service to Hispanics. In comparison, Telefónica relied on other regions for one-third of its proportionate subscribers and Telecom Italia for 60 per cent.

It is at this point, entering 2006, that it first became clear that Telecom Italia was willing to address the critical issue of whether it had any real future in the region. Hovering in the background, and presumed to be a willing purchaser of any assets that might be offered, stood América Móvil. However, it should be borne in mind that América Móvil by this point owned close to 100 per cent of the shares in all of its subsidiaries and hence was unlikely to pick up minority stakes, and that any developments in Brazil would undoubtedly be of grave concern to the regulator. Despite this, in practice, not only did the disposal of its wholly owned Venezuelan subsidiary by Telecom Italia not involve América Móvil, but the latter was alleged to have made an unofficial offer of $7.7 billion for TIM Participaçoes in October 2006.

Looking back at the four-year period 2005 to 2008 inclusive, the most striking factor is the disparity in the growth rates of proportionate subscribers. Telefónica began with 47 million and ended up with 88 million (+135 per cent), whereas América Móvil began with 87 million and ended up with 170 million (+95 per cent) and Telecom Italia began with 19 million and ended up with 29 million (+56 per cent). So, while América Móvil massively increased the difference between itself and Telefónica in absolute terms (from 40 million to 82 million proportionate subscribers), Telefónica was the better performer in relative terms.

In that context, it is worth noting the following. First, América Móvil's home market, Mexico, accounted for 42 per cent of its Latin American proportionate subscribers in 2006 but only 33 per cent in 2008. Brazil had remained roughly constant, so it is to Peru (up from 2.3 per cent to 4.2 per cent) in particular and to faster rates of growth outside Mexico in general that one must turn for an explanation. It may be added that it entered the Dominican Republic in 2006, although this did not yield large numbers of subscribers (albeit many more than were gained from its entry into Jamaica in 2007).

For its part, despite its excellent performance in Latin America, Telefónica relied on the region for 66 per cent of its proportionate subscribers in 2005 but for only 52 per cent in 2008. This was accounted for in particular by its entry into new markets in other regions discussed elsewhere in this book. Furthermore, Telefónica did not rely on a single market in the region to anything like the same extent as América Móvil. Fairly evidently, this has something to do with its home base being outside the region, and it is always tempting to assume that the answer is to be found in Brazil, given that this has long been the focus of media attention. In practice, Brazil diminished rather than increased in relative importance during the period – down from 20 per cent to 16 per cent of Latin American proportionate subscribers – as did Chile (down from 10 per cent to 7.5 per cent). The really interesting developments were in Peru (up from 7 per cent to 14 per cent) and Mexico (up from 12.5 per cent to 17 per cent).

Telecom Italia's sluggish overall performance slightly disguised the fact that it was increasingly relying on Latin America for subscribers due to its departure from Turkey, combined with a poor performance in Italy itself, but aside from Argentina it was not making much real progress in Latin America. It relied on Brazil for 85 per cent of its proportionate subscribers in the region and it was América Móvil that was scooping up a disproportionate share of the new subscribers there.

8.3 DEVELOPMENTS 2009–2013

Given the emergence of the 'big two' during the preceding years, the period since the beginning of 2009 has, not surprisingly, been relatively peaceful from a structural viewpoint. In early 2009, América Móvil was present in 14 countries in the region (plus Jamaica and Puerto Rico). It had obtained a licence in Panama in May 2008, along with Digicel (TeleGeography, 2008), and their networks were launched in March 2009 to compete with well-entrenched incumbents Telefónica and Cable &

Wireless. In practice, whereas Digicel has almost caught up with C&W, América Móvil has remained remarkably unpopular in Panama, as shown in Table 8.1.

In January 2011, América Móvil obtained a licence in Costa Rica along with Telefónica – there were no offers for a third licence – and it launched its network in November. The problem was that there had previously been a monopoly provider, ICE, and it was able without too much difficulty to keep its existing customers together with a significant share of new subscribers in a rapidly growing market. Once again, therefore, América Móvil's performance has been surprisingly poor, as shown in Table 8.1.

The most interesting developments during 2011 involved Jamaica, one of the largest islands in terms of subscribers. In March 2011, América Móvil entered into an agreement with Digicel (based in Jamaica) to acquire the latter's stakes in El Salvador and Honduras in return for América Móvil's network in Jamaica. This network had only a few hundred thousand subscribers, whereas Digicel was the dominant operator – Cable & Wireless provided only modest competition – and in its turn, América Móvil was already present in El Salvador and Honduras, so no new entries or exits were involved.

Unfortunately for the participants, the competition regulator in El Salvador ruled in May 2011 that for the transfer to progress, América Móvil would have to give up 20 MHz of spectrum. América Móvil at the time had roughly 2 million subscribers and would be acquiring roughly 1 million from Digicel. Although the combined network would be roughly the same size as Millicom,[1] and Telefónica also had a presence comparable to that of Digicel, the regulator wanted to create the capacity to ensure that no diminution in competition would occur and was reluctant to lose Digicel, which was operating as a price cutter in the market. The competition regulator reiterated its decision on appeal in September 2012 (*The Gleaner*, 2012), at which point América Móvil saw no further point in proceeding and withdrew that part of the deal with the agreement of Digicel.

The transfer of the América Móvil network in Jamaica to Digicel was completed in December 2011. This had initially been subject to the condition that the two networks should be run independently of one another, but the restriction was removed shortly before the transfer, whereupon Digicel declared that it would be merging the networks in February 2012. At this point, assuming the América Móvil subscribers had chosen to transfer, Digicel would control over 3 million subscribers, six times as many as its now sole competitor, Cable & Wireless Communications (LIME) (Ictpulse, 2012). Subsequently, various appeals against the decision to merge the networks were lodged, but the reality was that the two networks were now one and could not realistically be split again due to the single network infrastructure.

As for Honduras, Millicom was the clear market leader but the transfer would create an effective duopoly as Hondutel was only just keeping its head above water. Despite this, the transfer went surprisingly smoothly and was completed simultaneously with that of the Jamaican network.

For its part, Telefónica has entered only one country since 2009, namely Costa Rica in 2011. Licensed alongside América Móvil, it nevertheless performed considerably better, acquiring twice as many subscribers during 2012 but still remaining no better than one-seventh the size of ICE. However, the real story in recent times has concerned developments in Brazil, which, in turn, have involved Telecom Italia.

As previously noted, Telecom Italia was already the laggard among the original 'big three' in Latin America by 2008, with a single holding of consequence in Brazil. In November 2010, it finally concluded a bitterly fought arbitration process involving the government of Bolivia, which in 2007 had nationalised Entel in which Telecom Italia held a stake. It received $100 million in compensation, considerably less than it had originally wanted (Petersen, 2010). It disposed of its minority holding in Etecsa Cuba in February 2011 and its even smaller stake in Paraguay provided only a few hundred thousand proportionate subscribers. In effect, therefore, Telecom Italia depended on Brazil and (to a much smaller extent) Argentina for its continued presence in the region but it agreed to sell its Argentina (and hence also the Paraguay) network to Fintech in November 2013, subject to regulatory approvals.

The situation at the end of 2012 is set out in Table 8.1. The table also lists Millicom. The curious thing about Millicom is that each of its six holdings produced roughly the same number of proportionate subscribers – varying from 2.9 million in Colombia, where its stake, at 50 per cent, is the lowest of the six, to 4.4 million in Guatemala. Millicom is comfortably the market leader in Paraguay, where it is sole owner of the network – it trades as Tigo – as well as in Guatemala and Honduras, where it holds at least half of the equity. But it is absent from Argentina, Brazil and Mexico, and its grand total of 20 million proportionate subscribers in the region does not even begin to compare with that of Telecom Italia, let alone the 'big two'.

Table 8.1 shows market shares rather than the number of subscribers, although one can be readily turned into the other using the data on overall subscriber numbers (subject to the qualifications set out in the paragraphs that follow the table). As of the end of 2012, América Móvil had 237 million proportionate subscribers in Latin America (plus

1.2 million in Puerto Rico). Since all 17 networks were to all intents and purposes wholly owned, this figure was also its gross subscriber number. With 176 million gross subscribers at the end of 2012, Telefónica would not appear to have been far behind, but it had only 153 million proportionate subscribers, roughly one-third less than América Móvil. But at the beginning of 2009 the equivalent figures had been 170 million and 89 million, so at that point Telefónica had trailed América Móvil by almost one-half.

During the four-year period 2009 to 2012 inclusive, Telefónica's proportionate subscribers achieved an overall growth of 130 per cent compared to 40 per cent for América Móvil, and it is useful to analyse why this happened. The most obvious factor is that Telefónica had bought out its partner in Vivo. Largely as a result, its proportionate subscribers grew by a touch under 200 per cent. Meanwhile, because it controlled its network throughout the period, even though it was smaller in absolute terms than Vivo, América Móvil managed only an overall growth of 68 per cent. But this is clearly no longer a repeatable scenario. It is possible, but unlikely, that Telefónica will increase its current stake of 74 per cent in Vivo other than marginally in the near future, and it is possible, but unlikely, that Vivo's growth will significantly outstrip that of América Móvil. However, because of the involvement of parties not yet discussed, ongoing events in Brazil make speculation somewhat hazardous, which is why the matter is addressed separately in Section 8.4 below.

As for the rest of Telefónica's networks, these performed patchily. The odd good performance such as in Panama, where subscriber numbers doubled (but from a fairly small base) during the period, was more than balanced by the absence of any growth at all in Guatemala and Venezuela and, more importantly, a poor performance in Mexico (+25 per cent) and Peru (+18 per cent), the big successes during the previous period. Evidently, the pressures of competition, combined with rapidly maturing markets with much higher penetration rates, were taking their toll, as were adverse movements in exchange rates. That was clearly the main explanation why rumours first began to circulate in July 2012 that Telefónica was considering some kind of share sale in some or all of the Latin American networks in order to reduce its €57 billion debt mountain. At the time, Argentina and Venezuela were specifically identified as unattractive businesses, even though the former had 20 million subscribers.

However, developments elsewhere in the world – particularly in Europe – meant that nothing further happened until April 2013, when Telefónica announced that it would be selling 40 per cent of its networks in Central America – comprising El Salvador, Guatemala, Nicaragua and Panama – to Corporación Multi Inversiones for $500 million (Baigorri, 2013). The networks would be combined into a new subsidiary before the sale, but at the end of the day the sum involved was of little consequence.

But if Telefónica was struggling, the same was likely to be the case for América Móvil. As previously noted, its overall subscriber growth during 2009 to 2012 inclusive was a disappointing 40 per cent, and the two biggest markets indicate the reason why. In Mexico, subscriber numbers grew by only 25 per cent, and, although Brazil did better at 68 per cent, the combined improvement was 42 per cent – effectively the norm. The same was true for Argentina, whereas the third-largest market, Colombia, grew by a mere 13 per cent. This sluggish performance links with the discussion in Chapter 7 on Europe because América Móvil now had to address the issue that progress in Latin America was no longer a given. The fact that only one major competitor remained was obviously beneficial, but that competitor was doing rather better overall yet itself struggling to make good profits, so the halcyon days of low penetration rates and modest expenditures on new technology were now firmly consigned to the past.

It is appropriate to end this section by looking not at the big players on the mainland but at the two most significant players on the islands, as these have formed part of the above narrative. For its part, Cable & Wireless is worth only a passing mention because it does not appear to obtain even 1 million proportionate subscribers from any of its holdings in Latin America and the Caribbean. In terms of gross subscribers it exceeds that figure only in Panama, which is of interest because Digicel is also present there. As Digicel is a private company, it is able to choose not to disclose any more information than is strictly necessary. Evidently it has between 13 and 14 million gross subscribers distributed between 30 to 35 networks worldwide, but primarily in the Caribbean. From this it can be deduced that few of its networks are of any significance and, aside from Haiti, where it is the dominant operator, it is the relatively large presence in El Salvador, Honduras and Jamaica that caused it to become involved in regional restructuring. In practice, this ended up as an exchange of subscribers in El Salvador for subscribers in Jamaica, so in effect it has become largely irrelevant to any discussion of developments on the mainland since its network in El Salvador is only the fourth-largest.

Unfortunately, it should be noted that there are unresolved issues with some of the data in Table 8.1. Both América Móvil and Telefónica provide a good deal of detailed information about subscriber numbers in their annual reports, but not every country in Latin America is individually specified. For example, Telefónica lumps together Costa Rica, El Salvador, Guatemala, Nicaragua and Panama as Central America. According to the annual report for 2012, Telefónica had a total of 176.7 million mobile subscribers in Latin America. Of these, all bar 9.2 million are accounted for by the individual countries that are specified, and these have been reallocated by one reputable source into Costa Rica (0.7 million), El Salvador (1.9 million), Guatemala (2.5 million), Nicaragua (1.3 million) and Panama (2.8 million).

As for América Móvil, it lumps together Costa Rica, Dominican Republic, El Salvador, Guatemala, Honduras, Nicaragua, Panama and Puerto Rico as Central America and the Caribbean. By deducting the total for the specified countries from the overall total for the region, one is left with 20.7 million subscribers. It is possible to cross-check the two operators' data with a number of sources such as www.fosigrid.org or TheGlobalEconomy.com, and ultimately compatibility issues remain only in Honduras and Nicaragua and Panama, where different sources give wildly different subscriber numbers.

The situation in Honduras is difficult to unravel. A reasonable estimate appears to be 7.8 million. It is possible that part of the problem lies in the sale of Digicel's network to América Móvil, which already operated its own network in Honduras.

In the case of Nicaragua, Telefónica's own spokesman declared in August 2012 that it had 2.8 million subscribers and the regulator has stated that the country total for 2012 was 6.8 million. Other, private, sources such as www.fosigrid.org appear to offer totals in the region of 6.0 million. However, this is not such a big problem as there are only two service providers and their respective market shares can be taken to be as shown in Table 8.1.

In the case of Panama, the problem appears to lie in the fact that many more numbers have been issued than are active. The website www.fosigrid.org puts the number of active subscribers at 4.6 million, but that seems much too low given that Digicel, a relatively recent licensee, has 1.5 million. The figure of 6.3 million in Table 8.1 seems to be a reasonable estimate. +Móvil, controlled by Cable & Wireless and the state, has been losing customers at a great rate and is now lagging well behind Telefónica.

Table 8.1 Selected Latin America: percentage market[1] held by direct subsidiaries, 31 December 2012

	Networks (number)	Subscribers (million)	América Móvil[2]	Telecom Italia[3]	Telefónica[3]	Millicom
Argentina	4	58.3	35	7	30[3]	–
Belize	2	0.4	–	–	–	–
Bolivia	3	9.3	–	–	–	32
Brazil[5]	7[5]	264.3	24	18 [26]	21 [29][3,4]	–
Chile	4	26.1	23	–	38	–
Colombia	3	47.6	63	–	17 [25]	6 [12]
Costa Rica	3	6.0	6	–	12	–
Cuba	1	1.6	–	–	–	–
Dominican Rep.	4[7]	10.2	55	–	–	–
Ecuador	3	16.7	70	–	29	–
El Salvador	4[8]	8.9	32	–	21	35
Guatemala	3	15.9	35	–	16	28 [49]
Guyana	2	0.6	–	–	–	–
Honduras	4	–[9]	21	–	–	34 [51]
Mexico	4	100.0	70	–	19[13]	–
Nicaragua	2	–[10]	70	–	30	–
Panama	4	6.3	4	–	44	–
Paraguay	4	7.0	8[11]	4 [32][12]	–[3]	55[11]
Peru[6]	3	29.2	43	–	51	–
Suriname	3	0.7	–	–	–	–
Uruguay	3	4.9	16	–	37	–
Venezuela	3	30.2	–	–	34	–

Notes:

1. Calculated to nearest round number. The figures in brackets represent the overall market share held by the network in which the operator owns a less than 100 per cent stake.
2. América Móvil owns all the shares in all bar a few of its networks, where its stake is in every case nearly 100 per cent. It is the only operator listed in the table to have a presence in Puerto Rico, where it had a 31 per cent market share at the end of 2012.

3. Telefónica also has an indirect stake in Telecom Italia, so its direct plus indirect market shares in these countries are larger than those stated in the column by the equivalent of 10.3 per cent of the Telecom Italia figure.
4. Telefónica also has an indirect stake in Portugal Telecom, so its direct plus indirect market share in Brazil is a further 0.4 per cent larger than shown in the table.
5. Full nationwide coverage is not the norm. However, the 'big four' (including Oi) dominate the market. According to the regulator, there were 271 million GSM subscribers at the end of 2013 and the market shares were Telefónica (28.5 per cent), TIM (27.1 per cent), América Móvil (25.3 per cent) and Oi (18.5 per cent).
6. A fourth network, Viettel, is to become operational in July 2014.
7. Orange Dominicana is being acquired by Altice Group in 2014H1. Altice Group has also acquired another operator hence the number of networks will fall to three.
8. There is a fourth operator, Intelfon, but with fewer than 100 000 subscribers it does not offer direct competition for the three incumbents.
9. Estimates vary enormously, from 7.4 million to 9.2 million (see below).
10. Estimates vary enormously, from 4.3 million to 6.0 million (see below).
11. Respectively 7 per cent and 54 per cent at end 2013.
12. In November 2013, Telecom Italia agreed to sell its stakes in these networks to Fintech, subject to regulatory approval. An initial stake in Telecom Argentina was transferred in December 2013.
13. In February 2014, Telefónica initiated merger talks with Iusacell, which would create a virtual duopoly. Further developments are contingent on impending reforms of the telecommunications market.

8.4 BRAZIL

For a number of reasons, not least the sheer size of its mobile market, Brazil tends to dominate discussion about mobile provision in Latin America. As noted in Table 8.1, Telefónica not only controls its own network, Vivo, in Brazil but also has an indirect stake in TIM Participaçoes effected through a joint venture known as Telco. As of end 2012, Telefónica owned 46.2 per cent of Telco, which in turn owned 22.4 per cent of Telecom Italia. Hence Telefónica's indirect stake was 10.35 per cent. However, this percentage was of the voting shares only, and if the non-voting shares are included the percentage was lower.

In September 2013, Telefónica sought to raise its stake in Telco to 66 per cent, and this triggered an adverse response from the Brazilian regulator. As is only too common, media reports have significantly misrepresented the various figures involved. The first point to bear in mind is that in 2009 Telefónica owned half of a joint venture with Portugal Telecom, which in turn owned 59.4 per cent of Vivo. In July 2010, Telefónica agreed to buy out its partner, subsequently acquiring some of the minorities in March 2011 and raising its stake to 62.3 per cent. In March 2011, Vivo was merged with Telesp and Telefónica emerged with a 73.8 per cent overall stake in Vivo.

Meanwhile, Telecom Italia owned 69.8 per cent of TIM Participaçoes, which was worth roughly $11.5 billion at the end of 2013. As can be determined from the above, Telefónica at that point accordingly owned a doubly indirect stake of roughly 7 per cent in TIM Participaçoes. If Telefónica were to raise its stake in Telco to 66 per cent, this indirect stake would itself rise to over 10 per cent, and this appeared to be a step too far for the Brazilian regulator, especially given that there was the possibility that Telefónica might end up owning Telco in its entirety.

The regulator's response was to demand that Telefónica either restore its situation in 2010 when it owned 29.7 per cent of Vivo and a smaller stake in Telco or, if that was impractical, to find a way of exiting from TIM Participaçoes (Edgerton and Galvao, 2013). Given that Telefónica intended to take control of Telco, which in turn effectively controlled Telecom Italia, it certainly appeared to be feasible in due course for Telefónica to force Telecom Italia to sell its Brazilian subsidiary. As a result, intense speculation broke out in the Italian press that Telefónica was looking to set up a joint venture with rivals América Móvil and Oi with a view to taking over TIM Participaçoes and later breaking it up. However, this was utterly refuted by Telefónica in January 2014 (Telecom.paper, 2014), adding that it was considering legal action against the regulator for imposing an unreasonable ultimatum. For its part, Telecom Italia continues to insist that TIM Participaçoes is a strategic asset.

In March 2014, Assicurazioni Generali – holder of a 19.1 per cent stake in Telco – announced that it intended to sell the 4.3 per cent indirect stake in Telecom Italia in June. Telefónica would have been expected to acquire this via its existing option to buy, so this may precipitate some kind of resolution to the present stand-off between Telefónica and the Brazilian regulator.

It should also be noted that, after it sold off its stake in Vivo, Portugal Telecom took a 25.3 per cent stake in Oi, the fourth major network with 50 million subscribers in 2013 – that is, most of the market not provided for by the other operators in Table 8.1. Although Telefónica's residual stake in Portugal Telecom is tiny, and hence it obtains very few proportionate subscribers indirectly from Oi, it is extremely rare for an operator to have an interest, however small, in three networks in the same country. However, it is also extremely rare to find a country where the market is almost wholly accounted for by four operators with such similar market shares – the market share for Vivo in its entirety, as shown in Table 8.1, is 29 per cent and for TIM in its entirety is 26 per cent.

If one examines the four-year period 2009 to 2012 inclusive, one finds that the overall market grew by 53 per cent, while the network subscriber growth rates were TIM (75 per cent), Vivo (50 per cent), América Móvil (50 per cent) and Oi (40 per cent). This suggests that TIM is gaining market share exclusively from Oi, but whether the various regulatory developments will affect this in the future remains an open question. In that context, one additional development not discussed previously is the application to the regulator in October 2013 by Portugal Telecom and Oi to merge their holdings. Not surprisingly, given the overall market structure, the regulator gave the go-ahead in January 2014 without imposing any restrictions (Lennighan, 2014). However, although the merger had the backing of the largest shareholders because it was set up such that new shares would be issued in the new entity and used to pay off their outstanding debts, this meant that the stakes held by the minority shareholders would be diluted without compensation (TeleGeography, 2014). The Brazilian securities regulator has ruled in their favour, so the merger is not yet a done deal. The remaining issue is whether, assuming the merger is completed, this will improve the performance of 'new' Oi and hence affect the structure of the market.

8.5 LTE

A few remarks about LTE – which is currently what is meant in almost all cases where 4G is referred to in the region – are in order. The rolling out of LTE is an expensive process that may strain the coffers of the smaller networks with limited access to external funds, although it is less of an issue for the 'big two' because they can achieve economies of scale across their various networks. But LTE is also a driver of competition – by and large, if your main competitors are launching, then so must you.

Latin America has so far been something of a backwater where LTE is concerned. If one considers the countries where launches have occurred, then the situation at the end of 2013 was as follows:

- Argentina, Belize, Cuba, El Salvador, Guatemala, Guyana, Honduras, Jamaica, Nicaragua, Panama, Suriname: no launches
- Bolivia (1): Entel (12/12)
- Brazil (4): América Móvil (12/12); Oi (10/13); Telefónica (04/13); TIM (07/13)
- Chile (2): América Móvil (06/13); Telefónica (11/13)
- Colombia (2): Millicom (12/13); Telefónica (12/13); UNE-EPM (06/12)
- Costa Rica (1): ICE (06/13)
- Dominican Republic (2): Orange (01/13); Tricom (03/13)
- Ecuador (1): CNT (11/13)
- Mexico (2): América Móvil (11/12); Telefónica (11/12)
- Paraguay (2): Copaco (02/13); Personal (02/13)
- Peru (1): Telefónica (01/14)
- Uruguay (1): Ancel (12/11)
- Venezuela (1): Digitel (07/13).

Not only was there a small number of launches by end 2013 – 20 out of 260 worldwide is a modest performance for an entire region – but almost all of these occurred in December 2012 or later. Furthermore, the limited presence of América Móvil (three appearances) and Telefónica (five appearances) is striking. They have gone head to head in Brazil and Mexico presumably because they simply could not afford to let each other gain too much of an advantage – a factor that also seems to have applied to the other similarly sized competitors in Brazil – but elsewhere it is almost as though each is happy not to have to spend too much money provided the other does the same.

Since the beginning of 2014 there have been five further launches: América Móvil in Colombia (February), Uruguay (March) and Costa

Rica (April); Telefónica in Peru (January); and Entel in Peru (March). Hence, while América Móvil is making some progress, it can hardly be described as the pace setter for the region, despite its overall status.

It is possible that the opening up of the 700 MHz band will bring about some structural changes, but even if incumbents are subjected to caps on their overall spectrum holdings, new entrants are likely to need owners with deep pockets if they are to succeed in acquiring even modest market shares via this route (Curwen and Whalley, 2013: ch. 7). In essence, it is claimed that regulators in the region have made/are making four basic errors in trying to effect structural change via LTE licensing. The errors are to set aside spectrum for new entrants; allocate spectrum to state-owned bodies; impose spectrum caps on incumbents; and fail to re-farm 2G spectrum.

Overall, therefore, whereas the advent of LTE was expected to galvanise the mobile sector and result in significant structural changes in a good number of markets, this has yet to come about in any meaningful way other than in three or four countries.

8.6 CONCLUSIONS

As has become clear, the present situation in the region is relatively stable because only two operators of any consequence remain. There has not been any suggestion that a new investor of any size would single out the region as attractive from an investment perspective for over a decade. Historically, Latin America was in essence a colonial preserve, of interest to operators from the USA that regarded it as their backyard and to the European operators that spoke the relevant languages – Italian, Portuguese and Spanish. The withdrawal of the Baby Bells effectively left room for a major indigenous operator to emerge in the form of América Móvil, which branched out from its highly profitable home market of Mexico at a time when networks and licences were available quite cheaply, and it also had the advantage that, as a pure regional specialist, it did not allow itself to be distracted by the somewhat different challenges posed in other regions, especially those in a much more advanced stage of development.

It may be argued that the emergence of América Móvil went hand in hand with the emergence of AT&T because it held a minority stake in the former and hence did not need to look for regional assets independently, but, as previously noted, this was a passive investment with no control over the running of the networks.[2] It may further be noted that this left AT&T in something of an awkward position because it could not reasonably expect to set up an independent presence in a country where América Móvil was present, nor would it be possible to increase significantly its stake in América Móvil because the controlling shares are tightly held within the family of Carlos Slim.

As we now know, AT&T was interested in the region without it necessarily being its primary focus because it had Telefónica in its sights, allegedly offering to acquire that operator in May 2013 before the Spanish government made its displeasure clear. One must presume that a successful takeover would have forced AT&T to shed either its stake in América Móvil or (perhaps most of) Telefónica's Latin American assets, which would have been a fascinating development. However, it seems that AT&T must now content itself with what it already has, although it could still conceivably become involved in the unravelling of the situation in Brazil.

So far, its main rival in the USA, Verizon Communications, has done nothing to suggest that it plans to return to the region, and the sums involved in the payoff to Vodafone for its minority stake in Verizon Wireless more or less rule out international forays on any scale in the immediate future. With other international investors ruled out by the language problem, restructuring can really only proceed by shuffling the existing cards in the deck. On the whole, the most sensible approach is that adopted by América Móvil and Digicel, namely to swap assets that allow each party to build scale in countries where it already operates. The problem with this strategy is, first, that the number of possibilities is very limited and, second, that there is no guarantee that regulators will play along, as demonstrated in El Salvador.

Naturally, it may be argued that a country with fewer than 20 million subscribers does not need more than three networks, and there are still a few of them with four in the region, but regulators are entitled to interpret local market conditions as they see fit. What they cannot do, however, is make the region attractive to inward investment. In the light of the ever-shrinking role of Telecom Italia, combined with the gradual reduction in its overall commitment by Telefónica, it is fair to say that the region has never been less attractive to inward investors than now. Indeed, it may be reasonable to argue that if América Móvil is spending so much time and energy in acquiring assets in Europe, then it must itself have some doubts as to the wisdom of investing further in the region other than to consolidate its existing position.

NOTES

1. According to data compiled by the authors, but the competition regulator appears to have regarded América Móvil as the dominant operator even before the provisional acquisition of Digicel.
2. Although it did allow AT&T to maintain some kind of international presence at a time when it was otherwise withdrawing from overseas markets.

REFERENCES

Baigorri, M. (2013), 'Telefonica sells stake in Central America assets to cut debt', www.bloomberg.com, 30 April.

Curwen, P. and J. Whalley (2008), *The Internationalisation of Mobile Telecommunications: Strategic Challenges in a Global Market*, Cheltenham, UK and Northampton, MA, USA: Edward Elgar.

Curwen, P. and J. Whalley (2013), *Fourth Generation Mobile Communication: The Path to Superfast Connectivity*, London: Springer.

Edgerton, A. and A. Galvao (2013), 'Telefonica told by Brazil to engineer Tim sale or undo deals', www.bloomberg.com, 5 December.

The Gleaner (2012), 'Digicel to stay put in El Salvador', http://jamaica-gleaner.com, 19 October.

Ictpulse (2012), 'Has the Digicel–Claro merger been good for Jamaica?', www.ict-pulse.com, 17 October.

Lennighan, M. (2014), 'Portugal Telecom, Oi merger gets antitrust OK in Brazil', www.totaltele.com, 14 January.

Petersen, L. (2010), 'Bolivia settles bitterly-contested arbitration with Telecom Italia', www.iareporter.com, 25 November.

Telecom.paper (2014), 'Telefonica "not part of" any deal to sell off TIM Brasil', www.telecom.paper.com, 6 January.

TeleGeography (2008), 'Digicel and America Movil bag PCS licences', www.telegeography.com, 8 May.

TeleGeography (2014), 'Minority shareholders could upset Oi-PT merger', www.telegeography.com, 13 January.

9. Structural and strategic adjustment among mobile operators in North America

9.1 INTRODUCTION

Although this book is concerned with operators' international ventures, there is one region – if that is indeed the correct term for what is in effect a single country for our purposes, given that Canadian operators are fixated on the home market and the other parts of the region are too small to warrant analysis – that needs to be handled somewhat differently from the others.

The USA is one of the world's largest single markets and currently contains four large national operators – of which two are much larger than the others – and a number of relatively small regional/local networks. However, simply indicating who they are and listing their international assets would be highly misleading because the USA has undergone internal restructuring on a massive scale – indeed, some restructuring is ongoing while this chapter is being written – and this has, in turn, significantly altered the way in which the evolving tableau of operators has viewed the balance of advantage between domestic and overseas holdings.

For this reason, it is necessary to start by outlining the process of structural evolution before commenting on how this has affected operators' attitude towards ownership of overseas assets. In the process, the role played by foreign companies as strategic investors within the USA is put into perspective.

9.2 THE ERA OF THE BABY BELLS

Readers will be familiar with AT&T as one of the two largest mobile incumbents in the USA. So presumably it is the same AT&T with which they were familiar roughly a decade ago. In fact it is not, so it is necessary to begin the discussion about restructuring in the USA by explaining briefly how this has come about.

In early 1982, the chairman of the then AT&T announced that it had settled its long-running dispute with the Department of Justice (DoJ). This settlement, which technically altered AT&T's 1956 Consent Decree with the consequence that it was called the Modified Final Judgement (MFJ), required AT&T to divest its interests in 22 Bell Operating Companies (BOCs) that provided local telephone services. In return, AT&T would be freed from its Consent Decree obligations, thereby enabling it to enter new markets – for further details see Curwen and Whalley (2008: ch. 8).

As a result of the divestiture, AT&T was left as a provider of long-distance, manufacturing and research services. The regional BOCs (RBOCs), better known as the Baby Bells, had amalgamated into seven operating companies, each of which was valued at between $14 billion and $20 billion compared to the $35 billion market value of AT&T. US West covered the most states (14), while Pacific Telesis covered the fewest (2) and the other Baby Bells covered between five and eight. Initially, the MFJ restricted the Baby Bells from expanding into new markets, but these restrictions were gradually removed and interest began to be expressed in the provision of mobile services – or, as they were normally referred to in the USA – cellular services.

In the course of time, the Baby Bells underwent further restructuring such that by early 2006 only three remained – for a schematic representation of the restructuring process, see Curwen and Whalley (2008: 193). The first of these, US West, did not engage in M&A activity with other Baby Bells. However, in April 1998 it sold its domestic cellular operations to AirTouch Communications for $5.9 billion and these eventually became part of Verizon Wireless. After further restructuring of its other operations, including the hiving-off of MediaOne in June 1998, the residual company became the subject of takeover bids from Global Crossing and Qwest International in June 1999. The latter was successful, with a bid of $51.3 billion and US West became Qwest Communications.

In September 2004, a company known at the time as CenturyTel set up a relationship with the then Cingular Wireless. In 2005, it began a wireless voice and data service, and subsequently expanded via the purchase of other local operators. In April 2010, just before changing its name to CenturyLink to reflect the prior purchase of Embarq, CenturyTel agreed a merger with Qwest Communications via a share exchange that left CenturyTel with 51 per cent of the combined entity, which emerged in April 2011. However, despite its great size overall, this company is not important in relation to mobile provision in the USA.

The same cannot be said of the remaining Baby Bells. Shortly after the Telecommunications Law of 1996 was passed, Bell Atlantic entered into a $25.6 billion 'merger of equals' with another Baby Bell known as Nynex. As a result, the merged entity became a 50 per cent shareholder in PCS PrimeCo., a cellular joint venture that eventually became part of Verizon Wireless. In July 1998, Bell Atlantic merged with GTE (not a Baby Bell) in a deal valued at $67 billion, and on completion of the merger changed its name to Verizon Communications.

The origins of the present-day AT&T are more complicated, involving as they do the other four Baby Bells and the original AT&T. In late 1993, Pacific Telesis grouped together its non-regulated subsidiaries in PacTel Corporation, which it subsequently spun off into a company whose name was changed to AirTouch Communications. Meanwhile, in 1994, Southwestern Bell also changed its name, in this case to SBC Communications, and in April 1996 it entered into a $16.5 billion 'merger of equals' with what remained of Pacific Telesis. During 1998, it effectively acquired SNET and in May it merged with the third Baby Bell, Ameritech, in a deal worth $68 billion.

In 2000, SBC Communications and BellSouth merged their domestic cellular operations to form Cingular Wireless, a 60–40 joint venture. In October 2004, Cingular Wireless paid $41 billion in cash to acquire AT&T Wireless, one of the companies that had emerged from the restructuring of 'old' AT&T just after the turn of the millennium – it opened the bidding at $34 billion but was forced into a competitive auction with Vodafone (Curwen, 2005). 'Old' AT&T subsequently re-organised itself so as to concentrate on the corporate long-distance market. However, in early 2005, SBC Communications announced its intention to merge with 'old' AT&T in a deal valued at $16 billion. After the merger was approved in November 2005, SBC changed its name to AT&T – the 'new' AT&T. In March 2006, 'new' AT&T announced that it would be merging with BellSouth in a deal valued at $67 billion, plus $17 billion of debt taken over and the deal was authorised by the FCC in December (FCC, 2006).

At this point the era of the Baby Bells was finally over and the two main protagonists in the present-day US mobile market, Verizon Communications and AT&T, had emerged to do battle.

9.3 RESTRUCTURING CONTINUES APACE

To open this section, we should recall how much has changed during the past decade. At the end of 2003, there were 160 million mobile subscribers in the USA, equivalent to a 54 per cent penetration rate, which lagged by some margin that ruling in Europe. The 'big six' operators at that time – AT&T Wireless, Cingular Wireless, Nextel Communications, Sprint PCS, T-Mobile USA and Verizon Wireless – accounted to roughly 80 per cent of the subscribers and a further 10 per cent was accounted for by eight regional operators: Alltel, Centennial, Dobson Communications, Leap Wireless, Qwest, Rural Cellular, US Cellular and Western Wireless.

One year on, and AT&T Wireless was about to be taken over by Cingular Wireless subject to some wireless assets being sold to T-Mobile USA, leaving a 'big five', all of which could reasonably be referred to as national rather than regional operators. Post-merger, Cingular Wireless and Verizon Wireless, each with more than 40 million subscribers, were significantly larger than the other three. This is where the restructuring process is picked up in Table 9.1.

In practice, the 'big five' structure did not last for long since Sprint PCS reacted by making an ill-advised bid to merge with Nextel Communications, which used a technology known as iDEN rather than the CDMA used by Sprint. The post-merger entity would be worth over $60 billion and have almost as many subscribers as Verizon Wireless and Cingular Wireless. This new company was to be known as Sprint Nextel, reflecting the fact that the former Sprint shareholders held 51 per cent of the combined equity. However, even at the time there were good reasons to question the underlying logic of the merger, such as the incompatible technologies and the marked differences in management styles that presaged severe culture clashes.

Technically, this created a 'big four', but with T-Mobile USA lagging the other three by a significant margin it could reasonably be claimed that the era of the 'big three' had now dawned. But did this finally represent a structure that presaged a period of relative equilibrium?

Evidently not, although the next move did not directly involve the transfer of mobile assets. By the end of 2004, as noted above, AT&T had been reduced to a corporate long-distance operation and it appeared to be too highly valued to attract a predator. Nevertheless, SBC Communications chose to make a bid worth $16 billion, paying with its own shares. Interestingly, SBC's junior partner in Cingular Wireless, BellSouth, had itself attempted to buy AT&T in 2003 and the FCC had previously stated

Table 9.1 Main M&A activity involving US-based operators, January 2004 to April 2014

Bidder	Target	Date	Value, $billion
Cingular Wireless	AT&T Wireless	February 2004	41.0
T-Mobile USA	Cingular Wireless assets	May 2004	2.3
Sprint	Nextel	December 2004	35.0[1]
Alltel	Western Wireless	January 2005	4.4 + 1.5 debt
SBC Communications	AT&T	January 2005	16.0
Sprint Nextel	Alamosa Holdings	November 2005	3.4 + 0.9 debt
AT&T (SBC)	BellSouth	March 2006	67.0 + 17.0 debt
Verizon Communications	Vodafone USA	May 2006	48.0 + 6.0 debt
Goldman Sachs/TPG	Alltel	May 2007	27.5
AT&T	Dobson Communications	July 2007	2.8 + 2.3 debt
Verizon Communications	Rural Cellular	July 2007	0.8 + 1.9 debt
MetroPCS	Leap Wireless	September 2007	5.5 + 2.0 debt
Verizon Wireless	Alltel	June 2008	5.9 + 22.2 debt
CenturyTel	Qwest	April 2010	10.6 + 11.8 debt
AT&T	T-Mobile USA	March 2011	39.0
Dish Network	MetroPCS	August 2012	4.0
T-Mobile USA	MetroPCS	October 2012	[2]
SoftBank	Sprint Nextel	October 2012	20.1[3]
Sprint Nextel	Clearwire	December 2012	2.2[4]
Dish Network	Clearwire	January 2013	4.9
Dish Network	Sprint Nextel	April 2013	25.5[5]
AT&T	Leap Wireless	July 2013	1.2 + 2.8 debt

Notes:

1. Technically a 'merger of equals' but designed so as to give Sprint shareholders 51 per cent of the merged company.
2. See section on MetroPCS.
3. Bid for a 71 per cent stake. Successful after bid altered to $21.6 billion (with a larger cash element) for a 78 per cent stake.
4. Bid for the outstanding 51 per cent of voting rights.
5. Bid for a 78 per cent stake, causing SoftBank to raise its own offer to $21.6 billion.

that it would not look favourably on a Baby Bell acquiring its former parent, but in the event things had moved on to the point at which objections to the takeover could not be sustained.

But 2005 was notable also for the beginning of a restructuring process among the second tier of regional operators, initially involving Alltel. The list of both current as well as defunct, merged and acquired wireless operators in the USA can be found at http://en.wikipedia.org/wiki/List_of_United_States_wireless, but for our purposes only a handful have had a major role to play in the restructuring process and it is accordingly convenient to clarify this role before returning to the nationwide networks.

Alltel

In January 2005, Alltel (with roughly 8.5 million subscribers) made a takeover bid for Western Wireless (with roughly 1.5 million subscribers) worth $4.4 billion plus $1.5 billion of debt taken on. This still left Alltel only half as large as T-Mobile USA, but it did have a roaming agreement with Sprint Nextel that gave both operators near-nationwide coverage for voice and data.

In December 2005, Alltel became a wireless-only operator, having spun off its fixed-wire operations, which were then sold off. At its peak, Alltel provided wireless services in 34 states, covered a greater land area than any other network and achieved nationwide coverage via roaming agreements with both Sprint Nextel and Verizon Wireless, which in return achieved coverage in rural areas. However, by early 2007, Alltel appeared to be intent on getting itself taken over by one of its bigger brethren, although none of these showed much enthusiasm at the time. Surprisingly, given that (a) the operator was financially in good shape and (b) its share price had risen by one-third between December 2006 and May 2007, it was a Goldman Sachs/Texas Pacific Group consortium that acquired the company for $27.5 billion in cash, in the process outbidding two other private equity groups.

However, in June 2008, Verizon Wireless bid $28.1 billion, consisting largely of debt taken over, to relieve the consortium of its holding, with the FCC approving the acquisition subject to Verizon divesting assets in 105 Alltel markets. In May 2009, AT&T agreed to purchase 79 of these asset packages, including licences, network infrastructure and 1.5 million subscribers. The remaining 26 packages, including 800 000 subscribers, were bought in April 2010 by Atlantic Tele-Network, which continued to use the Alltel brand. In January 2013, AT&T acquired these Alltel assets for $780 million in cash – the deal was completed in September – and at the time of writing (June 2014) is in the process of incorporating them into its existing networks as part of AT&T Mobility.

MetroPCS

In September 2007, MetroPCS, at the time the USA's largest provider of non-contracted, flat-fee unlimited mobile services, made a hostile take-over bid for Leap Wireless, its main like-for-like competitor, based on an exchange rate of 2.75 MetroPCS shares for each Leap Wireless share. If accepted, this would have created a nationwide entity owned 34.6 per cent by Leap shareholders, but the Leap board turned down the offer flat on the grounds that it was too low. However, this did not prove to be a particularly wise move as the Leap share price subsequently went into a severe decline – hardly surprising, given that its only logical partner was MetroPCS and that it was enduring a run of annual net losses that has yet to come to an end.

Although an eventual link-up between the two companies continued to be the subject of much speculation, it was destined never to happen. Dish Network (see below) made an unsuccessful bid worth $4 billion for MetroPCS in August 2012 (Ramachandran, 2012). A further interesting development was the aborted (at literally the final hour) deal whereby Sprint Nextel would buy MetroPCS for $8 billion, including debt taken over (Telecom.paper, 2012). Given Sprint Nextel's difficulties, discussed elsewhere, it comes as no great surprise that they sought to link up with a large regional operator using cdma2000 technology, but for now it is not known why the deal fell through.

In the event, it was T-Mobile USA that ultimately won the day after following up an initial agreement in principle in October 2012 with a revised deal, tabled in April 2013 (Wood, 2013a), whereby it would merge MetroPCS with T-Mobile USA and hand over to MetroPCS shareholders 26 per cent of the merged company plus $1.5 billion in cash. The new operator would be known as T-Mobile US – a subtle but crucial distinction – but it remains the fourth-largest operator in the USA. The new company formally came into being in May 2013. In June, MetroPCS became an MVNO selling T-Mobile US 2G and HSPA+ services – it had already launched LTE on its own behalf – pending a switch-over of its CDMA subscribers to GSM, which is expected to reach completion by the end of 2015 (Gryta, 2012).

Leap Wireless

As noted above, Leap Wireless did not prosper after it turned down the offer from MetroPCS. By the second half of 2012, Leap's long-term debt had risen to $3.2 billion and it seemed that it would be left to flounder on its own. It accordingly came as something of a surprise when AT&T moved in July 2013 to offer $1.2 billion for Leap, roughly twice what it was worth before the bid (Hillen, 2013), as well as taking on its $2.8 billion of debt. However, Leap was sitting on $4.8 billion of tax credits that could be set against future profits, and these would become available to AT&T. The proposed deal was (not surprisingly) met with great enthusiasm by Leap shareholders and the transfer of assets officially took place in March 2014 (Cellular-news, 2013).

Others

In July 2007, AT&T – that is, the 'new' AT&T that emerged after SBC Communications took over BellSouth and hence took full control of Cingular Wireless – made a takeover bid worth $5.1 billion for Dobson Communications. Not to be outdone, Verizon Wireless responded by making a friendly takeover bid worth $2.7 billion for Rural Cellular (with 716 000 subscribers) and the latter was fully integrated in January 2009. Finally, in September 2007, T-Mobile USA attempted to bulk up by acquiring SunCom Wireless, which both strengthened T-Mobile's position in the south-east and gave it an entry into Puerto Rico and the US Virgin Islands. Excluding all of the above, this has left US Cellular as the only independent regional operator with more than 1 million subscribers.

9.4 THE 'BIG FOUR'

The previous section has detailed the manner in which certain of the second-tier regional operators have become part of the restructuring process involving the 'big four' nationwide operators, and it is to these that our attention must now turn. However, much of the history of AT&T and Verizon Wireless in their various forms has already been recounted above, and hence further discussion will be left until consideration is given to their international activities. This section is accordingly given over to an analysis of Sprint Nextel, but also reviews T-Mobile USA's ultimately doomed takeover bid by AT&T.

Sprint Nextel

In April 2004, the two previously separated parts of Sprint, namely Sprint FON (fixed-wire) and Sprint PCS (mobile) were re-integrated as Sprint Corporation. At the time, Sprint PCS had roughly 20 million subscribers in mainland USA plus a small number in Puerto Rico and the US Virgin Islands. In December, it instigated (for tax reasons) a 'merger of equals' with Nextel Communications valued at $35 billion, although in reality it was more akin to a takeover bid as Sprint shareholders ended up with a controlling stake. At the time, Nextel had 15 million subscribers in the USA but also owned a minority stake in four subsidiaries in Latin America held by Nextel International. The merger, which created Sprint Nextel, was completed in August 2005.

The change in its structure meant that Sprint was also obliged to pay in excess of $10 billion over a period of years to acquire outstanding stakes in most of its affiliates such as Alamosa Holdings (which cost $4.3 billion in November 2005) – see http://en.wikipedia.org/wiki/Sprint_Corporation and Curwen (2007).

At the time of the merger the two companies had a combined market value of over $60 billion and in March 2006 Sprint Nextel's value peaked at $76 billion. But disillusion rapidly set in, causing Sprint Nextel in 2008 to write down $29.7 billion of the value of Nextel on its books. One major contributory factor was the incompatible technologies used by the two companies – whereas Sprint provided conventional CDMA-based services in the 1900 MHz (PCS) band, Nextel used a proprietary technology known as iDEN in the 850 MHz band. This, in turn, meant that most new mobile subscribers with good credit histories preferred to opt for AT&T and Verizon Wireless, and Sprint Nextel found itself suffering a net loss of high-value contract customers while gaining subscribers on cheap pre-paid plans.

By the end of 2008 its 48 million (mainly pre-paid) subscribers had fallen well behind the total for its two main rivals, and in practice there has been remarkably little improvement during the subsequent five years. Sprint Nextel seemed unable to adopt a strategy that would restore its image. For example, as noted below, it became involved with Clearwire and adopted WiMAX as the technology of choice to deliver high-speed data services at a time when its three main rivals were all opting for LTE.

In October 2011, Sprint Nextel announced that it would not be selling any devices using Clearwire's WiMAX technology after 2012, although it would continue to support such devices as were already in circulation. Rather, it would be speeding up the roll-out of Network Vision, which would combine its various technologies (now including LTE) using the

1900 MHz band. The commercial launch took place in 15 cities in July 2012, with nationwide coverage to be achieved during 2013 (by which point it expected to have re-farmed the 850 MHz spectrum previously used by its terminated iDEN service). Sprint Nextel's credit rating was badly affected – the fact that it was going it alone in the 1900 MHz band meant at the very least that equipment vendors would not be rushing to provide compatible devices – and elicited the common response that it appeared to have relegated itself to the status of permanent also-ran behind AT&T and Verizon Wireless.

Not surprisingly, Sprint Nextel now sought a solution to its problems by acting as a predator while at the same time appearing in its weakened state to be a target. On the one hand, it was said to be monitoring the T-Mobile USA bid for MetroPCS with great interest, mulling over whether to rebid for MetroPCS or, more likely, to bid for a post-merger T-Mobile/MetroPCS entity. On the other hand, it was said to be a potential target for Japan's SoftBank, and if SoftBank were indeed to bid for a majority stake, then the situation would become somewhat complicated. In practice, in October 2012, SoftBank bid $20 billion for a 70 per cent stake, which was not expected to create regulatory problems. Sprint Nextel immediately responded by raising its stake in Clearwire (see below) so that, in effect, SoftBank was acquiring control of both companies (Osawa, 2012).

Further complications arose in April 2013 when Dish Network made an (allegedly informal) unsolicited bid for 68 per cent of Sprint Nextel worth $25.5 billion – equivalent to $7 per share, of which $4.76 consisted of cash, considerably more than the prior offer by SoftBank. SoftBank, in response, refused to raise its own offer, claiming that it expected its offer to be accepted on 1 July without further amendment. However, it soon changed its mind when shareholders reacted negatively, adding extra cash to its offer so as to raise it to $21.6 billion. Curiously, this meant that it would now end up with 78 per cent of Sprint Nextel, thereby reducing the value of the entire company from $28.9 billion to $27.8 billion (Gabriel, 2013).

T-Mobile USA

T-Mobile USA was the wholly owned subsidiary of Germany's T-Mobile and seemingly stuck in fourth place among the 'big four'. Although it was an important source of proportionate subscribers – the 33 million at the end of 2009 was second only to the 35 million in Germany – this number was exactly the same at the end of 2012 and parent Deutsche Telekom (operating as T-Mobile) was unwilling to invest on a scale that would resolve the problem internally. This indicated that the network should either be sold, or, failing that, bulk up by acquisition of other networks and licences. Its takeover of MetroPCS has been noted previously, but this was in effect the fallback solution in the face of a failure to deliver a sale.

Given that T-Mobile USA used GSM, there could realistically only be one possible buyer, which was the other GSM operator among the 'big four', AT&T. In March 2011, AT&T duly made a takeover bid worth $39 billion in cash and shares. However, in August the DoJ filed to prevent the merger and in November the FCC indicated that it would be conducting a full inquiry. As a result, the bid was called off at considerable cost to AT&T, which had agreed to hand over to T-Mobile USA $3 billion in cash and licences worth a further $1 billion if the deal fell through (Moritz and Rahn, 2011).

The completion of the MetroPCS takeover has created a new entity, T-Mobile US, which is only 74 per cent owned by T-Mobile, so the subscribers gained have been roughly balanced out by the subscribers lost. As a result, T-Mobile has not significantly improved its financial situation in the USA and there are no longer any feasible acquisitions worth pursuing. Furthermore, Sprint is in a much stronger financial position. This means that T-Mobile US has only two ways forward. The first is to accumulate licences in low bandwidths in order to improve its network and attract more subscribers, which is the tactic underlying the deal to pay $2.4 billion in cash together with licences worth $950 million, agreed with Verizon Wireless in January 2014 (Lennighan, 2014). The other is to be taken over.

But this presents a major difficulty. In the first place, there is no longer a GSM operator able to do a deal. Second, even if a CDMA operator is willing to have a shot at a takeover, it faces huge regulatory problems. Since these would be insuperable for Verizon Wireless, that only leaves Sprint and until recently Sprint (in the guise of Sprint Nextel) was also a non-starter. What has changed is that Sprint Nextel has itself been taken over by SoftBank and it somewhat surprisingly indicated that it intended to make a bid for T-Mobile US in early 2014. This is more conveniently discussed in Section 9.7 below.

9.5 SATELLITE

A unique feature of the USA is the opportunity to launch a wholesale LTE service using mobile satellite spectrum, and this has led to certain owners of satellite spectrum becoming involved with the 'big three' in the ongoing restructuring process. Three companies in particular are involved, namely LightSquared, Dish Network and Clearwire.

LightSquared

Harbinger Capital Partners set up LightSquared using licences owned by subsidiary SkyTerra. LightSquared partnered with Inmarsat in order to create the necessary contiguous bandwidth to supply terrestrial services and subsequently joined up with rural access start-up OpenRange to provide services in areas neglected by the 'big three'. However, widespread concerns arose over the potential for LightSquared to interfere with GPS signals (Fitch Ratings, 2011) because LightSquared hoped to use the 1545–1555 MHz (upper 10 MHz) band while GPS operates in the 1559–1610 MHz band. These concerns were first expressed by regulators in early 2011, but the FCC went ahead anyway with provisional approval for the non-satellite usage. As things stood at the end of August 2011, the Department of Defense standard of September 2008 effectively granted GPS a 4 MHz guard band, LightSquared had offered a 23 MHz guard band and the GPS industry was asking for a 34 MHz guard band, which would leave LightSquared with only 24 MHz of usable spectrum.

LightSquared had agreed in July 2011 to pay $9 billion over an 11-year period for Sprint Nextel to manage the roll-out of its wholesale network. However, LightSquared's prospects appeared to have been damaged irreversibly by the National Executive Committee for Space-Based Positioning, Navigation and Timing (PNT ExComm), which claimed that there could be interference with aircraft safety systems and that there was no solution to this problem (Cellular-news, 2012a). In late February 2012, the FCC responded to a letter from the National Telecommunications and Information Administration (NTIA), which contained a summary of tests conducted so far, by announcing that it intended to revoke LightSquared's provisional permission to proceed. This, in turn, was expected to affect Sprint Nextel's agreement with LightSquared and it was duly scrapped by Sprint Nextel in late March (Cellular-news, 2012b).

LightSquared's financial situation began to look increasingly precarious, given that it was obliged to repay large sums of interest on loans during the course of 2012, and in late February it announced that it would be dismissing 45 per cent of its staff. Chapter 11 bankruptcy rapidly became the inevitable outcome – a process that allows a company to continue in operation while it restructures its affairs. A debtor-in-possession loan in June 2012 meant that the company could at least remain in business until November 2013. The total losses accumulated during the year commencing May 2012, when it was declared bankrupt, amounted to $661 million.

In August, a court case was brought by LightSquared shareholder Harbinger Capital Partners against Dish Network chairman Charlie Ergen, claiming that he had illegally used a third party to acquire large quantities of LightSquared debt at a significant discount, thereby gaining effective control over its debt renegotiation process, without declaring any interest in that third party. The case was dismissed on the grounds that the suit should have been instigated by LightSquared rather than Harbinger. LightSquared chose to respond with a second suit, this time naming both Dish Network and Ergen as co-defendants on the grounds that Ergen had bought debt on behalf of Dish rather than himself, thereby violating the Chapter 11 process.

Subsequently, it appeared that a stalking-horse bid for LightSquared had been lodged by Dish affiliate L-Band Acquisition with a superficial value of $2.2 billion in cash but worth $3.3 billion in total after allowing for certain debts taken on. A private-equity firm, Centerbridge Partners, then intervened by tabling a provisional offer of $3.3 billion in cash and $5 billion in total, but withdrew soon afterwards, citing uncertainty over the eventual date for regulatory clearance to be given for LightSquared to roll out its network. Although Dish Network duly dropped its own bid in early January 2014, it was alleged that it might still be obliged to proceed as a result of agreements made with its own lenders.

In February 2014, with its cash reserves due to run out in mid-March, LightSquared lodged a proposal, supported by a significant number of major stakeholders, whereby it would receive $1.65 billion in debtor-in-possession financing plus $1 billion in exit financing. As a major debt holder, Ergen may have to ratify the deal.

Dish Network

Dish Network spent $2.8 billion buying up 40 MHz of contiguous S-band spectrum (2000–2020 MHz plus 2180–2200 MHz) – the S-band is known in FCC parlance as AWS-4. This was acquired from bankrupt DBSD and TerreStar, subject to regulatory approval by the FCC and the resolution of Chapter 11 proceedings. The S-band is considered to be superior to the L-band used by LightSquared because it can be used twice – once terrestrially in what is known as an Auxiliary Terrestrial Component and once via a satellite. However, whereas LightSquared had managed to circumvent the existing restriction on dual use via a waiver from the FCC (but run into interference problems), Dish Network struggled to secure a satisfactory solution. Nevertheless, its main problem remained a cash shortfall to fund a full terrestrial roll-out together with a shortage of spectrum. It had begun talking in vague terms of a link of some kind with either Clearwire or Sprint, but many commentators argued that since the takeover of T-Mobile USA by AT&T had failed to come to fruition, Dish Network could be interested in merging its spectrum with that of T-Mobile USA – it had made an unsuccessful bid worth $4 billion for MetroPCS in August 2012 before T-Mobile USA interceded and subsequently held informal discussions with T-Mobile USA about the possibility of a merger – or even offering to be bought by AT&T (Gabriel, 2012).

However, a link-up with AT&T seemed unlikely given the latter's petition to the FCC in February 2012 that Dish Network should have the same roll-out obligations as those imposed on LightSquared. Dish Network considered this to be a wholly unreasonable requirement on the grounds, first, that it intended to operate as a retailer rather than a wholesaler (which had ramifications for the prospects of 4G in the USA) and, second, that it intended to move straight to LTE-Advanced, a launch of which would not be feasible before 2015.

The situation in November 2012 appeared to be that the FCC was prepared to sanction full terrestrial rights, but only if onerous conditions were accepted. In effect, the FCC wanted Dish Network to disable 25 per cent of its uplink spectrum and impair a further 25 per cent in order to accommodate the potential future use of adjoining H Block spectrum by Sprint Nextel. As this 5 MHz block of spectrum was yet to be licensed, Dish understandably considered the proposal to be unreasonable and likely to cause yet further delays in its roll-out. The official FCC clearance finally came in mid-December with the expected conditions attached, together with a requirement that Dish Network build out at least 70 per cent of its network within five years. The coverage condition was subsequently reported as 40 per cent of the population within its footprint within four years and 70 per cent within seven years.

As noted above, LightSquared was in need of a concrete proposal to take it out of Chapter 11 in November 2013. Thwarted in the execution of its original plans, Dish Network now intervened to place on the table a so-called 'stalking-horse' bid for LightSquared spectrum designed to set in motion an auction for LightSquared. The bankruptcy court accepted the $2.22 billion bid in October 2013 and set the auction date as 25 November. Although Dish Network withdrew its bid in early January 2014, it was alleged that it might still be obliged to proceed as a result of agreements made with its own lenders.

Clearwire

The trend in the USA where 4G is concerned is unquestionably to proceed via LTE to LTE-Advanced. However, it is of interest to ask whether WiMAX has a future as a competitor for LTE. Until recently, it was arguable that only the USA was witnessing the roll-out of a network that might be seen in this light. When Sprint took over Nextel, it ended up with the great bulk of the spectrum in the 2.5 GHz band in the largest 100 markets in the USA. Given the existence of other users in the band, it appeared unlikely that Sprint Nextel could use it to roll out any kind of nationwide service, but the FCC imposed a condition for sanctioning the takeover of Nextel in the form of a requirement that the merged company offer services in the 2.5 GHz band to at least 15 million citizens by 2009 and to 15 million more by 2011.

In March 2007, AT&T sold its entire 2.5 GHz spectrum to Clearwire in order to satisfy one of the conditions attached to its purchase of BellSouth. This meant that, in effect, Clearwire and Sprint Nextel now had the 2.5 GHz field to themselves, although it is worth noting that while Clearwire now had access to extensive holdings of spectrum, these were highly fragmented. In July, Sprint Nextel and Clearwire announced that they now intended to build a joint nationwide WiMAX network branded as Xohm. Each would build out its own network and these would then be linked via a roaming agreement. By January 2008, Sprint Nextel was said to be desperately seeking partners in Xohm. In May came the announcement that the new partners would be Comcast, Time Warner Cable, Bright House Networks, Intel and Google. They would get a combined 22 per cent stake in a new network branded as Clear, with Clearwire itself holding a further 27 per cent and Sprint Nextel the residual 51 per cent. The deal was sanctioned by the FCC in November 2008 and closed in December. Sprint Nextel chose to provide services over the Clear network acting as an MVNO branded as 'Sprint 4G'. The initial launch was in Baltimore at the end of September 2008.

In March 2010, Clearwire appeared to be allowing for the possibility of at least a partial switch of technology in that it was one of the advocates of the opening up of the 2.5 GHz band for TD-LTE in the USA. Meanwhile, the rush to expand its WiMAX network to match the coverage of Verizon Wireless meant that it was running short of money. Although Sprint Nextel appeared to be unwilling to provide any additional finance, it duly signed a new long-term agreement in April 2011 that specified wholesale prices and guaranteed minimum payments to Clearwire. Subsequently, Clearwire announced that it would not be selling off any of its spectrum. In June 2011, Sprint Nextel reduced its voting rights in Clearwire to 49.9 per cent to allay shareholders' fears that it would be liable for debts incurred by Clearwire if the latter were deemed to be a subsidiary company.

In August 2011, Clearwire confirmed that it would be overlaying its WiMAX network with LTE-Advanced-ready technology in the main urban areas, provided it could secure the necessary funding. It reported that it had 7.7 million (mostly wholesale) subscribers, adding that it remained the largest wholesaler in the USA and that Sprint Nextel would not be alone in needing to utilise Clearwire's spectrum at some point.

In early December, Sprint Nextel agreed, subject to certain conditions, to pay Clearwire $926 million in return for unlimited 4G WiMAX retail services during 2012 and 2013. In addition, Sprint Nextel agreed to pay $350 million, via a series of pre-payments over a two-year period, in return for LTE capacity, provided that Clearwire achieved specified roll-out targets by June 2013. Finally, Sprint Nextel agreed to provide additional equity finance in the event of a further share issue, which was needed because, in early December, the owners of SpectrumCo, a group of cable operators including Clearwire investors Time Warner and Comcast, sold their AWS licences to Verizon Wireless for $3.6 billion, subject to regulatory approval – AT&T could not bid because of its ongoing merger negotiations with T-Mobile USA. This meant that they would cease to re-sell Clearwire's 4G services in March 2012. The deal received conditional approval from the DoJ and FCC in August 2012, and Verizon Communications agreed to sell on spectrum in the 700 MHz Lower A and B blocks to T-Mobile and/or other interested parties.

In August 2012, the Clearwire share price fell below $1, equivalent to a near-50 per cent fall during the year so far, in the process forcing Sprint Nextel to take an impairment charge. At 11 million, subscriber numbers appeared to be healthy, although the great majority were wholesale customers provided by Sprint Nextel.

The SoftBank takeover bid for Sprint Nextel prompted the latter to raise its voting rights in Clearwire from 49.8 per cent to a small majority (without affecting its majority equity stake). This meant that SoftBank would gain control of two companies, each with significant spectrum holdings. In December, it was alleged that Sprint Nextel would proceed with a plan to acquire full control of Clearwire, causing the latter's share price to rise to $2.68. SoftBank expressed its support for a bid of $2.97 per share for the outstanding voting rights in Clearwire; this was accepted by Comcast and Intel, although, inevitably, other shareholders set out to contest the bid in the courts. However, all these plans were thrown into disarray when Dish Network made an unsolicited bid for the whole of Clearwire in January 2013 – its motives were questionable since, given its majority voting rights in Clearwire, Sprint Nextel could forestall any takeover bid.

There were further developments in February 2013 when Clearwire announced that it needed additional funding, and hence that it intended to draw on funds made available by Sprint Nextel as part of its attempt to take full control of Clearwire. Sprint Nextel had offered to buy $800 million of bonds from Clearwire by way of ten monthly instalments, a facility that Clearwire had not taken advantage of during January and February because Dish had stated that its own offer was contingent on Clearwire not drawing on the facility. However, Clearwire did draw down the instalments, beginning in March, without eliciting any response from Dish Network.

In the event, it was Sprint Nextel that blinked first when presented with the likelihood that Clearwire's shareholders would reject its initial bid. In late May, it raised its offer for the outstanding Clearwire voting rights to $3.40 each, thereby valuing the whole of Clearwire at $10.7 billion. It declared this to be its 'best and final offer'. Within days, Dish Network responded by raising its own bid to $4.40 per share in cash, contingent on acceptance by at least half of the minority shareholders (equivalent to one-quarter of the voting shares) and the offer of at least three seats on the board. This raised the possibility that it would end up as a minority shareholder in a company controlled by Sprint Nextel.

Analysts were divided about the relative merits of the two offers for Clearwire. The main benefit of the Dish Network proposal was that it would create a portfolio of 230 MHz of spectrum across a range of bandwidths. This would enable it to provide a quad-play offer in competition with AT&T, Verizon Communications and Comcast. However, there would be a need to invest heavily over a period of years to create the requisite network of cell sites, especially in urban areas, and to integrate the two companies. Given that Fitch had awarded Dish Network a BB– debt rating in 2012, indicating little confidence in its existing prospects, taking on further debt to finance the takeover was not seen by many as a positive move, even though it already had $9.5 billion in cash and marketable securities at its disposal.

Sprint Nextel now raised its offer to $5 per share, valuing the whole of Clearwire at $14 billion, nearly three times its value in October 2012 before the bidding war began. In response, 9 per cent of Clearwire voting shares were irrevocably pledged to Sprint Nextel, meaning that, as board members had switched their support from Dish Network to Sprint Nextel, the latter now had pledges for half the independent voting shares. Sprint Nextel had emerged victorious, but it will be some time before we know whether the cost can be justified.

9.6 INTERNATIONAL ACTIVITIES

As discussed elsewhere, operators in the USA have occasionally attracted the attention of overseas investors, predominantly in recent times. However, the initial role of this section is to consider the actions of domestic operators in regard to investment in mobile assets overseas.

As noted in Curwen and Whalley (2008), the Baby Bells did not record their international operations in detail, so information is patchy about their overseas investments and the revenues and subscribers that they yielded. What seems clear is that the Baby Bells were in general expansionist-minded during the period before the turn of the millennium, but subsequently spent the period until the late 2000s divesting themselves of most of their overseas holdings. This was primarily the result of regulatory conditions imposed in the course of M&A activity, combined with a desire to concentrate resources on building up domestic networks at a time of rapid technological change. It may also be noted that the overseas assets never provided more than a modest return on the sums invested.

AT&T

It would be too complicated to reiterate here all the events during the period of consolidation among the Baby Bells. We take up the story where 'new' AT&T is concerned in 2003. At this point there is an operator in existence known as AT&T Wireless and during the course of the year it sells off its stakes in Far EasTone in Taiwan and EuroTel Praha in the Czech Republic, such that at the end of the year it has 30 million gross subscribers and 24.8 million proportionate subscribers. The difference is accounted for by the following holdings: Bermuda (60 per cent), Canada (34.3 per cent), India (32.9 per cent), Puerto Rico (100 per cent) and Slovakia (24.5 per cent). However, at some point during 2003 or 2004 it also held licences and/or stakes in Anguilla, Antigua and Barbuda, Barbados, Cayman Islands, Curaçao, Dominica, French West Indies, Grenada, Jamaica, St Kitts and Nevis, St Lucia, St Vincent and The Grenadines, and the US Virgin Islands. These yield very few proportionate subscribers and all bar the US Virgin Islands are sold off to Digicel, a privately owned operator, in June 2005.

During 2004, as noted previously, the ownership of AT&T Wireless is transferred to Cingular Wireless. The minority owner, BellSouth, had, like AT&T Wireless, built up a considerable portfolio of overseas assets, almost all of which are located in Latin America – regarded as the USA's 'backyard' at the time of their acquisition. During 2004, stakes in the following countries were sold: Colombia, Ecuador, Guatemala, Nicaragua, Panama, Peru, Uruguay and Venezuela. This left BellSouth with stakes in Argentina and Chile, which together yielded 3.2 million proportionate subscribers as well as Puerto Rico (0.2 million) and, rather oddly, Israel (0.9 million). The stakes in Argentina and Chile were sold in January 2005 and the stake in Israel in September 2005. Hence, by the end of 2005 only the stake in Puerto Rico remained.

The acquisition by Cingular Wireless of the residual stakes held by AT&T Wireless meant that they were transferred to SBC and BellSouth, and when SBC took over BellSouth it became their sole owner in the guise of 'new' AT&T. At this point, therefore, all that remained of a considerable overseas empire were stakes in Puerto Rico and the US Virgin Islands – with one exception.

The only holding that can truly be called foreign is a stake in América Móvil, the Mexico-based wireless conglomerate controlled by the family of Carlos Slim. This stake was originally acquired by Southwestern Bell in 1990 as part of the privatisation of Telmex, and although the share structure of América Móvil is complicated, the proportion of the overall equity held by AT&T in recent years has hovered around the 9 per cent mark – it was re-adjusted back to almost exactly that figure in December 2013 after a share buy-back operation by América Móvil.

After adjusting for the size of its stake, AT&T gained 17.8 million proportionate subscribers in 2009, 19.8 million in 2010, 21.5 million in 2011 and 24.5 million in 2012. In other words, América Móvil currently provides roughly one-quarter as many customers as AT&T's domestic operations. América Móvil operates in 16 countries in Latin America and is increasingly involved in Europe (widely defined), where it has a total of 14 holdings, so it is possible to argue that AT&T is highly internationalised. However, it has no operational control whatsoever at América Móvil, despite being the second-largest shareholder, so in effect its stake is an investment.

This raises the issue as to whether AT&T now intends to branch out into new fields on its own behalf. In 2007, AT&T made an unsuccessful bid for a licence in Qatar. According to a report in the *Financial Times* in 2008, AT&T was keen to expand in emerging markets such as India and the Middle East, and identified Telekom Malaysia International (now Axiata) as a specific potential target. However, in June 2010, AT&T denied a rumour that it was interested in a stake in India's Reliance Communications and, given the current problems in India, it is highly unlikely that it remains interested in pursuing an acquisition.

AT&T also expressed an interest in bidding for a 3G licence in Pakistan during 2012, but it looks very much as though its interest has switched during 2013 from the Indian sub-continent to Europe. This is, perhaps, surprising, given that it already has indirect access via América Móvil – strictly doubly indirect, given that América Móvil itself is present indirectly via KPN and Telekom Austria. Nevertheless, in March 2012, AT&T allegedly expressed an interest in a licence in Romania and in January 2013 it claimed that it was looking to do deals in Germany, the Netherlands and the UK.

It accordingly came as something of a surprise when, in June 2013, AT&T allegedly made a friendly takeover bid for Telefónica worth $95.3 billion, plus $69.5 of debt taken over – the attractions of such a move to AT&T are set out in Caplinger (2013). However, it is more likely that AT&T approached the Spanish government with a set of possible options, ranging from taking a 29.9 per cent stake – thereby avoiding the need to bid for the entire company – to acquiring Telefónica's UK network or its Latin American operations. This move appears to have been triggered by a combination of Verizon Communications agreeing to buy out Vodafone's stake in Verizon Wireless and relatively cheap share prices for telcos in Europe resulting from recessionary conditions. The government claimed (predictably) that it had not encouraged, blocked or influenced any kind of bid. As for Telefónica itself, it consistently denied that it had discussed a potential takeover by AT&T, but, irrespective of whether that was the case, AT&T was frankly unduly optimistic in expecting a government in Southern Europe to welcome a major foreign shareholder, even though, given Telefónica's indirect controlling stake (via Telco) in Telecom Italia, the Spanish government could be said to have behaved in a somewhat hypocritical manner if it did, indeed, discourage AT&T's approach.

An interesting aspect of the alleged offer was the sum involved, which, at $165 billion, would have been somewhat difficult to finance and would have saddled AT&T in turn with huge debts on its balance sheet. However, if AT&T realistically believes that such a sum is within its compass, then a massive range of potential acquisitions becomes feasible. In September 2013, it was said to be interested in taking a stake in Telecom Italia, but as that company is indirectly controlled by Telefónica, it is unclear what AT&T hoped to achieve other than, perhaps, to annoy Telefónica and the Spanish government.

More realistically, AT&T could make a move on EE, the UK network jointly operated by T-Mobile and Orange, or, alternatively, it could decide (bravely) to make a takeover bid for Vodafone now that it is formally shorn of its US holdings and hence is no longer subject to competition issues in the USA. If so, the asking price is likely to be in the region of $140 billion – it is difficult to be precise because of the size of payments

Vodafone is making to shareholders, the inevitable takeover premium and the general state of Vodafone's portfolio.

There are several good reasons for AT&T to make a move for Vodafone, in the process turning itself into the most internationalised operator in the world. In the first place, it is technically available in the sense that neither the UK government nor Vodafone's shareholders will raise objections – which was certainly not the case in Spain and the negative response to a bid for T-Mobile or Orange would be overwhelming. Second, AT&T evidently believes that it could raise the necessary finance. Third, the company is highly profitable and has consolidated tax losses waiting to be put to good use. Fourth, the outlook in Europe is likely to improve as LTE becomes established, and AT&T has already established its own network in the USA, so has experience to fall back on (Thomas, 2013).

However, the outlook is not uniformly positive. In the first place, it can be argued that AT&T is already large enough to achieve the necessary economies of scale in matters such as equipment purchase. Second, AT&T has no experience in operating in Vodafone's overseas markets, although it would have the option to divest some of these at a later date. Third, the tax losses would not be available to a company headquartered outside Europe. Fourth, as a pure mobile operator, AT&T would probably feel uncomfortable with Vodafone's recent forays into the fixed-wire market but would be unable to unravel any completed deals. It should also be noted that the loss of yet another iconic British company would not go unopposed, and that there is a good deal of regulatory uncertainty in Europe in general and in many of Vodafone's individual markets. Furthermore, the deal would have to be authorised in every individual market in which Vodafone operates. Buying Vodafone is a very far cry from taking a small non-controlling stake in América Móvil.

As of April 2014, the potential deal is clouded in uncertainty. In January, AT&T was forced to issue a statement denying any immediate intention to bid (Gabriel, 2014), but this did not rule out the possibility of a bid at a later date – AT&T has merely been prevented from acquiring more than a 30 per cent stake for six months.

Verizon Communications

Verizon Communications was the majority owner of Verizon Wireless. As such, it continued to operate a substantial international footprint until 2001, when it set in motion an abrupt exit. Beginning in 2002, all its overseas mobile holdings were sold, bar a 23.2 per cent stake in an Italian network majority-owned by Vodafone, the minority partner in

Verizon Wireless. The last to go were stakes in the Dominican Republic and (interestingly) Puerto Rico (sold to América Móvil in 2006), followed by Venezuela (sold to the state in 2007). In recent years the network in Italy has yielded roughly 6 million proportionate subscribers, but its continued presence as part of the Verizon mobile portfolio has remained something of an anachronism.

In recent years, Verizon Communications has made some rather desultory attempts to move back into the international arena. In 2007, it made an unsuccessful attempt to acquire a licence in Qatar and it made a further foray into the region when it became involved in the equally fruitless pursuit of a licence in Iraq. In 2013, it turned its attention to Canada, where it made a provisional offer for Wind Mobile in June and appears to have done the same in respect of Mobilicity after Telus had its own takeover bid rejected by the regulator (*The Globe and Mail*, 2013). However, between them the two networks muster fewer than 1 million subscribers, so this so far unsuccessful incursion north of the border cannot realistically be said to be the precursor of a new overseas empire. Indeed, Verizon Communications specifically ruled out a move into Canada in September 2013 (Wood, 2013b). Furthermore, the cost of acquiring Vodafone's stake in Verizon Wireless has hugely reduced the fund pool that might be used to buy overseas assets.

Inward Investment

There is very little to add here. As noted above, Verizon Communications had sold off most of its overseas assets by the end of 2007, and since 2009 the 'big three' US-controlled operators have owned between them little more than modest networks in Puerto Rico and the US Virgin Islands. No attempt has been made by an overseas operator to take a stake of any size whatsoever in Verizon Wireless or AT&T. Sprint Nextel is another matter altogether, but this is best addressed within a wider context covering recent events, since the successful takeover bid by Japan's SoftBank has also involved other companies that have formed part of the previous narrative.

9.7 RECENT EVENTS

As is evident from the above discussion, the restructuring driven by the need to provide high-speed data transfers in the USA became unusually complicated during 2013 and had unique characteristics because of the involvement of satellite. It is accordingly sensible at this point to bring the various protagonists together to form an overview of events during the year.

As of mid-June 2013, the position was as follows.

- T-Mobile USA was merged with MetroPCS to form T-Mobile US.
- Dish Network had offered to buy Clearwire.
- Sprint Nextel had offered to buy Clearwire minorities.
- Dish Network had offered to buy a majority of Sprint Nextel.
- SoftBank had offered to buy a majority of Sprint Nextel.
- Dish Network had made a ‘stalking-horse’ offer to buy Light-Squared spectrum.

The total amount being committed came to comfortably in excess of $50 billion, but whereas that might indicate huge synergies there for the taking via any number of possible combinations between the protagonists, it might also be seen as a sign that there is considerable uncertainty as to what might be termed a sustainable longer-term equilibrium structure for spectrum holdings in the USA.

Towards the end of June the first sign of matters moving towards some kind of resolution came in the form of public recognition by Dish Network that it was no longer an active bidder for Sprint Nextel. Sprint Nextel shareholders overwhelmingly voted in favour of the SoftBank offer and the FCC gave its approval in early July. The takeover was formally completed on 10 July and Sprint Corp. was merged into Sprint Nextel. This created ‘new’ Sprint, which became the parent company of Sprint Nextel, with Sprint Nextel its wholly owned subsidiary – in the process changing its name to Sprint Communications.

Dish Network also withdrew its offer for Clearwire (Telecom.paper, 2013a), in effect admitting that the SoftBank/Sprint Nextel/Clearwire link-up was a done deal. The FCC duly approved the Sprint Nextel/Clearwire deal, which was duly completed on 8 July.

It was felt in some quarters that Dish Network would now turn its attention to some kind of link-up with one or more of T-Mobile US, Leap Wireless, nTelos Wireless and US Cellular. However, Leap Wireless was promptly removed from the equation when AT&T made a takeover bid in July worth $1.2 billion in cash, plus taking on $2.8 billion of debt (Telecom.paper, 2013b). The deal included spectrum in the PCS and AWS bands covering 137 million people, which would complement AT&T’s existing holdings. The assets were transferred in March 2014.

While all of this suggests that Dish Network may remain independent, it cannot be entirely disregarded as a player in the US mobile market.

This is partly because Dish Network was set to gain from draft rules published at the end of June by the FCC in respect of the 1900 MHz H Block (1915–1920 MHz paired with 1995–2000 MHz) in 176 Economic Areas (TeleGeography, 2013). This was to be auctioned (via Auction 96) in mid-January 2014 (but postponed), and there would be no interference issues with Dish Network's S-band spectrum. Hence Dish Network was almost certain to emerge victorious, especially as it had agreed to pay the reserve price of $1.56 billion as part of a complex deal with the FCC that would allow it to convert part of its 2 GHz AWS-4 spectrum from uplink to downlink. In early March, Dish Network was duly declared to be the winner at the reserve price (Wood, 2014).

However, the real surprise emerged at the year-end, when SoftBank announced that it was in advanced negotiations with T-Mobile to take a stake of up to 70 per cent in T-Mobile US – if 70 per cent, then the cost would be roughly $19 billion at the ruling share price (Lawson, 2013), although T-Mobile was talking in terms of nearly $30 billion in early 2014. The formal takeover bid would be forthcoming, if at all, in 2014H1 and would in effect make SoftBank the clear majority owner of both Sprint and T-Mobile US because it intended to use Sprint as the vehicle for financing the takeover via the US bond market (MarketWatch, 2013). At the end of 2013, T-Mobile had 46.7 million subscribers, up by 10.4 per cent compared with 2012.

9.8 CONCLUSIONS

The battleground for mobile operators in the USA is that for high-speed data, which in essence currently means LTE and LTE-Advanced. Achieving an internal structure that facilitates this has preoccupied all US operators, both large and small, and has induced them in recent years to concentrate almost exclusively on investment in their domestic networks. Because the introduction of new technology has been so successful – the USA has amassed far more LTE subscribers to date than any other country – it has very recently drawn in an apparently well-funded foreign predator.

The only fairly unequivocal statement that can be made is that Verizon Wireless made most of the running in relation to LTE during 2013 – in late June, it announced that its LTE network was substantially complete after reaching 500 markets and that its thoughts were now turning to re-farming both its PCS spectrum and AWS spectrum. However, it can be argued that Verizon Wireless has been far less preoccupied with restructuring than its main rivals because it has remained under the direct control of its majority (now only) owner, Verizon Communications, and hence has not been directly affected by its parent's own bid for the stake held by Vodafone.

AT&T has largely kept up with Verizon Wireless in terms of overall subscribers, although it is playing catch-up for the time being in relation to LTE. The core issues accordingly relate to the other two nationwide incumbents, Sprint and T-Mobile US. For its part, Sprint is showing signs of a fight-back now it is owned by SoftBank. In October 2013, it announced an LTE service branded as Sprint Spark, which would be capable of a 50–60 Mbps peak downlink with immediate effect, combining Carrier Aggregation for speed, 8T8R for coverage, MIMO for capacity and TD-LTE for spectral efficiency. The intention was that TD-LTE in the 2.5 GHz band would be combined with paired spectrum in the 800 MHz and 1900 MHz bands, with tri-band devices available from November.

However, SoftBank's takeover bid for T-Mobile US may derail this progress because, if successful, it would force Sprint to face issues that it failed to address successfully on a previous occasion when it sought to merge its activities with those of Nextel. Once again there is an issue of incompatible technologies – Sprint uses CDMA while T-Mobile US uses GSM – although it may be argued that most 2G customers of both networks will eventually use LTE. Once again there is a need to bring together two sets of management with different cultures. While it is true that SoftBank will be overseeing the process, SoftBank has no prior history of tackling a merger process within an overseas mobile market and there is the not so minor matter of simultaneously sorting out an ongoing role for Clearwire.

Of course, it may never be necessary to address these issues. The regulatory implications of a SoftBank takeover bid for T-Mobile US are unclear. What it would do, in effect, is create a third national operator of a size that would much more closely match that of Verizon Wireless and AT&T, and hence, in principle, set the scene for tougher competition between three more or less equals. But there would be very little left of the regional structure that had previously provided competition in most markets, given that both Leap Wireless and MetroPCS have already been absorbed by the 'big four',[1] and the FCC has made clear on previous occasions – in particular, when it rejected the proposed merger between AT&T and T-Mobile USA – that it wanted to preserve at least four meaningful competitors in every market. By and large, three national operators does seem too few for a market the size of the USA, so it would come as something of a surprise if either the FCC or DoJ were to take a positive view of this development. So far, both organisations have done little other than to express considerable misgivings about the matter (Reuters, 2014).

Given that Verizon Communications has at least temporarily used up its financial firepower in completing the purchase of Vodafone's stake in Verizon Wireless and that both Sprint and T-Mobile US are involved in convoluted tie-ups of unknown outcome, this effectively leaves only AT&T in a position to contemplate a major move back into the international arena.

NOTE

1. It is of interest that, in April 2014, Verizon Wireless provisionally acquired the (admittedly modest) spectrum held by Cincinnati Bell. This appears to be the latest part of a process whereby the 'big three' are mopping up small regional operators.

REFERENCES

Caplinger, D. (2013), 'Will Telefonica be AT&T's answer to Verizon?', www.fool.com, 6 November.

Cellular-news (2012a), 'Blow to LightSquared as government agency says GPS interference cannot be solved', www.cellular-news.com, 16 January.

Cellular-news (2012b), 'Sprint Nextel scraps $9 billion network sharing deal with LightSquared', www.cellular-news.com, 20 March.

Cellular-news (2013), 'Leap shareholders approve $1.2 billion takeover bid from AT&T', www.cellular-news.com, 1 November.

Curwen, P. (2005), 'Consolidation in the USA: does bigger mean better?', *info*, **7**(5), 61–3.

Curwen, P. (2007), 'Anatomy of a merger: Sprint Nextel', *info*, **9**(4), 86–8.

Curwen, P. and J. Whalley (2008), *The Internationalisation of Mobile Telecommunications: Strategic Challenges in a Global Market*, Cheltenham, UK and Northampton, MA, USA: Edward Elgar.

FCC (2006), 'FCC approves merger of AT&T Inc. And BellSouth Corporation', *FCC News*, 29 December.

Fitch Ratings (2011), 'The road to 4G', www.fitchratings.com, 3 August.

Gabriel, C. (2012), 'Dish can play all sides in US 4G fall-out', www.rethink-wireless.com, 17 January.

Gabriel, C. (2013), 'SoftBank lures Sprint investors with more cash', www.rethink-wireless.com, 11 June.

Gabriel, C. (2014), 'AT&T will not bid for Vodafone any time soon', www.rethink-wireless.com, 27 January.

The Globe and Mail (2013), 'Verizon's interest in Canadian market deeper than admitted', www.theglobeandmail.com, 20 September.

Gryta, T. (2012), 'Spectrum needs at heart of any T-Mobile/MetroPCS deal', www.totaltele.com, 3 October.

Hillen, B. (2013), 'AT&T purchase of Leap Wireless brings Cricket under new ownership', www.slashgear.com, 12 July.
Lawson, S. (2013), 'SoftBank and DT near a deal on the Sprint/T-Mobile merger', www.totaltele.com, 25 December.
Lennighan, M. (2014), 'T-Mobile US to pay $2.4 billion for Verizon spectrum', www.totaltele.com, 6 January.
MarketWatch (2013), 'Softbank may buy U.S. bonds to buy T-Mobile', www.marketwatch.com, 25 December.
Moritz, S. and C. Rahn (2011), 'AT&T pulls its $39 billion bid for T-Mobile', www.bloomberg.com, 20 December.
Osawa, J. (2012), 'SoftBank deal for Sprint turns on spectrum', http://online.wsj.com, 21 October.
Ramachandran, S. (2012), 'Dish frustrated in dream of breaking into wireless', www.totaltele.com, 20 November.
Reuters (2014), 'Update 2 – FCC chief tells Sprint chair he is sceptical on T-Mobile deal', www.reuters.com, 4 February.
Telecom.paper (2012), 'Sprint backs out from MetroPCS at last hour', www.telecompaper.com, 7 March.
Telecom.paper (2013a), 'Dish withdraws Clearwire tender offer', www.telecompaper.com, 27 June.
Telecom.paper (2013b), 'AT&T to acquire Leap Wireless', www.telecompaper.com, 14 July.
TeleGeography (2013), 'FCC sets wheels in motion for 1900MHz H block auction', www.telegeography.com, 28 June.
Thomas, D. (2013), 'Should AT&T buy Vodafone?', www.ft.com, 3 December.
Wood, N. (2013a), 'T-Mobile, MetroPCS merger closes', www.totaltele.com, 19 March.
Wood, N. (2013b), 'Verizon rules out Canada move', www.totaltele.com, 3 September.
Wood, N. (2014), 'Dish spends $1.56bn in US spectrum auction', www.totaltele.com, 3 March.

10. Internationalisation as at end 2013

10.1 INTRODUCTION

Previous chapters have described internationalisation among mobile telecommunications service providers on a regional basis. The allocation of countries to particular regions was organised so as to bring out the strong relationship between specific operators and specific groups of countries. However, it is now time to move on the analysis in order to investigate internationalisation on a worldwide basis. For this purpose it is necessary to refer to the regional breakdown first encountered in Table 5.1. In particular, this splits up the previously wide definition of Europe used in Chapter 7, with the countries covered there allocated to a number of regions. Dividing Europe into West and East may no longer accord with the concept of a 30-country European Union, but it does recognise the fact that when operators first moved into Eastern Europe they did so in the knowledge that governance there was significantly different from that in Western Europe, and that the economies of scale and scope might not be as great as they would be if restricting themselves to one cohesive part of Europe.

It is also the case that the Middle East can now be considered as a region in its own right. It was not useful to do so in previous chapters because the focus was on how Middle Eastern operators were looking to expand outside their domestic region, whereas operators based outside the region showed little or no interest in entering it.

The discussion below is structured as follows. The first section outlines the methodology adopted and is followed by a section corresponding to each of the dimensions of internationalisation that are deemed to provide some insight in the context of the mobile industry. These are respectively the number of countries in which an operator has a stake in a network, the number of proportionate subscribers it controls in each country and the location of these subscribers from a regional perspective. These data sources are then brought together in a section that identifies which operators can be said to have been the most internationalised at the end of 2012. Conclusions are then drawn as appropriate.

10.2 METHODOLOGY

In an ideal world, it would be possible to access a wide range of data on a comparable basis across a significant sample of operators with international reach. But when it comes to data on mobile operators, the world is far from ideal. For example, while it is possible to extract financial information on some international operators in a manner that allows for a direct association between individual networks and their performance, this is mostly not the case either because networks are lumped together in regions or because they are simply not referred to at all, particularly if they are not direct subsidiaries.

As financial comparisons cannot usefully be made, three[1] alternative measures of internationalisation are used in this chapter. These are:

- The number of countries in which the mobile operator has invested.
- The number of proportionate subscribers.
- The dispersion of these subscribers by region.

To be included in the analysis, the internationalising mobile operator must be present in at least five countries including the domestic market and have a minimum of 20 million proportionate subscribers. Although both requirements are individually fairly modest, the combination produces a sample set of between 20 and 30 operators that can be compared on an ongoing basis over a period of many years, which is extremely useful for analytical purposes. It should also be borne in mind that the subscriber requirement is more onerous than it seems since it does not refer to gross subscribers but the gross number in each country adjusted to take account of the equity held in the relevant network by an operator. Operators themselves mostly prefer to announce gross figures as these tend to be much larger and hence make the operator appear to be more important than it is, or alternatively they may cite 'consolidated' subscribers where consolidation may be taking place while holding an equity stake of, say, just over 50 per cent.

A final point to note at this stage is that a decision has had to be made whether to use a parent company or merely its mobile subsidiary as the source of data. Historically, this was an awkward issue at times because, when telcos privatised all or part of their mobile subsidiaries, they occasionally held on to some mobile assets or, alternatively, they subsequently bought mobile assets via the parent company rather than the subsidiary. In practice, this problem now relates only to NTT and its subsidiary DoCoMo – see note 15 to Table 10.1.

10.3 NUMBER OF COUNTRIES

There is no widely accepted method for counting countries in which major operators have a presence. Indeed, only the authors are known even to have attempted to address this issue in the public domain, and their methodology is therefore definitive only by default. The methodology is as follows.

To begin with, the term 'countries' is used here to encompass any areas such as islands or territories that have independent networks and/or their own licensing regulators. While it is possible to identify a wide range of entry models that are available to an internationalising company, the crucial criterion in this chapter is that, to be included as a country presence, the investment must involve an operational mobile network. As a consequence, potential entry arrangements such as international roaming or brand franchising are excluded as they do not provide a sufficient degree of investment.

Where the mobile operator has multiple investments in the same country, only a single presence is recorded, even though the proportionate subscribers for each network are aggregated. Subject to this proviso, each equity stake, regardless of its size, is recorded as a single presence in a country. That said, the size of the stake is adjusted for when calculating the relevant number of subscribers to be allocated. Furthermore, a presence is recorded only when a network has been rolled out and is attracting subscribers – the mere acquisition of a licence is not counted (although it is noted). In contrast, the launching of an MVNO is counted because it generates subscribers and hence income, although again this is noted wherever relevant.

The use of holding companies greatly complicates the process of determining the precise number of countries in which an operator is present. If an operator has invested in a holding company, should this be counted as a single investment or should each of the holding company's subsidiary networks be counted? The approach adopted here is to count all the subsidiary networks – in other words, both directly held and indirectly held networks. Furthermore, given that, broadly speaking, it is not possible to determine the precise degree of control that each operator exercises over its networks, it is assumed that every network should be counted unless either the mobile operator has expressly stated that the holding is purely for investment purposes or the stake is below 10 per cent, generates a relatively small number of subscribers and creates incompatibilities if it is counted. Although it is possible to argue that a minority subsidiary of a minority-owned operator should not be counted as a country presence due to the size and indirect nature of the stake, it may reasonably be counter-argued that the strategy being pursued is precisely to gain a presence in a spread of countries by this method rather than to attempt to buy direct stakes in tightly held companies at exorbitant prices. Furthermore, a large number of subscribers may be generated, as was the case, for example, with Vodafone's 3 per cent stake in China Mobile (now sold on).

However, although the above works satisfactorily in the great majority of cases, there are inevitably certain exceptions. Hence, although 28 operators are listed in the main section of Table 10.1, AT&T is excluded because, despite a 9 per cent stake in América Móvil, which in principle provides it with wide coverage in Latin America, this is best regarded as a financial investment. As with the other main US-based operators, AT&T – somewhat confusingly not, in practice, the same company as a decade earlier – retreated from its overseas ventures after 2000, a strategic decision that the table reflects via the superficially surprising absence of US-based operators. The case of Vodafone is more awkward since it held a few indirect stakes via Bharti Airtel until 2009, which provided a very small proportion of its total subscribers. This stake was largely sold off once Vodafone became the majority shareholder in Vodafone Essar, but when Bharti Airtel acquired most of Zain Africa in 2010, the number of indirect stakes held by Vodafone suddenly rose sharply even though each provided on average only an insignificant number of subscribers. As a result, these stakes have been excluded as financial investments together with the other indirect stakes previously credited via Bharti (see note 10 to Table 10.1) – and in any case Vodafone is disposing of its residual stake in Bharti Airtel. In contrast, the additional Bharti Airtel stakes are credited to Singapore Telecom because it has a large minority stake in Bharti Airtel and helped to finance its expansion.

The stakes held by the likes of Russia's Altimo or Singapore's Temasek Holdings can more readily be excluded as financial investments on the grounds that they do not actually operate any networks, although Altimo is very active as a shareholder, as its quarrels with Telenor and TeliaSonera in Russia bear witness, so there is an element of ambiguity in that case. In contrast, Hutchison Whampoa is included because its subsidiaries operate its networks – confusingly, there is no single holding company for telecommunications assets (see Chapter 4) – even though it has (more important) shareholdings in a number of other industrial sectors.

This methodology is somewhat complicated and requires at times the addition of a great many notes to tables, as illustrated in Table 10.1. Although these are frequently omitted in other publications, particularly those originating from consultancies, such an approach gives the impression that the methodology being used is not subject to multiple qualifications, which cannot be the case in practice. The number of networks assigned to individual operators by other parties that publish data often differs from what appears in this chapter due to their use of an undisclosed alternative method of counting, but experience suggests that differences are just as likely to be accounted for by basic errors in research practices. For example, there is a (perhaps natural) tendency for researchers to rely heavily on the annual reports of operators. Unfortunately, this is not to be recommended, as not only do many operators neglect to list their indirectly held assets, but they sometimes even fail to acknowledge those that are directly held but modest in size – for example, France Télécom/Orange never appears to list the 42 countries indicated in Table 10.1 in a single public domain source.

Table 10.1 also raises issues relating to the identification of networks. Many discrepancies exist between the original names recorded as network licensees and those currently holding those licences. This complicates the data collection process, and reflects the extensive M&A activity that has occurred throughout the mobile telecommunications industry as well as operators changing their names for a variety of reasons. Operators themselves can be the source of confusion by applying for licences using the names of existing or newly created subsidiaries or via joint ventures/consortia.

It is by no means uncommon to find multiple names being used simultaneously in the public domain in relation to a single network. These names may refer, for example, to the majority owner of the network, to a minority owner whose name is nevertheless used because it is better known than that of the other owners, to the historic name of the original network, to the brand name of the main service(s) or to a marketing name that it is hoped will enhance brand recognition, which may involve another operator without an equity stake. Furthermore, even if a name may appear to have been used consistently, this may disguise a somewhat different reality – for example, the current AT&T Wireless is a quite different organisation from that which existed under the same name in 2009 – see Chapter 9.

There are 28 operators in the main section of Table 10.1. AT&T is also present but does not have any changes listed because it has not been active as either buyer or seller over a considerable period, choosing to concentrate on developing LTE in its home market. However, it may be about to become acquisitive in its own right – for example, it has been associated, if only by rumour, with an interest in buying Vodafone in its current form. Digicel is included simply because it controls so many networks and is discussed in note 24.

Table 10.1 Presence in countries, end 2012 and annual changes [2012/2011, 2011/2010, 2010/2009, 2009/2008, 2008/2007]

Company	Number of countries	Number of countries by region						
		Western Europe[1]	Eastern Europe[2]	Middle East[3]	Asia Pacific[4]	Latin America[5]	North America[6]	Africa[7]
France Télécom	42 [–1, +2, +2, +1, –1]	10	2	2	1	6	1	20
América Móvil	32 [+8, +6, 0, +1, 0, 0]	10	4	–	–	16	2	–
Telefónica[8]	31 [0, 0, –2, 0, +1]	8	–	–	4	15	–	4
Telenor	28 [–2, +7, 0, +3, 0]	5	11	–	7	–	1	4
Singapore Telecom[9]	28 [+1, 0, +16, +1, +1]	2	–	–	9	–	–	17
Vodafone[10]	28 [0, –4, –3, +1, +2]	12	1	2	4	–	1[25]	8
Bharti Airtel[11]	22 [+1, 0, +17, +1, +1]	2	–	–	3	–	–	17
MTN[12]	22 [+1, 0, 0, 0, 0]	1	–	3	1	–	–	17
TeliaSonera[13]	21 [–1, +1, –1, 0, +3]	9	9	1	2	–	–	–
Etisalat	20 [–1, +1, +1, +2, +1]	–	–	4	4	–	–	12
VimpelCom[14,16]	18 [–2, +9, +1, +2, 0]	1	8	–	4	–	1	4
Deutsche Telekom	17 [–2, 0, 0, 0, +5]	11	5	–	–	–	1	–
Millicom	13 [0, –1, 0, –3, 0]	–	–	–	–	6	–	7
Qatar Telecom	13 [0, 0, 0, +1, 0]	–	–	6	5	–	–	2
Hutchison Whampoa	13 [+1, –1, 0, –1, –1]	6	–	1	6	–	–	–
NTT DoCoMo[15]	10 [0, 0, –1, +1, +1]	–	–	–	9	–	1	–
Tele2	10 [0, +1, +1, –1, –4]	7	3	–	–	–	–	–
Orascom[16,14]	9 [–1, –2, –1, +2, –5]	–	–	–	3	–	1	5
Saudi Telecom	9 [–1, –1, 0, +2, +1]	–	1	4	3	–	–	1
Axiata[17]	8 [0, 0, 0, 0, –1]	–	–	1	7	–	–	–
Portugal Telecom	8 [0, +1, –2, 0, 0]	1	–	–	2	1	–	4
Telekom Austria	8 [0, 0, 0, 0, 0]	4	4	–	–	–	–	–

Turkcell[18]	8 [0, +1, 0, 0, +1]	1	6	1	–	–	–	–
Vivendi Universal[19]	8 [0, 0, 0, +1, 0, +2]	1	–	–	–	–	–	7
Zain[11]	8 [0, +1, –14, 0, +2]	–	–	5	–	–	–	3
Maroc Télécom[20]	5 [0, 0, 0, +1, 0]	–	–	–	–	–	–	5
Mobile TeleSystems[21]	5 [–1, 0, 0, 0, 0]	–	5	–	–	–	–	–
Vodacom[22]	5 [0, 0, 0, 0, 0]	–	–	–	–	–	–	5
Total	449	91	59	30	74	44	9	142
AT&T[23]	33	10	4	–	–	16	3	–
Digicel[24]	32	–	–	–	6	26	–	–

Notes:

1. Includes the 27 EU current member states plus residual Western Europe.
2. Includes the Balkans, Belarus, Moldova, Russia and Ukraine.
3. Includes the Arabian Peninsula and Turkey.
4. Includes Australia and New Zealand.
5. Central and South America plus most of the Caribbean.
6. Mainly constitutes Canada, Puerto Rico and the USA.
7. Defined geographically to include Egypt.
8. Telefónica took over the networks of O_2 during 2006.
9. Because SingTel is a significant minority shareholder in Bharti Airtel, with 32 per cent, the stakes acquired by Bharti in 2010 are credited in this case as indirect holdings.
10. There are issues with this total because in early 2010 there were technically a further 15 stakes held indirectly via Bharti Airtel, even though the stake in Bharti itself had been reduced to a small residual of less than 5 per cent, which could more appropriately be treated as a financial investment without control (as with AT&T's stake in América Móvil), especially given the stake in Vodafone Essar, also in India. The inclusion of Bharti's stakes had not previously created a distortion of any consequence, but now appeared to do so because of Bharti's acquisitions in 2010. At the same time, the removal of Bharti's stakes would include the other four Bharti stakes previously credited to Vodafone. If Bharti is included, the total becomes 49 (because Vodafone is already in India). Vodafone also owns a majority stake in Vodacom, which is counted. It should be noted that Network Partnership Agreements are not included within the Vodafone total as no shareholding is involved.

11. Zain was formerly The Mobile Telecommunications Co. In May 2005, it acquired 85 per cent of Celtel, a pan-African operator with 14 networks. In 2010, it sold 15 networks in Africa (essentially Celtel) to Bharti Airtel. It also has a management contract in Lebanon.
12. MTN, which is based in South Africa, took over the networks of Investcom during 2006.
13. TeliaSonera's home market is defined as Finland plus Sweden.
14. In 2011, VimpelCom acquired a majority stake in Orascom Telecom Holding as well as Wind in Italy. It sold two of its inherited stakes in Africa in 2013.
15. The use of DoCoMo is so pervasive that it is included here although it is technically a majority-owned subsidiary of NTT. NTT is unusual in that it has two stakes in mobile networks that are separate from those held by DoCoMo but only one is in a different network. The total number of countries for NTT is accordingly 11.
16. Orascom Telecom Holding (OTH) was based in Egypt. In 2010 it was 50 per cent owned by Weather Investments, which also owned two stakes in Western Europe, the first of which, in Italy, it bought in 2006 and the second of which, in Greece, it acquired in 2007 but sold in 2010. In 2011, it became a majority-owned subsidiary of VimpelCom. Its previous owner, now renamed Wind Telecom, acquired a minority stake in VimpelCom as part of the merger (treated as a financial investment) as well as retaining its existing stakes in Egypt and North Korea. In September 2013, OTH was renamed Global Telecom Holding (GTH).
17. Before 1 April 2009, the company was known as Telekom Malaysia International, which had been split off from Telekom Malaysia.
18. North Cyprus is included within Turkey.
19. There is a problem with Poland. After a long-running dispute, Deutsche Telekom eventually consolidated the whole of PTC despite Vivendi's continuing claim to be part-owner, and so the stake is not included here. However, DT made a large payment to Vivendi in 2011 in respect of PTC, indicating that Vivendi's claim was at least partly legitimate, so the number of stakes in Europe could be taken to be two and the total could be taken to be nine.
20. Vivendi Universal owns 51 per cent of Maroc Télécom. Its only other networks comprise SFR in France and its small networks in the French overseas islands of La Réunion and Mayotte.
21. MTS had its licence revoked in Uzbekistan in 2012.
22. A majority stake is owned by Vodafone.
23. AT&T operates directly only in the USA, US Virgin Isles and Puerto Rico, but also has a stable stake below 10 per cent in América Móvil, which is treated here as a financial investment.
24. Digicel claimed to have 32 networks in 2010, but uses 'French West Indies' as a generic term with coverage on three islands. It is difficult to trace precise numbers for 2012, so Digicel's 2010 figure is used. It is included here for illustrative purposes, although it should be borne in mind that (a) it is privately held and does not release full data and (b) the average number of subscribers per country is very low (due to its numerous Caribbean networks) at roughly 350 000, which renders it incompatible with the operators in the main table.
25. Sold in March 2014.

As for the 28 operators, these predictably show a very diverse range of activity, especially bearing in mind that all changes are net – that is, they net out purchases and sales. The largest movements warrant a brief mention, as follows:

- América Móvil acquired a net six additions in 2011 and eight additions in 2012.
- Bharti Airtel acquired 17 additions in 2010.
- Singapore Telecom acquired 16 additions in 2010 via Bharti Airtel.
- Telenor acquired seven additions in 2011 via VimpelCom.
- VimpelCom acquired nine additions in 2011.
- Zain disposed of 16 holdings and gained two in 2010.

As can be seen, Singapore Telecom and Telenor made significant acquisitions indirectly via their stakes in other operators. The two direct gains were made by Bharti Airtel, which bought Celtel from Zain, and by VimpelCom, which bought most of Orascom Telecom Holdings (now Global Telecom Holdings). As for América Móvil, it took a small stake in KPN in 2011 (much enlarged in 2012) and a large minority stake in Telekom Austria in 2012.

Partly due to the above, Table 10.1 shows some signs of clustering. The top ten operators account for 274 of the 449 entries, equivalent to 61 per cent, whereas the second ten account for 120 entries, equivalent to 27 per cent, and the remaining eight account for 55 entries, equivalent to 13 per cent. It could be argued that a line could also be drawn beneath the twelfth entry since these 12 would account for 309 entries – 69 per cent of the total – while representing only 43 per cent of the operator sample.

It is not appropriate to read too much into this clustering at this stage, because Table 10.1 does not provide information about whether the networks are large or small, directly held or indirectly held and so forth. However, the existence of multi-country and, in particular, multi-regional operators is obviously very important when considering the overall structure of the mobile sector and when assessing how that is likely to develop in the future. This is because they are likely to have a major say in resolving inappropriate market structures such as the overprovision of networks, which is a matter addressed at various points in this text. It is reasonable to argue that a structure that was appropriate for a given time period and a given degree of competition may change considerably over time, and that operators should not be held to account for what turns out years later to be an inappropriate structure. But that argument has a potentially fatal flaw, which is the assumption that the underlying structure made sense both operationally and financially in the first place.

Although operators can be expected to claim that they acquire overseas assets in accordance with some master plan, the reality is that assets cannot readily be bought unless they are available at the right (albeit often somewhat inflated) price, and if that does not apply, the temptation may be to settle for whatever is available, even if second-best. While it is technically the case that an operator can make hostile takeover bids for assets that have not been flagged up for sale, this is not all that common in practice. Given the potential for strong objections by either shareholders or regulators or both, most takeovers are quietly negotiated on an unofficial basis over quite lengthy periods and in reality it is only a distinct minority that result in an outcome acceptable to all of the parties with some kind of stake in the outcome.

Furthermore, a coherent strategy for expansion overseas strictly requires a sequence of purchases in the right place and at the right price, yet operators have no means of predicting what will become available and when. Hence such a sequence does not tend to happen and, in any event, factors such as changes at the top of companies invariably lead to reassessments of previous strategies – anything that has gone wrong is, naturally, claimed to be the fault of a previous CEO – and a desire to rearrange the portfolio of assets. The financial health of an operator is also highly relevant in respect of its trading of overseas assets, so coherence may have to be sacrificed on the altar of financial expediency.

Predictably, five of the top ten in Table 10.1 are based in Europe[2] – they are among the longest-established and have colonial associations – while US-based operators are non-existent, having largely withdrawn in recent times to their home market (Curwen and Whalley, 2008: ch. 8). Of the total sample, 13 are based in Europe (including Russia), with the others spread out fairly evenly in Asia, the Middle East and Africa. However, a matter of particular interest is the appearance of Bharti Airtel, Etisalat and MTN among the top ten. In the mid-2000s it appeared to be the case that Middle-East-based operators would become major international players, but the rise of Etisalat has been mirrored by the decline of Zain, with the latter providing a cautionary tale for its peer group (Curwen and Whalley, 2011). Meanwhile, African operators have, on balance, retreated rather than expanded over the past several years, in the process failing notably to expand in their home continent in the face of competitive bidding for assets and licences from operators based elsewhere.

It may be noted that Asia-based operators tended historically to have limited empires – among other matters, large differences in culture, language and religion between adjoining countries in Asia militate against the creation of a tight grouping of assets. Furthermore, although

DoCoMo, which has relatively large Asian holdings, aspired to expand beyond Asia, it was notably unable to make a success of these, and all of them have now been sold. Whether Bharti Airtel's African acquisitions in 2010 – also held indirectly by Singapore Telecom – indicate that this may be about to change is a moot point, especially given that the three Chinese operators have as yet engaged in little overseas activity.

10.4 NUMBER OF PROPORTIONATE SUBSCRIBERS BY REGION

Although the point has been made previously in this text, it should be mentioned once again that the counting of subscribers is a complicated business. This is evident if only because, for example, penetration rates in most advanced economies currently exceed 100 per cent – in other words, every single inhabitant has, on average, more than one device. But those below seven years of age and those above 70 clearly do not all own handsets, so the average for the 7–69 age group must be comfortably greater than one (and even higher if the poorest decile of the population is also excluded). The most obvious reason for this is that many individuals carry one device for work purposes and one for social purposes (and, in certain countries, possibly a further one for very private matters). However, since both devices are likely to be generating non-trivial amounts of revenue for operators, counting both is perfectly sensible.

Overestimation of subscriber numbers is almost inevitable for a number of reasons. For example, there exist what are, in effect, phantom subscriptions. These can arise because an individual has switched operators, and while the new operator has opened up a new subscriber account, the old operator has yet to delete the closed account. In addition, the minimum period during which an account lies dormant before it is automatically deleted is normally three months, but it can be considerably longer.

Number portability is of relevance in this respect because, where it is not permitted, there is a tendency for subscribers to hold on to an old SIM for some time in case calls and texts are still being sent to it by those as yet unaware of a replacement number. A different sort of portability is also applicable in so far as individuals travelling abroad for business or pleasure may buy SIM at their destination in order to reduce international roaming charges.

A further issue arises in countries where pre-paid is the norm rather than the post-paid contract. In the case of pre-paid, what is being counted is generally the number of SIM cards, but there is nothing to prevent an individual from acquiring multiple SIMs and switching from one to the other within a single handset. Indeed, there is an increasing number of handsets on the market that can hold two or occasionally three SIMs, which greatly eases the task of switching networks between calls (Garner, 2007; Wikipedia, 2012; and see www.multisim.co.uk). This is not particularly common in advanced countries because price differences do not warrant the effort of monitoring the cost of individual calls and fiddling with the handset, but it is very common in emerging economies because of significant cost differences between on-net (between two subscribers to the same network) and off-net calls and according to the time of day. It may also be the case that reception is much better for different operators in different places because none has full national coverage.

It is worth noting that multiple SIM use is primarily restricted to towns and cities where several operators' signals can be received. Clearly, there is little to be gained in more rural areas in emerging countries where there is often only one service provider. A consequence of this is that, although penetration rates in many emerging countries exhibit a rapidly rising trend line, this may be misleading in relation to mobile penetration in rural areas, which may have remained relatively constant.

Finally, an increasing number of mobile numbers is being assigned not to people but to machines (M2M), so that individuals can access services such as vending machines or communicate directly with appliances at their homes.

However, the biggest problem of all is that in some countries there is a tendency for operators deliberately to inflate their subscriber numbers and to do so in different ways, making comparability an issue. Although this problem is rarely significant in the USA and Europe, it is periodically significant elsewhere. The problem in Africa has already been noted in Chapter 6. In Asia, MobiFone (Vietnam Mobile Telecommunication Service) issued 24 million SIMs in 2009 but only 10 million were activated, while Pakistan blocked several million unverified SIMs in May 2011. In the Middle East there are examples both in Bahrain, where 400 000 SIMs were de-activated in August 2010, and in Iraq in January 2011, when Zain was fined for dumping 5 million SIMs on the market.

A somewhat different, but related, issue has arisen recently in respect of demands by certain governments that all handsets be registered by a specified date. Ostensibly, this is due to a desire to prevent terrorists from using untraceable handsets, although there is an alternative view that it is to bolster oppression by the state. Irrespective of the motive, the effects can be dramatic: for example, in Zimbabwe in April 2011, 1.4 million

Econet subscribers, 700 000 Telecel (Orascom) subscribers and 300 000 NetOne subscribers were cut off as a result of failure to register their handsets, thereby reducing the total from roughly 8.4 million to 6 million – equivalent to 28 per cent of all activated SIMs at the time (Cellular-news, 2011b). This exercise has been conducted elsewhere in Africa, for example in Tunisia, and a recent study in India by the TRAI revealed a total of 549 million active customers in January 2011 compared to the previous estimate based on operator numbers of 771 million. The study used a point-in-time count of active subscribers at particular cell sites known as a visitor location register (VLR – see http://en.wikipedia.org/wiki/Network_switching_subsystem), the virtue of this method being that any one subscriber can only be present at one VLR irrespective of the number of SIMs possessed. Not surprisingly, the main lesson from the study was that the relatively new entrants such as Etisalat, Uninor and Videocon had VLRs below 50 per cent, and hence were in a worse financial condition than they made out, although the opposite was true for established operators (Cellular-news, 2011a). Interestingly, the shortage of phone numbers has begun to make a difference, with Vodafone cutting off anyone not making calls for two months in order to reallocate numbers as from December 2011.

It is impossible to say precisely what effect the above factors have on subscriber numbers. Ideally, what is needed is a calculation of the number of 'unique' subscribers where all double-counting is eliminated, but there is no way of achieving this with any certainty. It is fair to say that the overestimation of subscriber numbers is not severe in advanced countries, whereas it can be a major issue in countries where regulatory oversight is weak and multiple SIMs are the norm. However, the increasingly widespread obligation to pre-register all SIMs before use at least means that the problem is steadily diminishing over time.

Table 10.2 shows the number of proportionate subscribers controlled by each of the 28 operators listed in Table 10.1. In addition, the table breaks down this figure by region. The most striking feature of Table 10.2 is arguably that it suggests strongly that there is no clear relationship between the process of internationalisation by mobile operators and the scale of their operations. This can be examined by inspection of Table 10.3.

Table 10.2 Proportionate subscribers by region, 31 December 2012

Company	Total subs (millions)	Number of subscribers by region[1] (millions)						
		Western Europe	Eastern Europe	Middle East	Asia-Pacific	Latin America	North America	Africa
Vodafone[2]	359.9	118.7	2.0	19.2	115.3[10]	–	44.2	60.5
América Móvil	276.5	13.7	2.2	–	–	237.4	23.2	–
Bharti Airtel	243.0	neg.	–	–	184.1[10]	–	–	58.9
Telefónica	238.4	69.2	–	–	12.0	157.2	–	neg.
Telenor	177.8	18.8	40.6	–	114.6[10]	–	neg.	3.8
Singapore Telecom	169.7	neg.	–	–	150.6	–	–	19.1
VimpelCom[3]	168.8	21.6	103.3	–	33.0	–	0.2	10.7
MTN	145.5	0.2	–	28.4	4.8	–	–	112.1
France Télécom[4]	145.2	78.0	2.4	2.5	neg.	3.8	neg.	58.5
Deutsche Telekom	126.0	90.7	1.9	–	–	–	33.4	–
NTT DoCoMo[5]	97.6	–	–	–	97.2[10]	–	0.4	–
Mobile TeleSystems[6]	96.7	–	96.7	–	–	–	–	–
Axiata	89.5	–	–	neg.	89.5	–	–	–
Orascom	85.7	–	–	–	63.3	–	0.4	22.0
TeliaSonera	80.7	25.2	35.5	13.8	6.2	–	–	–
Etisalat	64.8	–	–	12.1	23.3	–	–	29.4
Qatar Telecom[7]	60.4	–	–	11.0	37.6	–	–	11.8
Saudi Telecom	53.6	–	0.1	22.4	28.4[10]	–	–	2.7
Turkcell[8]	46.5	0.2	10.9	35.4	–	–	–	–

Hutchison Whampoa	44.4	19.2	–	1.0	24.2	–	–	–
Vodacom	41.7	–	–	–	–	–	–	41.7
Millicom International	39.0	–	–	–	–	20.3	–	18.7
Vivendi Universal	35.9	21.8	–	–	–	–	–	14.1
Tele2	33.3	8.4	24.9	–	–	–	–	–
Zain	33.0	–	–	18.6	–	–	–	14.4
Maroc Télécom	24.4	–	–	–	–	–	–	24.4
Portugal Telecom	23.0	7.5	–	–	0.6	12.5	–	2.4
Telekom Austria	20.8	11.6	9.2	–	–	–	–	–
AT&T[9]	132.4	1.2	0.2	–	–	21.0	110.0	–

Notes:

neg. = negligible.

1. Defined as per Table 10.1.
2. Network Partnership Agreements are not included within Vodafone's total. It agreed to sell its North American subscribers to Verizon Wireless in August 2013.
3. The data are for 'active subscribers' as published by VimpelCom. Hence direct comparability with other Russian operators that use other methodologies must be treated with caution.
4. Now known as Orange, the brand previously used for most of its mobile networks. Those French overseas territories where Orange is present are counted separately and not included within France.
5. DoCoMo is the mobile subsidiary of NTT. In this case, however, NTT has stakes in two networks independently of DoCoMo, and hence the subscribers of NTT are not simply a proportion of those belonging to DoCoMo. In 2012, NTT had roughly 66 million proportional subscribers, all in Asia.
6. MTS is majority-owned by Sistema, which also operates a network in India that was shut down during 2012.
7. Now known as Ooredoo.
8. The Turkcell listed here is the operator. However, the term 'Turkcell' is often used erroneously to refer to Turkcell Holding, which owns 51 per cent of the operator. The relevant subscriber numbers for Turkcell Holding are therefore 51 per cent of the above.
9. Of these, 107.9 million are direct subscribers in three North American countries. The rest result from an 8.9 per cent stake in América Móvil, which is treated as an investment rather than an attempt to gain control over international subsidiaries.
10. Biased upwards by overstatement of subscribers in India.

Table 10.3 Operators and their proportionate subscribers (millions)

Range	20–50	50–100	100–150	150–200	200–250	250–300	300+
No. of operators	9	9	3	3	2	1	1
Average	[1]	79	139	172	241	276	360

Note:

1. Not relevant as there is a minimum value of 20.

Altogether, the 28 operators account for 2935 million proportionate subscribers, which averages out at 105 million apiece. However, only ten operators have more subscribers than the average, whereas 18 have fewer. Given that in an even distribution each average figure would rise by 50, the gap of 69 between the 150–200 million and 200–250 million ranges is very marked. It is also the case that the largest ten operators were (for the time being, at least) in something of a league of their own at the end of 2012. Of these, five are based in Western Europe plus one in Russia, two in Asia, one in Latin America and one in Africa. Interestingly, during the five years beginning with 2008, all the European operators grew more slowly in proportionate subscriber terms than those based in Russia and elsewhere. This tends to confirm the point that the most acquisitive operators are no longer to be found in Europe (or indeed the USA).

Comparing the top ten companies in Tables 10.1 and 10.2 is informative, as several differences can be observed. In the first place, by no means all of the same operators appear in both tables (see discussion below). As noted in Table 10.1, this feature would have been even more marked if all of the countries where AT&T and Digicel could be said to operate had been included. However, the discrepancies can be found throughout the tables. Thus, for example, Etisalat, Millicom and Telia-Sonera are significantly more highly rated in terms of number of networks, whereas the reverse is true for Mobile TeleSystems.

10.5 PERCENTAGE OF PROPORTIONATE SUBSCRIBERS BY REGION

Table 10.4 reworks the data contained in Table 10.2 above so that the evident regional differences are highlighted. It is clear from the table that, for many mobile operators, their domestic market provided a significant number of their proportionate subscribers at the end of 2012, notwithstanding their internationalisation. Furthermore, it must be borne in mind that, for the mobile operators that are not listed, given the very small number and/or small size of overseas assets in every case, the domestic market is bound to represent the bulk of their proportionate subscribers. Expressed crudely, the interest here therefore lies mainly in those operators that rely upon their domestic market for less than half of their proportionate subscribers. There are 21 such operators listed in Table 10.4 but, interestingly, the 28 operators listed do not show a continuum of values, as one might expect.

Of the 28, four have a domestic market share of over 70 per cent, a surprisingly high figure for an international operator, although all four are narrowly clustered just above the 70 per cent mark. Curiously, the number of proportionate subscribers is highly variable between them, so it is difficult to generalise from these data. Furthermore, the next three in the list have a domestic market share of more than 60 per cent, but all seven have little in common except for two features – the first of which is that, aside from Vivendi Universal, which is anyway not a pure mobile operator and is currently divesting its interest in the sector, none is based in Europe. The second – albeit seemingly somewhat weaker – connection is that five of the seven are to be counted among the operators with interests in no more than seven overseas networks.

It may be argued that Bharti Airtel should be excluded because it is a major participant in the world's second-largest market, which is bound to result in a high domestic percentage, no matter how many overseas networks it invests in – currently over 20 – in which case there seems to be quite a strong association between having a large domestic percentage and a small number of overseas investments. However, it also has to be said that the home markets in question are almost all of a significant size, so a more complicated version of the above would be to the effect that an association between a large incumbent in a relatively large market and a small number of overseas investments is the main explanatory factor for a large domestic percentage.

Although the sample size is too small to warrant generalisations, it is undeniable that the gap between Vivendi Universal and Saudi Telecom is extremely large at almost 25 per cent. There is also a slightly surprising gap of 6 per cent between América Móvil and France Télécom. However, Curwen and Whalley (2008: table 9.4) reveals a larger gap in approximately the same place in the distribution, so this can be disregarded.

There are nine operators – excluding Millicom, which strictly has no home market – with a domestic percentage below 10 per cent. These do not appear to have anything much in common other than that in most cases – Hutchison Whampoa, Qatar Telecom, Singapore Telecom, Tele2

Table 10.4 Percentage of proportionate subscribers by region, 31 December 2012

Company	Total subs (millions)	Domestic as % of total	Percentage of subscribers by region[1]						
			Western Europe	Eastern Europe	Middle East	Asia-Pacific	Latin America	North America	Africa
Turkcell	46.5	76.3	0.4	23.3	76.3	–	–	–	–
Maroc Télécom	24.4	73.6	–	–	–	–	–	–	100
Mobile TeleSystems	96.5	73.5	–	100	–	–	–	–	–
Bharti Airtel	243.0	73.1	neg.	–	–	75.8[10]	–	–	24.2
Vodacom	41.7	68.8	–	–	–	–	–	–	100
NTT DoCoMo	97.2	62.5	–	–	–	99.6	0.4	–	–
Vivendi Universal	35.9	60.7	60.7	–	–	–	–	–	39.3
Saudi Telecom	55.6	36.4	–	0.1	41.9	53.0[10]	–	–	5.0
VimpelCom	168.8	33.3	12.8	61.3	–	19.5	–	0.1	6.3
Portugal Telecom	23.1	33.0	33.0	–	–	2.4	54.0	–	10.6
Deutsche Telekom	127.1	28.9	71.9	1.5	–	–	–	26.6	–
Telekom Austria	20.8	25.5	55.8	44.2	–	–	–	–	–
América Móvil	276.5	25.4	5.0	0.8	–	–	85.8	8.4	–
France Télécom[2]	145.2	19.3	53.7	1.7	1.7	neg.	2.6	neg.	40.3
MTN[3]	145.5	17.5	0.1	–	19.5	3.4	–	–	77.0
Axiata	89.5	14.0	–	–	neg.	100	–	–	–
TeliaSonera[4]	80.7	12.2	31.2	44.0	17.1	7.7	–	–	–
Etisalat	64.8	11.3	–	–	18.6	36.0	–	–	45.4
Tele2[5]	33.3	9.9	25.3	74.7	–	–	–	–	–

Telefónica	238.4	8.6	29.1	–	–	5.0	65.9	–	neg.
Zain[6]	33.0	6.7	–	–	56.4	–	–	–	43.6
Vodafone[7]	359.9	5.5	33.1	0.6	5.4	31.7[10]	–	12.3	16.9
Hutchison Whampoa	44.4	5.1	43.2	–	2.3	54.5	–	–	–
Qatar Telecom[8]	60.4	3.6	–	–	18.2	62.3	–	–	19.5
Singapore Telecom	169.7	2.2	neg.	–	–	88.8	–	–	11.2
Orascom	85.7	1.7	–	–	–	73.8	–	0.5	25.7
Telenor	177.8	1.8	10.6	22.8	–	64.5[10]	–	neg.	2.1
Millicom International[9]	39.0	n/a	–	–	–	–	52.0	–	48.0

Notes:

1. Defined as per notes to Table 10.1.
2. Now known as Orange, the brand previously used for most of its mobile networks. Those French overseas territories where Orange is present are counted separately and not included within France.
3. The company's home market is South Africa.
4. The company's home market is defined as Finland plus Sweden.
5. The company's home market is Sweden.
6. Zain's home market is Kuwait.
7. Network Partnership Agreements are not included within Vodafone's total.
8. Now known as Ooredoo.
9. The company is registered in Luxembourg but has no mobile assets in Europe.
10. Biased upwards by overstatement of subscribers in India.

and Telenor – the home market is relatively small and has no scope for further growth. However, this is manifestly not the case for Vodafone or Telefónica. Many are present in a large number of countries – Vodafone, Telenor, Singapore Telecom and Telefónica – although for others this is much less evident – notably Zain. Some have large numbers of subscribers, but others fairly modest numbers. Some are based in Europe, but as many are based randomly elsewhere.

As can be seen in Table 10.5, which summarises the data in Table 10.4, three of the 28 operators sampled had all their proportionate subscribers in a single region – Maroc Télécom and Vodacom in Africa and Mobile TeleSystems in Eastern Europe. The commonest decile was 70–80 per cent and all bar three operators had at least half of their proportionate subscribers in one region, the exceptions being Vodafone (33.1 per cent in Western Europe), Etisalat (45.4 per cent in Africa) and TeliaSonera (44.0 per cent in Eastern Europe).

Twenty-five operators were present in at least two regions. However, 18 of these had at least 90 per cent of their proportionate subscribers in only two regions and only two had less than 70 per cent – Vodafone (64.8 per cent in Western Europe and Asia-Pacific) and TeliaSonera (75.2 per cent in Eastern Europe and Western Europe).

Table 10.5 Number of operators: regions where present by decile

Percentage	1 region	2 regions	3 regions	4 regions
100	3	9	4	15
90–100	1	9	11	1
80–90	2	5	1	0
70–80	7	1	0	0
60–70	5	1	0	0
50–60	7	0	0	0
40–50	2	0	0	0
30–40	1	0	0	0

Sixteen operators were present in at least three regions. However, only Vodafone recorded a figure of less than 90 per cent. Unlike the other operators, therefore, it was able to record non-trivial percentages – over 5 per cent – in five regions and only TeliaSonera and VimpelCom were able to do so in four.

It may further be asked how many operators had a significant percentage – 20 per cent or more – of their proportionate subscribers in a second region. It was possible for this to be the case for 22 operators and in practice the entire sample would have qualified if that target were to be lowered to 19 per cent. However, no operator managed to achieve the target in three regions, although not merely Vodafone, TeliaSonera and Etisalat, but also Qatar Telecom, came close.

A review of the individual regions indicates that Asia-Pacific (nine out of 17) and Latin America (four out of six) were the only regions where a significant number of the operators present had more than half of their proportionate subscribers – in other words, they were regional specialists. Nevertheless, among the Asia-Pacific specialists Bharti Airtel (24 per cent in Africa), Hutchison Whampoa (43 per cent in Western Europe), Orascom (26 per cent in Africa), Saudi Telecom (42 per cent in Middle East) and Telenor (23 per cent in Eastern Europe) all had a significant dependence on a second region, so the only heavily specialised operators in the Asia-Pacific region were Axiata, NTT DoCoMo and Singapore Telecom. As for Latin America, the only true specialist was América Móvil, whereas Millicom (48 per cent in Africa), Portugal Telecom (33 per cent in Western Europe) and Telefónica (29 per cent in Western Europe) were essentially bi-regional.

The true Africa specialists were, predictably, Maroc Télécom, MTN and Vodacom. No other operators had in excess of 50 per cent of their proportionate subscribers located in the region. As for the Middle East, the only specialist could be said to be Turkcell, although it had 23 per cent of its proportionate subscribers in Eastern Europe so could be said to be bi-regional, as indeed could Saudi Telecom and Zain, where both had their proportionate subscribers divided roughly equally between two regions.

Neither Western Europe nor Eastern Europe showed any real tendency to be a home region for regional specialists and, interestingly, only Tele2 could be said to be an additional regional specialist, were the two regions to be treated in tandem. A significant number of operators present in either Western or Eastern Europe were bi-regional, but the two exceptions are worthy of note. Vodafone was very unusual in that it was present in five regions and had no more than one-third of its proportional subscribers in any one region. While VimpelCom could also claim to be present in five regions, its dependence on North America is trivial and hence it cannot be said to match Vodafone, which, admittedly, had a very modest dependence on Eastern Europe, albeit that a given percentage was equivalent to many more proportionate subscribers than was the case for VimpelCom. TeliaSonera, in contrast, was present in four while deriving 75 per cent of its proportionate subscribers from only two of them, and Telenor was also present in four but the equivalent figure was 88 per cent.

Otherwise, only Etisalat could be said to be truly tri-regional as it derived less than one-half of its proportionate subscribers from any individual region, whereas for Qatar Telecom the Asia-Pacific region was arguably too significant to permit it to be called tri-regional.

10.6 COMPARING INTERNATIONAL OPERATORS

In the sections above it has been noted that, if internationalised operators are listed numerically according to the criterion relevant to the discussion in hand, the list does not match that/those used in previous sections. This feature is worthy of further analysis.

Table 10.6 Most internationalised mobile operators at 31 December 2007

Rank order	Number of countries	Proportionate subscriber numbers	Percentage of proportionate subscribers outside the domestic market[1]
1	France Télécom	Vodafone	Telenor
2	Vodafone	América Móvil	Hutchison Whampoa
3	Telefónica	Telefónica	Singapore Telecom
4	MTN	Deutsche Telekom	Zain
5	TeliaSonera	France Télécom	Vodafone
6	Zain	Mobile TeleSystems	Orascom
7	América Móvil	Telenor	TeliaSonera
8	Telenor	Singapore Telecom	Qatar Telecom
9	Etisalat	Telecom Italia	Telefónica
10	Hutchison Whampoa	Orascom	Tele2
11	Millicom International	Sprint Nextel	Telekom Malaysia
12	Orascom	TeliaSonera	Telekom Austria

Note:

1. Excluding Millicom International, which has no subscribers in the country where it has its HQ.

It is of interest to start by comparing some of the indicators of internationalisation at the end of 2007 and the end of 2012. In most industrial sectors one would expect to see minimal change over a five-year period, but the mobile industry is very volatile. Tables 10.6 and 10.7 compare operators according to three measures: number of countries

Table 10.7 Most internationalised mobile operators at 31 December 2012

Rank order	Number of countries	Proportionate subscriber numbers	Percentage of proportionate subscribers outside the domestic market[1]
1	France Télécom	Vodafone	Telenor
2	América Móvil	América Móvil	Orascom
3	Telefónica	Bharti Airtel	Singapore Telecom
4	Telenor	Telefónica	Qatar Telecom
5	Singapore Telecom	Telenor	Hutchison Whampoa
6	Vodafone	Singapore Telecom	Vodafone
7	Bharti Airtel	VimpelCom	Zain
8	MTN	MTN	Telefónica
9	TeliaSonera	France Télécom	Tele2
10	Etisalat	Deutsche Telekom	Etisalat
11	VimpelCom	NTT DoCoMo	TeliaSonera
12	Deutsche Telekom	Mobile TeleSystems	Axiata

Note:

1. Excluding Millicom International, which has no subscribers in the country where it has its HQ.

where present; the number of proportionate subscribers; and the percentage of those subscribers located outside the home country. Several features are apparent:

- Orascom (taken over by VimpelCom), Zain and Hutchison Whampoa (due to divestments) and Millicom International (due to lack of growth) have disappeared from column 1.
- Orascom, Telecom Italia (due to divestments) and Sprint Nextel (no longer qualifying) have disappeared from column 2.
- Telekom Austria has been replaced by Etisalat in column three – Telekom Malaysia has become Axiata.
- Singapore Telecom now appears in all three columns (and MTN would have been 13th in column 3).
- América Móvil now occupies second spot in columns 1 and 2, and has moved up five places in column 1.
- Bharti Airtel now appears in the top ten in columns 1 and 2.
- VimpelCom now appears in columns 1 and 2.
- Deutsche Telekom now squeezes into column 1 but falls back six places in column 2.

- TeliaSonera has dropped four places in column 1 and disappeared from column 2.

The situations for Singapore Telecom and Bharti Airtel are closely aligned because the former is a major shareholder in the latter – in effect, Bharti Airtel drags Singapore Telecom in its wake in columns 1 and 2. However, they diverge significantly in column 3, since Singapore Telecom is high up the list due to its small home market whereas Bharti Airtel does not appear at all due to its enormous home market.

The most notable feature in many ways is the second position in columns 1 and 2 occupied by América Móvil, given that it is located in neither Europe nor the Asia-Pacific region. This is indicative of a degree of dominance that is likely to be keeping regulators on their toes – as it is currently in Mexico (Reuters, 2014). Telefónica's situation is similar in some respects, as Table 10.7 indicates, but the big difference is that, whereas América Móvil is dependent on Latin America for 86 per cent of its proportionate subscribers, that figure falls to 66 per cent for Telefónica, which also obtains a further 29 per cent from Western Europe.

Four of the top six entries in column 1 are European operators, but France Télécom/Orange drops sharply in column 3, to be replaced by Bharti Airtel. As noted, this implies that Orange's networks must often be sub-scale and that there is a need for restructuring.

A number of operators appear in all three columns – Singapore Telecom, Telefónica, Telenor and Vodafone. These four would accordingly appear to be the contenders for the title of most international operator, a matter that will be discussed in the next section.

10.7 CONCLUSIONS

It is now time to try to answer the question as to which of the operators covered in this chapter can be said to be the most international, and which operators can be considered to be the runners-up. It is of interest to note that the answer to these questions in Curwen and Whalley (2008: 235) was first Vodafone and second América Móvil, France Télécom, Hutchison Whampoa and Telefónica. Five years later, what has changed, if anything?

Vodafone clearly looks to have retained its pre-eminence if four relevant pieces of information are factored into the question – number of countries where present (28), total number of proportionate subscribers (360 million), number of regions where present (six) and percentage of proportionate subscribers located overseas (94.5). However, it is important to reiterate the point made in 2008, namely that Vodafone may be highly internationalised but it is by no means a global presence. At the end of 2012, it had a substantial number of proportionate subscribers in only four regions out of a possible seven and, significantly, its recent disposal of its stake in Verizon Wireless has reduced that to three out of seven, which is hardly global.

A further point that needs to be reiterated is that a superficially significant presence in a region may in practice reflect the accumulation of modest proportionate subscriber numbers in several countries. Clearly, especially where the stakes held are not controlling, this cannot be taken as truly indicative of a global reach. This is something of an issue for France Télécom (now Orange), since at the end of 2012 it had only 145 million proportionate subscribers spread over 42 networks – an average of 3.5 million compared to 12.9 million for Vodafone at that time. As suggested above, Orange has been effectively bi-regional for some years and hence is not even a strong contender for a runner-up position.

América Móvil looks a more impressive candidate since the four numbers comparable to those for Vodafone are 32, 277 million, four and 74.5. Alternatively, there is Telefónica, with 31, 238 million, three (effectively) and 91.4. However, given that América Móvil is essentially uni-regional whereas Telefónica is essentially bi-regional, neither can be said to be remotely a global presence. Orange, as noted, is arguably even less credible as a global presence, and Hutchison Whampoa no longer has a shred of credibility. Of the others, the most credible is arguably VimpelCom with 18, 169 million, four (effectively) and 66.7, but the latter percentage is not encouraging.

So the only conclusion to be drawn is that there is no such thing as a global operator and only Vodafone comes even vaguely close to such a description. This obviously implies that there is no correlation between sheer size and a global presence as things stand, and, as there is no prospect of any operator acquiring a significantly larger scale in the near future, this is not likely to change. This will probably not discourage operators from using the term 'global' to describe their aspirations – it has been a particularly popular term with up-and-coming operators based in the Middle East – but the evidence shows clearly that even highly internationalised operators tend to fall back into a bi-regional or occasionally a tri-regional structure.

It may be argued that this is a reflection of sound strategic thinking – after all, it is no easy task to exercise control over subsidiaries in far-away countries, especially those with significantly different cultural environments. On the other hand, as will become clearer in Chapter 11, many operators appear to be willing to acquire whatever happens to be available at 'the right price', so one should not perhaps overdo the sound strategic thinking aspect, especially as the current period appears to be associated with a good deal of portfolio readjustment. The only sound adage appears to be to stay in the comfort zone, but that often means limited growth prospects and hence few opportunities for glory for CEOs.

NOTES

1. Curwen and Whalley (2008) also contains a discussion of psychic dispersion. However, this did not provide enough useful data to warrant a repeat of the exercise here.
2. This is not, of course, necessarily the same thing as owned by Europeans, since many large mobile operators such as Vodafone are widely held by international investors.

REFERENCES

Cellular-news (2011a), 'India's active wireless subscriber data shows diverging trends', www.cellular-news.com, 18 March.

Cellular-news (2011b), 'Two million SIM cards disconnected in Zimbabwe', www.cellular-news.com, 20 April.

Curwen, P. and J. Whalley (2008), *The Internationalisation of Mobile Telecommunications: Strategic Challenges in a Global Market*, Cheltenham, UK and Northampton, MA, USA: Edward Elgar.

Curwen, P. and J. Whalley (2011), 'The restructuring of African mobile telecommunications provision and the prospects for economic development,' *info*, **13**(2), 53–71.

Garner, M. (2007), 'Multiple SIMs per user compared to market penetration', *Wireless Intelligence*, **26**(April).

Reuters (2014), 'Mexico regulator says Slim's America Movil dominant in telco', www.reuters.com, 7 March.

Wikipedia (2012), 'Dual SIM', http://en.Wikipedia.org/Wiki/Dual_SIM, 20 September.

11. Retrospect and prospect

11.1 INTRODUCTION

Chapter 10 has examined in some detail the situation in relation to the most important international operators at the end of 2012 and concluded with an attempt to identify the most internationalised among them by aggregating the data contained in the chapter. It was noted that no operator could reasonably lay claim to having a global presence irrespective of what they state officially in press releases or unofficially at the likes of conferences.

However, a few tasks remain in any attempt to take an overview of the mobile sector, and these are addressed in this chapter. For example, there has so far been no discussion of the operational performance of internationalised operators during the past several years – a discussion that must be conducted in terms of the acquisition of revenue-yielding subscribers for want of comparable financial data across the sample. Second, there is a need to bring together the information on recent acquisitions and divestments together with data on something that is less tangible, namely information on what operators would like to have achieved if they had been free to do so.

A third issue relates to licences. Competition is between networks that already exist but operators with licences provide potential competition that also needs to be taken into account. By and large, as noted previously, new entrants have rarely posed a significant threat to incumbents, and that conclusion held good during the heyday of 3G, which covered roughly the period 2000 to 2005. Now that 3G is to be found pretty much everywhere, operators have had time to adjust their strategies to take its effects into account. But the era of 4G – for now essentially LTE – is upon us, so operators, especially those with multiple networks, must factor in the costs and benefits, which are more diffuse than they were with 3G because of the much wider range of bandwidths that can be employed.

Finally, something needs to be said about the 'dark side' of the mobile sector, which involves bribery and corruption. Naturally, the mobile sector is far from unique in offering opportunities to procure an unfair advantage over rivals, but it has been too commonplace in recent years to ignore.

11.2 GROWTH RATES

It is of great interest to look back at the relevant sections in Curwen and Whalley (2008: 247–51), which discuss activity among the most prominent international operators during 2004 to 2007 inclusive. The most striking feature is the growth rates of proportionate subscribers – hereafter p/subs – during the years in question. The minimum rate for a listing was 50 per cent per annum, yet five operators exceeded this in 2004, an astonishing 11 in 2005, six in 2006 but a mere three in 2007. As the sections below will reveal, the comparable numbers were zero in 2009, three in 2010, one in 2011 and zero in 2012.

Naturally, there is a statistical feature that explains part of this away. Starting at 1000, it requires only an additional 1000 to produce a growth rate of 100 per cent, whereas starting at 1 million requires an additional 1 million and so forth. Hence, given that growth in the number of subscribers will necessarily slow down because populations are finite and the need for a second handset is clearly much less than for a first handset, it would be normal for there to be a declining trend in annual growth rates, particularly once a market reaches a nominal 100 per cent penetration ratio.

This obviously poses an awkward question, namely what should be taken as an equivalent to a 50 per cent growth rate – which had effectively disappeared by 2007 – when considering the period after 2008. Inspection of the data for the period 2009 to 2012 inclusive suggests that if an operator has grown its p/subs by more than 20 per cent in any given year, this should be treated as an exceptional performance.

It is also useful at this point to refer back to Table 10.1 because it contains a feature about which no comment has yet been made, namely changes in the number of countries where individual operators were present during the five-year period 2008 to 2012 inclusive. These appear in the brackets after the number of countries where present at the end of 2012. However, it must be borne in mind not merely that the figures are net – that is, entries less exits – but that a large plus or minus figure could involve more than one operator simultaneously where one is directly involved and the other(s) indirectly.

The net annual figures are as follows: 2008 = +9; 2009 = +13; 2010 = +21; 2011 = +20; 2012 = 0. In 2008 the only changes of any significance were the gain registered by Deutsche Telekom and the losses suffered by Tele2 and Orascom. In 2009 no significant gains or losses were registered, the plus figure representing an accumulation of small gains. In 2010 there were large gains registered by Bharti Airtel and Singapore Telecom and a large loss suffered by Zain. In 2011 there were significant gains registered by América Móvil, Telenor and VimpelCom, and a smaller loss suffered by Vodafone. Finally, in 2012 there was a single large gain made by América Móvil offset by a number of minor losses. These will all be discussed in more detail below.

However, it is also worth observing that there were significant changes in seven cases taking the period as a whole, as follows: Bharti Airtel = +20; Singapore Telecom = +19; América Móvil = +15; VimpelCom = +10; Telenor = +8; Zain = –11; and Orascom = –7. Combining the two sets of data leads to the conclusion that, first, there was a head of restructuring steam that built up in the period around the turn of the decade which suddenly fizzled out in 2012 and, second, that throughout the period relatively few operators were engaged in restructuring on other than a modest scale.

This is quite interesting because there is currently so much commentary about the need to restructure that the apparent equilibrium of 2012 does not seem to be sustainable. This will be discussed further in the concluding section.

11.3 ACTIVITY IN 2009

An increase in p/subs can occur either because an operator is acquiring new subscribers at existing networks or because it has acquired new networks. However, during 2009, a good many international operators barely grew their p/subs. For example, VimpelCom and T-Mobile managed less than 10 per cent growth, and Telefónica, TeliaSonera and Vodafone only just exceeded that percentage. That might be seen as reasonable purely as a measure of organic growth, but it was already quite clear that a notional 100 per cent penetration rate was not, in practice, a meaningful ceiling because of the increasing ubiquity of subscribers with multiple devices. Furthermore, there was a fair number of acquisitions overall, as shown in Table 11.1. The only noteworthy changes were as follows:

- Qatar Telecom's p/subs grew by 38 per cent to 32.9 million, primarily due to a 45 per cent increase to 21.4 million in Indonesia.
- Saudi Telecom's p/subs grew by 36 per cent to 35.2 million.

However, the massive growth of 334 per cent in Indonesia reflected the very low starting value, as was to a lesser extent true of the 93 per cent growth of p/subs to 5.7 million in India. In the home market, much the largest, p/subs grew by 29 per cent to 19.7 million.

- Axiata's p/subs grew by 29 per cent to 63.5 million, primarily due to an 85 per cent increase to 11.3 million in India and a 36 per cent increase to 8.3 million in Indonesia.

What is unusual about these three is that none of them is European and two are based in the Middle East, where for the most part home markets are relatively small, so growth must largely take place overseas. This was indeed a period when Middle Eastern operators including Etisalat were actively building empires – another on a much smaller scale was Batelco – but if they were not already aware of the pitfalls of over-expansion overseas relative to a modest home base, then the collapse of Zain was about to remind them in no uncertain terms.

11.4 ACTIVITY IN 2010

During 2010, subscriber growth was patchy but overall there was a distinct improvement compared to 2009. However, Deutsche Telekom saw its p/subs shrink by 5 per cent and Vodafone saw minimal growth, so size and growth were clearly not directly linked. The real casualty was Zain, which saw its p/subs shrink by 54 per cent from 55.7 million to 25.5 million essentially due to the sale of its Celtel subsidiary to Bharti Airtel and (hence indirectly to) Singapore Telecom. It would have been much worse for Zain had it not retained the Celtel network in Sudan, which provided 10.4 million p/subs. In contrast:

- Bharti Airtel grew its p/subs by 64 per cent. This was quite complicated because, although Bharti Airtel had 117.3 million p/subs in 2009, its three overseas networks accounted for only 1.3 million p/subs between them. In 2010, India accounted for 148.9 million p/subs while the other non-Celtel networks, now amounting to three in total, accounted for 3.9 million. The Celtel networks – which had not grown by much during the year – provided a further 39 million p/subs, representing a meagre 20 per cent of the total and not much more than the additional p/subs provided for by the 28 per cent increase in p/subs in India.

- Hutchison Whampoa grew its p/subs by 57 per cent, but once again this increase of 10.5 million p/subs was not across the board. In practice, the 200 per cent growth in Indonesia provided 6.1 million extra p/subs, the 200 per cent growth in Australia an additional 2.1 million and the 40 per cent growth in the UK a further 2 million. It can be deduced from this that not much happened at the other networks.
- VimpelCom grew its p/subs by 44 per cent from 63 million to 90 million. However, Russia – which accounted for roughly 50 million of these – barely grew. The dramatic growth in Ukraine (1115 per cent) from 2.0 million to 24.4 million p/subs was accounted for by the merger of the former Ukrainian RadioSystems and Kyivstar in April 2010 and a further 1.9 million resulted from a new network in Kyrgyzstan.
- Singapore Telecom grew its p/subs by 30 per cent to 132 million, of which 62 million were provided by its 32.1 per cent stake in Bharti Airtel – 65 per cent up from 37.6 million in 2009. In contrast, its other networks barely grew at all, with 24.4 million of the overall increase for the year of 30.6 million being provided by growth at Bharti Airtel.
- Etisalat grew its p/subs by 26 per cent. Interestingly, p/subs at some of its largest networks in the UAE, Pakistan and Saudi Arabia barely grew at all, whereas growth at others – Afghanistan (61 per cent), Sri Lanka (42 per cent) and Egypt (37 per cent) – was impressive, while Nigeria grew by 200 per cent albeit from a low base value.
- Qatar Telecom grew its p/subs by 25 per cent – a rare example of an operator appearing in two successive years even though it only just managed to qualify. The 33 per cent growth in Indonesia accounted for 7 million of the overall 8 million increase in p/subs.

Nevertheless, Etisalat did not improve on the previous year's growth figure by much, and Qatar Telecom, Saudi Telecom and Zain performed much worse, so it was not a good year overall for the current and former acquisitive operators based in the Middle East. Also, not only was it a poor year for operators based in Europe, but those operators that prospered did so partly because they did not have significant operations based in that region.

11.5 ACTIVITY IN 2011

During 2011, activity slowed down sharply compared to 2010. Thus:

- VimpelCom grew its p/subs by 79 per cent from 90 million to 161 million. Over a two-year period this represented growth of 157 per cent and is the only example of an operator growing by over 25 per cent in two successive years. However, the existing networks including Russia and Ukraine essentially stood still other than Uzbekistan (up by 32 per cent). A total of 86 per cent of the improvement was due to the acquisition of a majority stake in Orascom Telecom Holding (40 million p/subs) and Wind in Italy (21 million p/subs).
- Saudi Telecom grew its p/subs by 31 per cent. However, of the increase of 12.4 million p/subs, 8 million was accounted for by Indonesia, which grew by 155 per cent partly as a result of the increase in the stake held from 62.0 per cent to 83.8 per cent.
- Telenor grew its p/subs by 30 per cent. Of the 35 million increase in p/subs, 10.8 million were accounted for by the 131 per cent growth in India, while the minority stake in VimpelCom (somewhat reduced from 2010) accounted for an additional 15.3 million. This 43 per cent improvement was actually less than the equivalent 69.4 per cent recorded in 2010, which also witnessed an increase of 7.5 million p/subs in India, albeit from a very low base. However, these were offset by the loss of 12.4 million p/subs transferred from the independent entry for Ukraine (via Kyivstar) to VimpelCom – in effect simply a transfer with a net effect of near zero in terms of p/subs.

During the year, as noted above, there were substantial gains in countries where present registered not only by Telenor and VimpelCom but also by América Móvil. However, the latter came as a result of acquiring a minority stake in KPN, which brought with it a mere 1.8 million p/subs. In essence, therefore, organic growth was not a general feature of the year. Rather, the potential of countries in the Asia-Pacific region such as India and Indonesia to exhibit very rapid growth was more worthy of note.

11.6 ACTIVITY IN 2012

During 2012, reductions in p/subs appeared on a wider scale than in previous years. The only significant gain was as follows:

- Qatar Telecom – renamed Ooredoo in June 2013 – grew its p/subs by 26 per cent. The growth of 136 per cent in Tunisia reflected a large increase in the stake held rather than in the network's subscribers, and the same was true to a lesser extent in the case of Algeria and Iraq. Overall, gross subscriber numbers increased by less than 10 per cent.

In comparison, reductions in p/subs were in evidence at Mobile TeleSystems (as a result of its revoked licence in Uzbekistan – see below), Telefónica (down by 3 per cent despite growth in Brazil), TeliaSonera (down by 5 per cent, mainly as a result of a stake reduction in Russia, which cost it 11.2 million p/subs), Vodacom (down by 2 per cent largely due to a poor performance in South Africa) and Vodafone (down by 5 per cent largely due to a poor performance across Europe).

In may be noted that, once again, there was a large increase in countries where present by América Móvil. On this occasion it resulted from the acquisition of a minority stake in Telekom Austria, which brought with it nearly 5 million p/subs. The much-increased stake in KPN also resulted in an additional 10 million p/subs, but for an operator with 260 million p/subs elsewhere in the world such gains were of little consequence. Unfortunately, growth at these other networks was a mere 8 per cent, indicating a severe slowdown in subscriber growth in Latin America.

Looking across the most recent three-year period (2010 to 2012 inclusive) for which data are widely available at the time of writing, only four operators managed to average a p/sub growth rate in excess of 25 per cent: VimpelCom (43 per cent); Zain (39 per cent); Hutchison Whampoa (32 per cent); and Bharti Airtel (30 per cent). It may be noted that none of these is European, which simply reflects the slowdown in subscriber growth in the already saturated markets that are the norm across this region. The performance of France Télécom/Orange was notably poor, with average growth of 10 per cent despite a heavy presence in emerging countries, while Vodafone did even worse, although it exited from seven networks during this period. As for Telefónica, its presence primarily in Europe and Latin America was somewhat unfortunate as these two regions were not doing well and hence accounted for a poor average p/sub growth of 9 per cent.

Information on p/subs during 2013 is not yet readily available, so it is difficult to assess how operators performed during the year. However, data from América Móvil, Millicom and NII Holdings indicate that there has been a modest upturn in Latin America and that may turn out to be essentially the story of the year for most operators – a slight improvement across the board with some black spots. Growth rates of 25 per cent achieved organically look to be a thing of the past, since even in large emerging markets saturation must come about sooner or later, but they might still come about via M&A. As the section below demonstrates, M&A activity was not a characteristic of 2013, and most takeovers anyway take a long time to come to fruition due primarily to regulatory delays. But consolidation is very much on the minds of those involved in the sector, so it may be that 2014 and 2015 will be different.

11.7 ACTUAL AND WOULD-BE INTERNATIONAL TARGETS

The most sensible way of determining the international strategies of operators is to examine where they historically have been and are currently (end May 2014) located. For the most part, country entries and exits involve the taking and divesting of stakes, either majority or minority, in existing networks, but occasionally the preferred entry route is via the acquisition of a new licence, possibly offered for 3G or increasingly for LTE. It is also possible to examine statements made (mainly) by CEOs that indicate where an operator would/would not like to be located. Table 11.1 accordingly summarises the most recent data on actual entries and exits covering the period beginning in January 2009, but also includes data on transactions that have been agreed but have yet to culminate in the physical transfer of a licence or stake.

In contrast, Table 11.2 gathers together information on the countries that operators would like to enter or to have entered during the same period. It therefore covers a variety of possibilities such as an unsuccessful bid for a stake or licence, a withdrawn bid for a stake or licence, an expression of interest not subsequently pursued or the flagging up of a desire to sell out. Although there is evidence of considerable activity in these tables, it should be borne in mind that they encompass only 32 operators – albeit the most active internationally – and hence understate by some margin the total relevant activities on a worldwide basis.

One thing immediately stands out. Roughly 70 per cent of a similar sample of operators examined during 2006 and 2007 proved to have entered more countries than were exited, yet in this case it is only 12 of the 32 (37.5 per cent), with 14 exiting more countries than were entered. A turnaround on this scale would indicate that a rethinking of asset portfolios is under way.

A comparison of Tables 11.1 and 11.2 reveals that successful entries and exits are merely the tip of the iceberg. In practice, operators are constantly reviewing their options and are aware that many if not most of their aspirations and actual offers will not come to fruition. The fundamental reason is that bidding for stakes and licences is almost always a competitive process, and the most attractive assets may appeal simultaneously to a significant part of the sample set, not to mention many other parties not included therein. Plus, of course, international operators tend to announce periodic lists of 'it would be nice if something of interest to us came on the market in the following countries', although they do not really expect it to happen.

So far as licences are concerned, those providing spectrum essentially intended for 2G use are rarely made available – just occasionally a Myanmar comes along – and when they are it is customary for there to be a number of established incumbents that will clearly not make way for a new entrant without a fight, so these licences are not necessarily of interest to the larger international operators. New 3G licences no longer appear very often either – they may be combined with 2G licences – although some are still outstanding in surprisingly large countries such as Pakistan. While 4G/LTE licences are proliferating, they present the same problems for would-be new entrants as did the 2G/3G licences that preceded them, namely the need to reach agreement on roaming over an incumbent's 2G/3G network in order to achieve nationwide coverage.

All the entries contained in Table 11.1 that relate to the period 2009 to 2012 inclusive have been discussed on at least one prior occasion, including as part of the preceding sections, so detailed repetition would serve no useful purpose. However, the table also notes information relating to 2013 and even the first quarter of 2014. Hence a few brief comments are in order.

In the first place, five operators stand out as having pursued an expansionist strategy in recent years, namely América Móvil, Bharti Airtel, Etisalat, Telenor and VimpelCom. Singapore Telecom has also done so indirectly via Bharti Airtel, but that is not truly comparable. However, only Vodafone (voluntarily) and Zain (involuntarily) have undertaken major divestment programmes. Nevertheless, it is of interest that 2013 and even 2014 – that is, [13] and [14] – are quite prominent in the exit column but appear rarely in the entry column. Second, only one major transfer appears during those years, namely the pending transfer of multiple networks from Vivendi to Etisalat.

Table 11.1 Country entries and exits, 2009–2014[1,2]

Operator	Entries	Exits
América Móvil	Panama [9] Belgium, Costa Rica, France, Germany, Netherlands, Spain, Switzerland [11] Austria, Belarus, Bulgaria, Croatia, Liechtenstein, Macedonia, Serbia, Slovenia [12]	Jamaica [11] Belgium, Spain [13] Germany [14][6]
AT&T	–	–
Axiata	–	–
Bharti Airtel	Sri Lanka [9] Bangladesh, Burkina Faso, Chad, Congo-Brazzaville, DR Congo, Gabon, Ghana, Kenya, Madagascar, Malawi, Niger, Nigeria, Seychelles, Sierra Leone, Tanzania, Uganda, Zambia [10] Rwanda [12]	–
Deutsche Telekom	–	Serbia [12]
Etisalat	CAF, Sri Lanka [9] India [10] South Sudan [11] Burkina Faso, Gabon, Mauritania, Mali, Morocco [13][6]	India [12][4,5]
France Télécom	Armenia [9] Morocco, Tunisia [10] DR Congo, Iraq [11]	Switzerland [12] Austria, Portugal [13]
Hutchison	Israel [12]	Israel [9] Thailand [11]
KPN	–	Spain [13] Germany [14][6]
Maroc Télécom	Mali [9]	–
Millicom	–	Cambodia, Sierra Leone, Sri Lanka [9] Laos [11]
Mobile TeleSystems	–	Uzbekistan [12]
MTN	South Sudan [12]	–
NTT DoCoMo	India [9]	Malaysia [10]
OTE	–	Macedonia [9] Bosnia, Montenegro, Serbia [12] Bulgaria [13]
Portugal Telecom	Brazil [11]	Brazil, Morocco [10] Macau [13]
Qatar Telecom[3]	Palestine [9] Myanmar [13]	Saudi Arabia [14]
Saudi Telecom	Bahrain [10]	Romania [9] Indonesia [13]
Singapore Telecom	Sri Lanka [9] Bangladesh [10] Rwanda [12]	Pakistan [13]
Telecom Italia	–	Cuba [11] Argentina, Paraguay [14]
Telefónica	–	Isle of Man, Morocco [10] Cuba [11] Ireland [14]
Telekom Austria	–	–

Telenor	Cambodia, India, Vietnam [9] Kyrgyzstan [10] Algeria, Burundi, CAR, Italy, Laos, Namibia, Zimbabwe [11] Myanmar (L) [13]	Namibia, Vietnam [12] Burundi, CAR, Laos [13][6]
Tele2	Kazakhstan [10] Germany [11]	Russia [13]
TeliaSonera	Germany [11]	Philippines [10] Cambodia [12]
Turkcell	Germany [11]	–
Verizon	–	–
VimpelCom	Cambodia, Vietnam [9] Kyrgyzstan [10] Algeria, Bangladesh, Burundi, CAR, Italy, Laos, Namibia, Pakistan, Zimbabwe [11]	Namibia, Vietnam [12] Burundi, Cambodia, CAR, Laos [13][6]
Vivendi	Mali [9]	Burkina Faso, Gabon, Mauritania, Mali, Morocco [13][6]
Vodacom	–	–
Vodafone	Qatar [9]	China, Hong Kong, Pakistan [10] France, La Réunion, Mayotte, Poland [11]
Zain	Morocco, Sudan [10] South Sudan [11]	Burkina Faso, Chad, Congo-Brazzaville, DR Congo, Gabon, Ghana, Kenya, Madagascar, Malawi, Niger, Nigeria, Seychelles, Sierra Leone, Tanzania, Uganda, Zambia [10]

Notes:

1. To end April 2014, together with those agreed in 2013 with a potential completion date of 2014.
2. The relevant year appears in square brackets.
3. Renamed Ooredoo in 2013.
4. Licences revoked by Supreme Court.
5. Network declared bankrupt.
6. Provisional.

Table 11.2 Strategic interests, 2009–2014[1,2]

Operator	Interested in stake or licence (L)	For sale[3]
América Móvil	Serbia [10]	–
AT&T	Pakistan (3GL), Romania (L) [12] Germany, Netherlands, Telecom Italia, Telefónica, UK [13]	–
Axiata	Cambodia [9] Bangladesh, Myanmar, Sri Lanka [13]	Iran [13]
Bharti Airtel	Bhutan (L) [10] Cameroon (L), Mali (L), South Africa [12] Comium Côte d'Ivoire/Gambia/Liberia/ Mauretania /Sierra Leone [13]	–
Deutsche Telekom	Moldova [9] Croatia, Kosovo, Serbia, Ukraine [10] Cameroon (L) [12] Belgium [13]	UK [13]
Etisalat	Morocco, Nigeria [9] Syria (L) [10] Cameroon (L), Iraq (L), Turkmenistan [11] Libya [12] Bangladesh, Tunisia [13] Benin [14]	–
France Télécom	India [9] Cameroon, Mozambique, Serbia, Syria (L) [10] Kosovo, Morocco, Spain [12] Burkina Faso, Gabon, Mauretania, Myanmar (L), Vodacom [13]	Dominicana [13] Uganda [14]
Hutchison	Poland, Serbia [10] Italy [13][4]	Denmark [13]
KPN	–	–
Maroc Télécom	Cameroon (L) [12] Côte d'Ivoire [13]	–
Millicom	Costa Rica (L) [12] Myanmar (L) [13]	–
Mobile TeleSystems	Azerbaijan [9] Telekom Austria [10] Czech Rep. [13]	–
MTN	Angola (L), Ethiopia (L), Kenya [9] India [10] DRC, Iraq (L), Zimbabwe [11] Morocco [12] Mauretania, Myanmar (L) [13]	–
NTT DoCoMo	Cambodia [9] Pakistan [10] Pakistan (L) [12] Myanmar (L) [13]	India [13]
OTE	–	Albania, Bulgaria, Romania [12]
Portugal Telecom	Mali, Zambia [9] Cameroon (L), Mozambique (L) [12]	–
Qatar Telecom[1]	Syria (L) [10] Libya, Morocco, Pakistan (L), Syria [12]	–
Saudi Telecom	Morocco [9] Syria (L) [11] Morocco [12]	–
Singapore Telecom	Bangladesh [9] Myanmar (L) [13]	–
Telecom Italia	–	–
Telefónica	Cuba, Georgia (L), Nigeria [9] Mexico (L) [10] Poland, Serbia [11]	–
Telekom Austria	Kosovo [10]	–
Telenor	Bangladesh [9] Italy [10] Poland [11] Romania [12] Croatia [13]	–

Tele2	Albania [9]	–
TeliaSonera	Kyrgyzstan, Poland [11] Denmark [12]	–
Turkcell	Libya (L), Macedonia, Tunisia, Ukraine [09] Poland, Serbia, Syria [10] Iraq (L) [11] Belarus, Bosnia, Bulgaria, Kosovo [12] Tunisia [13] Pakistan (L) [14]	–
Verizon	Iraq (L) [11] Canada [13]	–
VimpelCom	Sri Lanka [9] Serbia, Zambia [10] Myanmar (L) [12] Azerbaijan, Bangladesh, Belarus, Moldova, Turkmenistan [13]	Cambodia, Zimbabwe [12]
Vivendi	Algeria [09] Brazil, Mexico (L) [10]	–
Vodacom	Angola (L) [9] Malawi, Nigeria [10] Sierra Leone, Zambia [12] Zambia (L) [14]	DRC [11]
Vodafone	Brazil [10] Italy [11] Myanmar (L) [13] Uganda [14]	Australia, New Zealand [10]
Zain	Algeria, Iran (L), Palestine, Syria [9] Mozambique (L) [10]	–

Notes:

1. To end April 2014. This table is best regarded as illustrative rather than comprehensive. Although all the table entries have been discussed in the media or on telecoms websites, some of the items may best be regarded as rumours rather than hard facts and some have officially been denied (which may or may not reflect the reality).
2. The relevant year appears in square brackets.
3. These entries are sometimes ‘floated’ in the media just to see if there is any interest ‘out there’.
4. A mooted merger between Hutchison and Telecom Italia.

Necessarily, the table does not identify cases where takeovers are occurring within a country, such as Telefónica's proposed acquisition of KPN's E-Plus network in Germany since that is not a new entry as such, nor does an exit technically occur if a small residual stake is retained. But these have been discussed in detail in previous chapters, so overall the table summarises the current position satisfactorily.

Table 11.2 is a more speculative affair for several reasons. In the first place, it reflects only what has been released into the public domain and noted by the authors. Second, it is sometimes difficult to distinguish rumour from fact. Third, operators often deny that anything is happening or simply refuse to comment as a matter of internal policy, but there have been numerous examples of denials being followed by confirmation that the denied events did indeed take place.

As can be seen, there are far more entries in the left-hand column. This is not surprising because, as indicated above, many items are essentially speculative – in other words, they represent wish-list items without a licence or network being available. In addition, there are multiple entries where something concrete has been on offer but by definition there can be only one successful party. The list of items for sale also contains an element of speculation in that the operator may simply be fishing for a reaction, such as was the case for T-Mobile in the UK in 2013. As can be seen, there are a few stakes currently for sale but arguably not as many as might be expected. To some extent this reflects the state of play of the significant takeovers/mergers currently undergoing regulatory inspection. If regulators let them proceed, then one suspects that this will be followed by an upsurge of M&A activity. Meanwhile, there is little to gain by making public any negotiations, no matter how advanced, to rationalise asset portfolios.

11.8 THE EFFECTS OF LTE LICENSING

The licensing of 4G networks – which to all intents and purposes means the licensing of LTE networks – is a potentially important topic because it is evident that incumbents with 2G/3G licences must expect to acquire LTE licences if they are to remain competitive. This is not merely because LTE and LTE-A offer the potential to increase the speed of data transfers, but also because they provide incremental bandwidth needed to keep up with demand.

The flipside is that both the licences themselves and the roll-out of the networks are costly. To some extent the latter can be controlled: licences tend to fix minimum requirements for coverage and it is up to individual operators to decide whether to proceed at a faster pace. Ultimately, however, LTE networks will cover the bulk of advanced countries and as a matter of competitive expediency all major networks will need to provide a roughly comparable service. In emerging economies with widespread rural populations the issue is more complicated and it has to be borne in mind that 3G networks may still not be widespread – indeed, some have yet to be rolled out.

The technology has been discussed in detail in Chapter 2, which also contains a table summarising information on launches that are now proceeding at a rapid pace. What is missing, however, is any detail about the financial costs involved. It is not possible to be specific about roll-out costs because these tend to be subsumed in aggregate figures for investment over a given period, and it is unusual to attribute specific sums among 2G, 3G or 4G networks. However, information about licensing costs is available and a representative worldwide sample is presented in Table 11.3 – as ever, the notes need to be read with care as these relate to matters that are rarely discussed in public-domain documents.

The critical aspect is to compare like with like, which is complicated by the fact that, unlike 3G, which, especially in Europe, required the opening up of new spectrum bands and hence at times resulted in very high licence fees, LTE can be provided either by re-farming existing spectrum or by opening up new bands – primarily the so-called digital dividend bands or the relatively high 2.6 GHz band. In addition, the size of the country, the number of incumbents, the take-up of smartphones and other factors obviously affect what operators are willing to pay.

The standardised measure for comparing licence costs is dollars per MHz per head of population ($/MHz/pop), although that does not take account of inflation or variations in exchange rates – on the whole the former has not been significant since 2010, whereas the latter has occasionally made a noticeable difference. Even so, the final column can be taken as accurate enough for comparative purposes.

Table 11.3 Illustrative results of sales of spectrum with potential 4G usage

Country	Date	Band MHz	MHz sold[1]	Revenue million	Revenue $ million	$ per MHz per pop[11]
Austria	Sep 2010	2600	190	€40	52	0.03
Austria	Oct 2013	800/900/1800	280	€2 014	2 779	1.18
Belgium	Nov 2011	2600	155[2]	€77.8	104	0.06
Belgium	Nov 2013	800	60	€360	482	0.72
Brazil	Jun 2012	2600	120	R2 625	1 270	0.05
Chile	Mar 2014	700	70	CLP11 500	21	0.02
Colombia	Jun 2010	2600	50	P80 000	41	0.19
Colombia	June 2013	Multiple	190	COP770 000	399	0.04[12]
Croatia	Oct 2012	800	40[3]	€40	52	0.30
Denmark	May 2010	2600	200[4]	DKK1 010	168	0.15
Denmark	Sep 2010	900/1800	30	DKK12	2.1	0.01
Denmark	Jun 2012	800	60	DKK742	125	0.38
Finland	Nov 2009	2600	190	€3.8	5.7	neg.
Finland	Oct 2013	800	30	€108	146	0.89
France	Sep 2011	2600	140	€936	1 280	0.14
France	Dec 2011	800	60	€2 639	4 480	0.95
Germany	May 2010	Multiple	358	€4 383	5 960	[13]
Greece	Nov 2011	900/1800	70/40	€381	515	0.41
Hong Kong	Apr 2011	850/2000[1]	20	HK$1 925	244	1.78
Hong Kong	Feb 2012	2300	90	HK$481	61	0.10
Hong Kong	Mar 2013	2600	50	HK$1 540	198	0.56
India	Jun 2010	2300	60	$8 230	8 230	0.12
Italy	Sep 2011	800/1800/2600	240	€3 945	5 300	0.36[14]
Latvia	Jan 2012	2600	140[5]	LVL2 341	4 390	0.01
Lithuania	Oct 2013	800	60	LTL8.1	3.1	0.02
Macedonia	Jul 2013	800/1800	150	€31	39	0.12[15]
Mexico	Jul 2010	AWS	60	P5 248	409	0.07

Netherlands	Apr 2010	2600	130[6]	€2.6	3.2	neg.
Netherlands	Dec 2012	Multiple	360	€3 805	4 870	[16]
New Zealand	Oct 2013	700	80[7]	NZD176	140	0.40
Peru	Jul 2013	AWS	80	$258	258	0.10
Poland	Feb 2013	1800	50	€228	301	0.16
Portugal	Oct 2011	Multiple	299[8]	€372	510	0.17
Romania	Sep 2012	Multiple	485[9]	€682	880	0.08
Singapore	June 2013	1800/2600	270	S$360	284	0.20[17]
Spain	Jul 2011	800/900/2600	210[10]	€1 650	2 380	0.26[18]
Sweden	Mar 2011	800	60	SEK2 054	325	0.59
Sweden	Oct 2011	1800	70	€148	203	0.32
Switzerland	Feb 2012	Multiple	575	CHF997	1 090	0.24
Taiwan	Oct 2013	700/900/1800	270	T$118 700	4 034	0.64[19]

Notes:

neg. = negligible (less than $0.01/MHz/pop).

1. The two 850 MHz licences were sold but the 2 GHz licence was left unsold.
2. 15 MHz paired of the available 70 MHz of paired spectrum was not sold.
3. Only two of the available three (equal-sized) licences were sold.
4. Including 15 MHz in the 2010–2025 MHz band.
5. Comprising 70 MHz paired. A 50 MHz unpaired block was left unsold.
6. Comprising only the paired spectrum on offer as no bids were made for the unpaired spectrum.
7. One block of 5 MHz paired was left unsold.
8. Spectrum in only three of the available six bands (800 MHz, 1800 MHz and 2.6 GHz) attracted competitive bidding, no spectrum was sold in the 450 MHz and 2100 MHz bands and not all of the available spectrum was sold in the 900 MHz and 1800 MHz bands.
9. Of which 110 MHz was for a 14-month period only. All blocks in the 900 MHz and 1800 MHz bands were sold, but one block in the 800 MHz band and eight blocks in the 2.6 GHz band were left unsold.
10. 270 MHz was on offer, of which 50 MHz of unpaired 2.6 GHz spectrum and 4.8 MHz paired of 900 MHz spectrum, together with a regional licence in Extremadura, were left unsold.
11. Other sources tend to give slightly different values for entries in this column. It is generally not possible to work out why this is the case as they tend not to provide their underlying data. It may be noted that the euro is often used rather than the dollar.

12. The 35 MHz not sold comprised 15 MHz paired in the 2.6 GHz band and 2.5 MHz paired in the 1900 MHz band. The 45 MHz paired of AWS spectrum fetched $259.4 million, equivalent to $0.06/MHz/pop; the 30 MHz paired in the 2.6 GHz band fetched $99.4 million, equivalent to $0.04/MHz/pop; and the 40 MHZ unpaired fetched $40.2 million, equivalent to $0.02/MHz/pop.
13. The overall figure was $0.27/MHz/pop, but this disguised big variations: 800 MHz = $0.97; 1800 MHz = $0.1; 2.6 GHz = $0.03; 2 GHz = $0.09.
14. The overall figure was $0.36/MHz/pop, but this disguised big variations: 800 MHz = $1.10; 1800 MHz = $0.35; 2.6 GHz = $0.08 for FDD, $0.06 for TDD.
15. 60 MHz in the 800 MHz band and 90 MHz in the 1800 MHz band were sold but the licence fees related to dual-band bundles.
16. The complex mix of spectrum bands, paired/unpaired spectrum and licence lengths precludes the calculation of a meaningful average.
17. 150 MHz in the 1800 MHz band sold for $0.24/MHz/pop whereas 120 MHz in the 2.6 GHz band sold for $0.15/MHz/pop.
18. This figure applies only to the 190 MHz of nationwide spectrum won by the three incumbents. As in Germany, 800 MHz band spectrum sold for considerably more per MHz than 2.6 GHz spectrum.
19. The 90 MHz in the 700 MHz band sold for $0.43/MHz/pop while the 60 MHz in the 900 MHz band sold for $0.22/MHz/pop and the 120 MHz available in the 1800 MHz band sold for $0.93/MHz/pop.

A number of conclusions can be drawn:

- It is very unusual for a licence to cost more than $1/MHz/pop. In the case of Hong Kong in 2011, it reflected the scarcity of spectrum in the lower bands. In the case of Austria in 2013, the price was considered to be excessive and led to (so far unsuccessful) appeals.
- The reasonable upper limit is generally considered to be $0.5/MHz/pop. This has rarely been exceeded and almost always in relation to digital dividend 800 MHz band spectrum, which normally sells for more than spectrum in bands above 1800 MHz.
- However, there are, as ever, the odd exceptions, such as Lithuania in 2013.
- Spectrum in the 2.6 GHz band rarely sells for more than $0.1/MHz/pop.
- There does not appear to be any correlation between the amount of spectrum on offer and the prices paid.

Because most of the data refer to countries in Europe, it is not possible to make inter-regional comparisons, but prices appear to be relatively low in Latin America, which is understandable given the relatively low level of smartphone penetration and the difficulties of rolling out networks beyond the major cities. It is also not possible to make comparisons with the USA, where LTE is essentially being rolled out using re-farmed spectrum.

In virtually every case where any of the main international operators have bid for licences, they have been successful, although that does not imply that they have necessarily obtained their optimum portfolios of spectrum. It is fair to conclude that they have behaved conservatively in most cases, even though, irrespective of what governments may have said about meeting social objectives and the like, the primary objective of most auctions – markedly the most common form of licence assignment, unlike in the case of 3G, where beauty contests were common – has been to raise revenue. It is clear from the provisional reports covering 2013 so far available that international operators' finances have taken a short-term hit due at least partly to LTE licensing (and roll-out), but this seems to be far less of a worry for them than downward pressure on their margins as a result of regulatory alterations to the likes of roaming fees[1] and interconnection charges.

11.9 THE DARK SIDE OF INTERNATIONALISATION

It is not the intention of this section to suggest that there is always a dark side to the process of internationalisation, merely to indicate that it does exist on occasion and that doing business in certain parts of the world is not without its hazards.

Several long-lasting and one very recent example will suffice, although they are by no means unique. Two of the former relate to Turkey and the latter to Uzbekistan. The cases are clustered geographically for convenience and because there are some common protagonists, rather than because it is the only part of the world where these things happen.

One case in Turkey dates back to 1998 (Mansfield, 2014a) but has yet to be resolved. At the time, one of the major Turkish networks traded as Telsim and was 73.5 per cent owned by the Uzan family. In April 1998, Telsim affiliates entered into an agreement with Motorola to provide Telsim with funding for a GSM licence and the purchase of equipment needed to roll it out. Shares in Telsim were provided as collateral. However, unknown to Motorola at the time, Telsim and the Uzan family then entered into a similar deal with Nokia – which was itself unaware of the Motorola deal – pledging the same collateral. In total, the two vendors were owed roughly $4.8 billion.

Within a year, payments to Motorola ceased to be made. The Uzan family tried to buy time via requests to reschedule payments, although in practice the Uzans had made off with a large part of the funds provided by Motorola. Motorola and Nokia subsequently discovered the fraud and took the Uzans to court in the USA in January 2002 (Glovin, 2009). However, the situation was complicated by the fact that, beginning in 2003, the Turkish government was pursuing the Uzans for their part in the fraudulent collapse of a Turkish bank. The government duly seized certain Uzan assets by way of recompense, including their stake in Telsim. This was sold to Vodafone, which offered $4.6 billion in December 2005 and took control in 2006 (Curwen and Whalley, 2009).

There have been no issues with Vodafone's ownership as such, but Motorola is still seeking damages, in pursuit of which it entered a claim for $5.6 billion in a Hong Kong court in March 2014. Significantly, this was a sum that had already been sanctioned by courts in the UK and USA, indicating how difficult it is to establish where assets are held and the location of courts that have the authority to seize them or, indeed, to rule on whether they should be seized. In that respect, it is reminiscent of the tax cases against Vodafone discussed in an earlier chapter.

It may be argued that the above is more of a salutary lesson for vendors than for telcos, since ultimately Vodafone could be said to have benefited from what happened. However, that has not generally been the case, as another example taken from Turkey demonstrates. The problems began in 2005 when TeliaSonera, a shareholder in the operator Turkcell – itself the subsidiary of Turkcell Holding – tried to acquire a further stake from the controlling shareholder Çukurova. The agreed deal subsequently fell through when Çukurova pulled out and TeliaSonera successfully sought damages through the courts (Martin, 2011). Çukurova later borrowed $1.7 billion from Altimo[2] using a 13.7 per cent Turkcell Holding stake as collateral, only to refuse to repay the money when it ran into financial difficulties. Altimo sued and the case eventually came before the Judicial Committee of the Privy Council in Britain – demonstrating the curious nature of jurisdiction in cases involving internationally held assets, in this case stemming from the fact that Çukurova Holding is registered in the British Virgin Islands – with the Committee ruling that Çukurova had either to repay the money or forfeit its stake in Turkcell (Rapsinews, 2013). The extraordinary machinations surrounding this decision are dissected in Shirbon (2013), but as further extensions have been granted (TeleGeography, 2014a), the matter remains unresolved.

Further examples can be found in respect of the CIS countries after their separation from the USSR. For example, in December 2011, Russia's Alfa Group agreed to reserve 2.7 per cent of its stake in VimpelCom against an adverse outcome in a case involving the ownership of BiTel GSM in Kyrgyzstan. In 2005, a subsidiary of Russia's Mobile TeleSystems, MTS Finance, acquired from Nomihold Securities a 51 per cent stake (plus an option to purchase the residual shares) in Tarino Limited, believing it to be the legitimate owner – albeit indirectly via subsidiaries – of BiTel. Unfortunately for Mobile TeleSystems, an obscure Russian company called Rezervspetsmet, allegedly linked to Altimo, came forward to claim the stake and the claim was accepted by the Russian Supreme Court.

In late 2006, BiTel was transferred to Kyrgyzstan operator Sky Mobile, which was owned by Altimo, and Mobile TeleSystems was forced to write off both what it had paid and the cost of exercising the option (Mansfield, 2011) – a sum running to hundreds of millions of dollars. In May 2008, BiTel apparently evolved into Nurtelecom (or Nur Telecom) and applied for a new licence. Mobile TeleSystems is attempting to have its ownership of BiTel restored via the courts. Sky Mobile is now owned by VimpelCom, hence the possibility of its major shareholder Alfa Group needing to pay a forfeit if Mobile TeleSystems were to regain ownership.

There are currently also issues in Uzbekistan, where, at the beginning of 2012, three large networks were in operation. These were controlled respectively by Mobile TeleSystems (the former Uzdunrobita with 9.6 million subscribers), TeliaSonera (branded as Ucell with 7.7 million subscribers) and VimpelCom (branded as Buztel with 6.2 million subscribers). Two other, small, networks in the form of Rubicon Wireless and Uzmacom were also present.

Before the year was out, only two large networks remained in business because Mobile TeleSystems had its licence revoked in August – it was subsequently declared insolvent in January 2013, causing Mobile TeleSystems to take a write-off amounting to $1.1 billion. The reason given was that local staff had been engaged in criminal activities, but Mobile TeleSystems claimed that it had been subject to a 'classic shakedown' of foreign investors in the country, of which it had been but one of a number of victims. It currently has a case outstanding against the Republic of Uzbekistan before the International Centre for Settlement of Investment Disputes (TeleGeography, 2013).

However, while this network closure was necessarily excellent news for its rivals, they were themselves under investigation for malpractices such as bribery and money laundering (Kiselyova, 2014). For its part, VimpelCom – with a listing in New York – found itself being investigated by the US Securities and Exchange Commission (SEC) and the Dutch authorities. This case followed on from a 2012 Swedish investigation into TeliaSonera's operations in Uzbekistan, where it was alleged that in 2007 it had paid out $350 million for a 3G licence to a company – Gibraltar-based Takilant Ltd – known to be a front for Gulnara Karimova, the daughter of the country's president, Islam Karimov. She was herself placed under investigation by the authorities in Switzerland. In the latest development the US Department of Justice has also opened an investigation into TeliaSonera Uzbek Telecom Holding (TeleGeography, 2014b). Apparently the (alleged) modus operandi was that 100 base stations would be shut down and not switched on again until a 'charitable' payment had been made to a specified entity, with the party receiving the charity turning out to be one or more of the members of the ruling elite.

A final example not involving the CIS countries also dates back to 2005, when Hungary's Magyar Telekom, majority-owned by Deutsche Telekom, acquired a 76.5 per cent stake in Montenegrin Telekom in the course of its privatisation. A case was brought in 2011 – by the SEC since Magyar Telekom had a US listing post-privatisation – alleging that three Magyar Telekom directors paid bribes to government officials. This concluded in December 2011, with both Deutsche Telekom and Magyar Telekom being fined without formally admitting their guilt (OCCRP,

2012). In March 2012, the Montenegrin Parliament opened an investigation but took no action. However, the SEC has once again opened proceedings (Mansfield, 2014b), so there are evidently always opportunities to start again in new jurisdictions. The sums involved in this case are modest, but it illustrates the fact that licence awards are not necessarily above board and that this behaviour is not entirely the preserve of countries outside Europe – see, for example, Curwen (2012a, 2012b), which relate to licensing in India.

11.10 CONCLUSIONS

In essence, this chapter has been concerned with various matters that shed light on why internationalised operators seek to enter or exit overseas markets, although in the process it has not sought to replicate material discussed at length in previous chapters. The reasons why operators seek to internationalise are many and varied, ranging from a slow-growing or small home market to the hubris of CEOs. But, as has been noted on several occasions, the process has rarely been entirely rational. To expand, either licences or stakes must be acquired. A wish list can be drawn up, but items on it may either not become available, be competed for by other operators with a higher valuation in mind, fall foul of contractual problems or be denied by regulators.

In the circumstances, it is not surprising that history is replete with examples of operators grabbing whatever comes on to the market, even if synergies are hard to identify or justify. One way of mitigating this risk has been for operators to acquire relatively cheap minority stakes that can be written down without too much impact on the balance sheet if things go wrong. But, as has increasingly become clear, minority stakes rarely exercise control, and partners with controlling stakes may behave in ways that are unacceptable.

While operators enjoyed a relatively benign environment, this did not matter too much. But there are several circumstances that may cause problems. For example, a recession can adversely affect operating margins, regulators can exert downward pressure on key sources of profitability such as roaming charges, shareholders can become more vocal in demanding dividends and unethical/illegal behaviour can take place. On the whole, when these pressures combine there is a greater likelihood that structural change will follow.

In most developed markets – and emerging markets may well follow suit when they reach saturation point – it is now accepted, at least by all bar some regulators, that consolidation is the way forward because over the past year or two all the above factors have tended to crop up simultaneously. In essence, consolidation means acquiring majority stakes where only minorities exist, acquiring licences and stakes in countries that offer the opportunity to achieve synergies and, primarily, the disposal of non-core – often unprofitable – stakes.

Naturally, it may be difficult to determine what is core and what is not, and the fact is that CEOs will have their own opinions on the matter. However, with so many international operators reporting a downward trend in profitability during 2013, it is financial expediency that currently takes precedence. The offsetting factor is that there is necessarily a limited market for unprofitable minority stakes, and if the only feasible buyer is another shareholder, then the price is unlikely to be attractive. In emerging markets, particularly in Africa, governments are the ones that are desperate for funds and are seeking to sell off state-owned networks and to issue new licences, but there is invariably a gulf between what the government thinks these are worth and what a potential buyer is willing to pay.

In contrast, when the assets on offer are not tarnished by the 'non-core' label, they can elicit competitive bidding. Hence Vivendi Universal's decision to exit the mobile market has provided opportunities to buy that are being eagerly pursued. Unfortunately, however, the status of the potential buyer still enters the equation. Although the highest offer for Vivendi subsidiary SFR in France initially came from another incumbent, Bouygues Télécom, Vivendi was inclined to accept a lower offer from Altice Group in order to avoid regulatory issues. In the end, both bidders offered much the same amount overall but via different combinations of cash – Bouygues offered more – and stakes in the new merged entity, but the Altice offer still won the day (subject to regulatory approval).

Ultimately, subscribers demand better connectivity and higher data transfer speeds; new, costly licences/roll-outs are needed to satisfy subscribers; regulators are reluctant to condone what they see as discriminatory practices; and shareholders demand a better return on their investments. This is a potentially toxic combination, and in March 2014 a Bouygues Télécom executive was driven to compare their liking for their operating environment to the Japanese liking for Fukushima (Lennighan, 2014). According to his calculations, Bouygues provided a triple-play bundle for €20, of which €12–13 was paid to France Télécom for network access and roughly €7 to providers of content – on top of which there was the need to subsidise set-top boxes.[3]

Obviously, these figures do not necessarily hold good for other operators in France or for operators in other countries, but, as has been observed in previous chapters, especially that on Europe, which is leading the way in this respect, operators such as Vodafone that had previously espoused the mobile-only model have come to accept that quad-play is the way forward. If so, the price of a bundle needs to have a built-in profit margin and that must take account of the costs involved for a mobile operator to acquire the relevant fixed-wire assets.

As it happens, Vodafone can fall back on the massive cash reserves provided by the sale of its stake in Verizon Wireless,[4] but even these are finite and most other internationalised operators are short of cash. It is of interest that this has engendered a different response to that traditionally offered when such operators are unhappy with regulators that argue that markets are anti-competitive. The new mantra is that this does not take account of competition in the quad-play market, competition from disruptive MVNOs and competition from over-the-top (OTT) companies such as Skype that by-pass traditional channels via use of the Internet.

So, all are agreed: consolidation is the answer – but possibly not right now since there is also widespread agreement that empire-building will receive the thumbs-down from the financial markets – even Vodafone's use of its cash pile is eliciting nervous reactions. Because in-market consolidation makes economic sense, it is likely to receive a better response than cross-border forays, yet the concessions demanded of those engaged are currently so onerous that they seem to be dissipating most of the potential synergies – as in Austria and Ireland. This suggests that consolidation will be driven by operators making forays into new regions and hence avoiding regulatory censure, but there is only a limited number of operators out there like América Móvil, which has itself fallen foul of vested interests in its unsuccessful pursuit of KPN – unless, of course, US-based operators such as AT&T[5] and Verizon Wireless decide to make a move.

NOTES

1. For example, the European Parliament voted in April 2014 to end roaming charges in the EU by end 2015 (Davies, 2014).
2. Altimo – Alfa Telecom International Mobile – is the telecommunications investment arm of the Alfa Group. Altimo was established on 1 December 2005, before which it was known as Alfa Telecom. It is worthy of note that Altimo Holdings & Investments Ltd is registered in the British Virgin Islands while being majority-owned by CTF Holdings, which is registered in Gibraltar, which is in turn owned by Crown Finance Foundation, a foundation based in Liechtenstein. Or so it appears.
3. However, the Altice takeover of SFR might ease the pressure because Altice–SFR will be saddled with debt and hence end up ceding market share to Bouygues if it maintains these very low prices.
4. Opinions vary as to whether Vodafone will use the money wisely. For a somewhat sceptical view see Gabriel (2013).

5. Given that AT&T still believes in the mobile-only model whereas Vodafone and most other European operators have moved on to quad-play, AT&T cannot really delay its entry into Europe or it will acquire an operator with a large portfolio of fixed-wire assets. However, it could simply build up a portfolio of individual networks such as EE in the UK.

REFERENCES

Curwen, P. (2012a), 'Licence issue and transparency: A cautionary tale', *info*, **14**(1), 70–71.

Curwen, P. (2012b), 'Licence issue and transparency: Reprise', *info*, **14**(4), 78–9.

Curwen, P. and J. Whalley (2008), *The Internationalisation of Mobile Telecommunications: Strategic Challenges in a Global Market*, Cheltenham, UK and Northampton, MA, USA: Edward Elgar.

Curwen, P. and J. Whalley (2009), 'Merging incumbent European mobile operators: *veni*, *vidi*, non *vici*', *info*, **11**(4), 34–51.

Davies, S. (2014), 'It'll all be over by Christmas – for European roaming charges', http://cellular-news.com, 3 April.

Gabriel, C. (2013), 'Vodafone risks squandering its $130 bn US windfall', www.rethink-wireless.com, 2 September.

Glovin, D. (2009), 'Motorola, Nokia seek $3 billion Uzan World Bank claim', www.bloomberg.com, 13 October.

Kiselyova, M. (2014), 'Update 2 – Russia's Vimpelcom investigated over Uzbekistan', www.reuters.com, 12 March.

Lennighan, M. (2014), 'Bouygues Telecom likens French telecoms market to nuclear disaster', www.totaltele.com, 18 March.

Mansfield, I. (2011), 'Alfa Group freezes part of stake in VimpelCom pending court action', www.cellular-news.com, 7 December.

Mansfield, I. (2014a), 'Motorola takes action to seize assets in $5.6 billion lawsuit', www.cellular-news.com, 17 March.

Mansfield, I. (2014b), 'Second bribery probe into Montenegro Telekom privatization', www.cellular-news.com, 21 March.

Martin, S. (2011), 'TurkCell braces for new fight between corporate parents', http://emergingmoney.com, 19 October.

OCCRP (2012), 'Montenegro opens probe into Telekom privatization', https://reportingproject.net, 1 March.

Rapsinews (2013), 'British court orders Turkish Cukurova to pay Russia's Altimo $1.6 billion', http://rapsinews.com, 9 July.

Shirbon, E. (2013), 'Cukurova says Altimo thwarting efforts to recover Turkcell stake', www.reuters.com, 23 July.

TeleGeography (2013), 'Uzbek authorities offer to purchase assets of bankrupt MTS subsidiary', www.telegeography.com, 15 August.

TeleGeography (2014a), 'Cukurova secures extension to reclaim Turkcell stake', www.telegeography.com, 6 February.

TeleGeography (2014b), 'TeliaSonera, Vimpelcom and Karimova face anti-corruption probe', www.telegeography.com, 13 March.

Index